UNPUBLISHED
SCIENTIFIC PAPERS OF
ISAAC NEWTON

PLATE 1

Portrait of Isaac Newton about 1690, wearing his own hair.
By (or after) Sir Godfrey Kneller.

UNPUBLISHED SCIENTIFIC PAPERS

OF

ISAAC NEWTON

A SELECTION FROM THE

PORTSMOUTH COLLECTION

IN THE

UNIVERSITY LIBRARY, CAMBRIDGE

CHOSEN, EDITED AND TRANSLATED BY

A. RUPERT HALL

AND

MARIE BOAS HALL

History of Science and Technology Department,
Imperial College, University of London

CAMBRIDGE UNIVERSITY PRESS

CAMBRIDGE

LONDON · NEW YORK · MELBOURNE

Published by the Syndics of the Cambridge University Press
The Pitt Building, Trumpington Street, Cambridge CB2 1RP
Bentley House, 200 Euston Road, London NW1 2DB
32 East 57th Street, New York, NY 10022, USA
296 Beaconsfield Parade, Middle Park, Melbourne 3206, Australia

First published 1962
First paperback edition 1978

Printed in Great Britain at the
University Press, Cambridge

ISBN 0 521 29436 3   paperback

FOREWORD

The renaissance of Newtonian studies was only just beginning in 1959 when, following earlier forays into the Portsmouth Collection, we began to prepare material for the first issue of this book. The edition of Newton's *Correspondence*, now completed,[1] had not yet appeared in print and the first volume of D. T. Whiteside's magisterial edition of Newton's mathematical papers was still in its early stages.[2] At the end of that year I. Bernard Cohen, whose distinguished career in these studies was then opening, in offering to the History of Science Society at Chicago an account of Newton's manuscripts and of the then state of Newtonian scholarship, justly remarked that 'What is especially astonishing about the huge number of Newtonian articles and books is the general failure of the authors to consult the manuscript collections'.[3] Cohen was able to show that there was then a fresh interest in all aspects of Newton's life and writings, indicated by re-issues and translations of his works and a scattering of articles, and to indicate a number of projects in being, including his own splendid edition (with Alexandre Koyré) of the *Principia*, which happily came to fruition in 1972. But the vast legacy of Newton's papers was still largely unread, and when (in the first issue of this book) we tried to furnish a guide to studies based on Newton's manuscripts the result, incomplete indeed, was extremely modest.

The situation is very different today, and so we have removed those out-of-date pages from this reissue, which is otherwise identical with our original book. The contemporary student has available for his consultation the recent and very ample bibliography by Peter and Ruth Wallis, whose pages record the large and important scholarly output of recent decades as well as the work of earlier years.[4] Although some of the desiderata noted by Cohen in 1959 are still lacking,

[1] H. W. Turnbull, J. F. Scott, A. Rupert Hall and Laura Tilling, *The Correspondence of Isaac Newton*, Cambridge, 1959–77.

[2] D. T. Whiteside, *The Mathematical Papers of Isaac Newton*, Cambridge, 1967 – (seven volumes of eight now published).

[3] I. Bernard Cohen, 'Newton and Recent Scholarship', *Isis*, 51, 1960, 496.

[4] P. and R. Wallis, *Newton and Newtoniana, 1672–1975: A Bibliography*, Folkestone, 1977.

such as a worthy biography of Newton to replace those of Sir David Brewster (1855) or L. T. More (1934) and an edition of Conduitt's papers, no serious student of Newton now fails to pursue his thought into the manuscript arcana. Besides the *Correspondence* and the great editions of Cohen and of Whiteside (the latter, too, now almost finished) we have available John Herivel's study of the development of Newton's ideas in mechanics, along with which Cohen's *Introduction* to the *Principia* should be noted;[1] in optics the many articles by Zev Bechler, Henry Guerlac, J. A. Lohne, A. E. Shapiro, R. S. Westfall and others have likewise explored in numerous different ways the rich implications of the manuscripts, while other scholars including B. J. T. Dobbs, K. Figala, J. E. McGuire and P. M. Rattansi have delved into the deepest and most obscure recesses of Newton's preoccupation with alchemical and other esoteric learning, which would be almost unknown to us had Newton's private papers not survived in such vast quantity.

In making even so partial and summary a note of scholarly achievement since the first publication of this book, we cannot omit mention of two scholars both living in 1962 and now dead: Joseph Hofmann and Alexandre Koyré. The former was naturally led from Leibniz's manuscripts to Newton's; the recent English version of one of his principal books (first printed, in fact, in 1949) must always be at the elbow of any historian of the calculus dispute.[2] Posthumously, Hofmann has given a definite edition of Leibniz's early scientific and mathematical correspondence (1672–76) which is inevitably of deep interest to students of Newton.[3] Koyré's later years were almost wholly given to Newtonian researches and especially to the manuscripts, for he was one of the first to appreciate how the development of Newton's ideas could be traced through them. His collection of essays on Newtonian topics is an essential starting-point for understanding Newton's contribution to the

[1] John Herivel, *The Background to Newton's 'Principia'*, Oxford, 1965. I. Bernard Cohen, *Introduction to Newton's 'Principia'*, Cambridge, 1971. See also R. S. Westfall, *Force in Newton's Physics: The Science of Dynamics in the Seventeenth Century*, London and New York, 1971.

[2] Joseph E. Hofmann, *Leibniz in Paris, 1672–1676*, Cambridge, 1974, a revised version of *Die Entwicklungsgeschichte der Leibnizschen Mathematik während des Aufenhalts in Paris (1672–76)*, Munich, 1949.

[3] G. W. Leibniz, *Sämtliche Schriften und Briefe, herausgeben von der Akademie der Wissenschaften der DDR*, dritte Reihe, erster Band, Berlin, 1976.

revolution in scientific and philosophic ideas of the seventeenth century.[1]

The present reissue is an exact facsimile of the original, apart from the one omission already noted. Newtonian scholarship has advanced so greatly in almost twenty years that, if we were to try to improve our texts and historical comments, it would be necessary to make a new book, which is not possible. We therefore ask the indulgent reader to accept it 'warts and all' as it was in 1962, conscious as we are of many mistakes in it and things we would now express differently. Yet we are also pleased to find that many opinions we expressed—even on such points as Newton's non-use of fluxions in preparing the *Principia*—have stood the test of time. The prudent reader will know that he may find an ampler context and richer annotation in the *Mathematical Papers* of Whiteside, and the books of Herivel and Koyré, besides many specialist papers dealing *inter alia* with documents first printed in this book.

Of the documents included in this volume the most important subsequent re-editions (not always from the identical manuscript, however) are as follows:

I, 1 To resolve problems by motion. Whiteside, *Mathematical Papers of Isaac Newton* I, 400–48.

II, 2 The Lawes of motion. Herivel, *Background to Newton's 'Principia'*, 208–15.

II, 4 Gravia in trochoide descendentia. Herivel, 199–203; Whiteside, II, 420–31.

IV, 1 Early drafts of propositions in mechanics. Herivel, 257–74; Whiteside, VI, 30–91.

IV, 2 On motion in ellipses. Herivel, pp. 246–54; Whiteside, VI, 553–6.

VI Comments on Hooke's *Micrographia*. Sir Geoffrey Keynes, *A Bibliography of Dr. Robert Hooke*, Oxford, 1960, 97–108.

Keynes' book was published before ours, but after our manuscript was complete and in the Press. It should also have been noted in the original issue that V, 1 'Of educating youth in the Universities' was first printed by W. W. Rouse Ball in his *Cambridge Papers*, Cambridge, 1918.

[1] Alexandre Koyré, *Newtonian Studies*, Cambridge, Massachusetts, 1965.

What are the prospects for further extensive editions of Newton's vast legacy? Two obvious areas of his work suggest themselves: optics and chemistry. In optics, many early papers and drafts have indeed been published wholly or in part,[1] but there would be value in a convenient, thoroughly annotated collection of the whole mass including early drafts of the main *Opticks* itself and of the Queries. Far less has been printed of the sheets, covered with Newton's writing, devoted to chemical experiment and to alchemy, largely because most of these latter are transcripts from books, not original compositions. A competent scholar could, however, make an interesting selection from that part of this material which seems more directly related to Newton's own thought and experimentation. For the rest, it seems certain that the preponderance of Newtoniana still unedited is likely to remain so, though no longer wholly ignored by scholars as it was during the earlier part of this century.

A. RUPERT HALL

MARIE BOAS HALL

IMPERIAL COLLEGE, LONDON
April 1978

[1] Beginning with A. R. Hall, 'Sir Isaac Newton's Notebook', *Cambridge Historical Journal*, **9**, 1948, 239–50, through *Correspondence*, Vol. I, 1959 to the recent facsimile of the *Unpublished first version of Isaac Newton's Cambridge Lectures on Optics, 1670–72*, with an introduction by D. T. Whiteside, Cambridge, The University Library, 1973, and many documents in the same editor's *Mathematical Papers*.

CONTENTS

LIST OF PLATES

PREFACE

This is the first attempt to add to the published collection of Newton's scientific writings since 1838, when S. P. Rigaud printed a document that he entitled 'Propositiones de Motu'. Editorial neglect is much to be lamented: while the *Philosophia Naturalis Principia Mathematica* (1687) and the *Opticks* (1704) have been reissued (in English) more than once during the present century, the standard edition of Newton's writings is still the *Opera Omnia* collected by Samuel Horsley and published between 1779 and 1785. As for the immense mass of manuscript material that remained after Newton's death, although the most important part of it has been accessible in the Cambridge University Library for many years it has been little explored by scholars, even by the biographers of Newton.

Now, at last, Newton's correspondence is being collected and published for the first time in full. But there is still slight promise that England will honour the memory of its finest intellect in the way that Italy has honoured Galileo, France Descartes, Holland Huygens, or America Franklin. No complete publication of his writings is contemplated. As a very small step towards such a monument, and in the hope of encouraging others by making the Portsmouth Collection in the Cambridge University Library better known, we have selected for this volume a few documents from the great, still imperfectly sorted, number that it contains.

In making this selection we have been chiefly interested in tracing the development of Newton's ideas on the nature of matter; this was the starting-point from which the volume grew. In reading Newton's thoughts on this subject one penetrates as profoundly as may be into his comprehension of the ultimate structure of the physical universe, into the shadowy realm where physics, metaphysics and theology overlap in Newton's mind. We make no apology, however, for venturing beyond this theme in order to include some papers on mathematics and other topics. Far more remains untouched: we have printed nothing from the several working notebooks extant, nothing of Newton's chemistry, nor of his optics, nothing from his drafts on the controversies in which he was involved, nor from his personal papers. In particular we should draw attention to the volumes of mathematical papers, and those concerned with the writing and revision of the *Principia*; these eminently deserve further study, and at least partial publication.

Possibly it would be futile to print the whole of the Portsmouth Collection because, in spite of rumours reported by Newton's biographers, nearly all his papers have been preserved, and he kept so much. (The scattering of documents that occurred as a result of the Lymington sale in 1936 has been much less significant than many people imagine.) One finds, for instance, a copy of the following satirical advertisement of a pamphlet, naturally covered with Newton's indefatigable handwriting on the back:

A philosophical Essay upon Actions on distant substances, clearly explicating according to the Principles of the new Philosophy & Sr Isaac Newton's Laws of Motion all to actions [*sic*] usually attributed to Sympathy & Antipathy; as Taliacolius's nose in Hudibras, how it happened to fall of from the Gentleman's Face at Brussels when the Porter that owned it died 500 miles off at Boulogne? How Mothers mark their infants? with a receipt for to prevent Hare lips. Why when one person yawnes others do the same? Whence comes the aversion in many persons to Eat cheese &c? Why dogs bark at beggars, why whistling makes horses stale? Why corns shoot against change of weather. With an explication of the Loadstone, Amber, Jet, Glass and other Electric bodies, sympathetic powder, Amulets, Cramp rings, &c. Dedicated to ye Royal Society Tis given gratis at the Milliners shop right against the three Tuns in Cary-street, facing Lincoln's Inn fields Play-house Gate.

There is much else of trivial significance—pages of laborious computation whose purpose is indiscernible, detached sentences noting the opinion of some early Father on a doctrinal question, drafts in four or five different versions of passages in Newton's printed works—affording the searcher no excitement of discovery. But there is much else that does, and that deserves to be printed. In this volume we offer no more than a sample.

Newton's papers are not always easy to read, nor in perfect condition, nor complete. In our selection we have had to leave many lacunae, and fit some passages together without entire coherence. We have endeavoured to reproduce the documents as accurately as possible except that, for the most part, we have silently omitted words, sentences and occasionally whole paragraphs that were struck out by Newton and rewritten. In a few cases where a cancelled draft included matter of interest we have reproduced it. Contracted forms commonly used in Latin have been expanded without comment in the transcriptions, and we have not endeavoured to correct errors in Newton's Latin. Square brackets have been used occasionally where the sense seemed to require a different word from that written. Of course, no reproduction of a document other than a facsimile can be perfect: apart from the inevitable errors in transcription that have escaped our care, any printed version must appear considerably more neat and uniform than Newton's hurried script.

The translations of papers in Latin are our own, aided here and there by reference to Andrew Motte's English rendering of the *Principia*. Motte is a good guide to the seventeenth-century meaning of a Latin word not used in its classical sense, or not occurring in classical Latin. In some cases we have translated such a word by assuming that Newton had in mind the sense of the English one derived from it. We have preferred to use phrases that Newton either did use elsewhere, or that he might have used had he written in English rather than in Latin. In the translations we have aimed rather at literalness than elegance; to make a truly readable modern English version of Newton's heavy Latin prose, if that were at all possible, would involve so rewriting his sentences that all the spirit of the original would be lost. Newton thought in Latin as easily as in English, not infrequently slipping from one language to the other in the same note (see Plate II); hence the form of his argument and of his thinking was to some extent shaped by the structure of Latin prose. It would be frivolous to modify this more than need be.

The frontispiece is reproduced by gracious permission of Her Majesty the Queen from the portrait in Kensington Palace. We are grateful to the Syndics of the Cambridge University Press for undertaking publication of this complex book, and to the Press which has produced it so excellently.

PREFACE

Our thanks are also due to Mr J. B. Hall, of St John's College, for his care in checking both the transcription and the translation of the Latin manuscripts, which has saved us many mistakes.

A. R. H.

M. B. H.

UNIVERSITY OF CALIFORNIA
LOS ANGELES
August 1960

A NOTE ON THE HISTORY OF
NEWTON'S MANUSCRIPTS

Newton's niece, Catherine Barton, married (1717) John Conduitt, who succeeded Newton as Master of the Mint. Their only child, also named Catherine, married (1740) the Hon. John Wallop who became Lord Lymington when his father was created Earl of Portsmouth. Conduitt had the custody of Newton's papers after the latter's death in 1727, and collected many memorials of his predecessor. These papers all remained in the possession of Conduitt's descendants as a single collection for over 150 years. They were explored by Samuel Horsley when he prepared his edition of the works of Newton (1779–85) and were again made available to Sir David Brewster when writing his *Memoirs* (1855).

In 1872 the then Earl of Portsmouth entrusted the whole collection to the University of Cambridge, where its contents were arranged and catalogued by a committee consisting of H. R. Luard, Sir George Stokes, John Couch Adams, and G. D. Liveing. Their catalogue of the Portsmouth Collection, published in 1888, divides the materials into fifteen sections:

I   Mathematics   (MSS. Add. 3958–72)[1]
II   Chemistry   (MSS. Add. 3973–5)
III   Chronology
IV   History
V   Miscellaneous, chiefly theological
VI   Correspondence   (MSS. Add. 3976–86; copies in Add. 4007)
VII   Books   (MSS. Add. 3987–4004)
VIII   Miscellaneous Papers   (MSS. Add. 4005)
IX   Papers concerning Flamsteed   (MSS. Add. 4006)
X   Correspondence concerning Fontenelle's *Eloge*
XI   Conduitt's drafts for a Life of Newton
XII   Papers relating to Newton, after his death
XIII   Papers concerning Newton's family, and the Mint
XIV   Books and Papers not by Newton
XV   Complimentary Letters

[1] The numbers in brackets are the present shelf-marks of the Newton papers in the Cambridge University Library.

When the work of the committee was done, the Earl of Portsmouth presented the scientific part of the collection to the University of Cambridge, that is, the papers catalogued in Sections I and II, and VI to IX above: they remain in the University Library. The remainder of the Collection was returned to Lord Portsmouth, including a considerable number of original letters of which, however, copies had been made.

It was this latter portion of the Collection that was offered for sale by Messrs Sotheby's of London in 1936, on the instructions of Viscount Lymington. It was composed largely of Newton's annotations from alchemical authors, of his theological extracts and writings, and of the Conduitt and other personal papers. There was also a mass of material on Newton's administration at the Royal Mint, which was presented to the Mint by Viscount Wakefield. Lord Keynes was the principal buyer at this sale; his acquisitions (apart from some gifts he made to Trinity College) are now in the Library of King's College, Cambridge. Thanks to his energetic and generous actions the complete dispersion of Newton's papers was prevented, and although many were scattered, a large part of the original Portsmouth Collection can now be consulted in Cambridge, at the University Library and King's College.

NEWTON MANUSCRIPTS IN THE CAMBRIDGE UNIVERSITY LIBRARY PRINTED IN THIS VOLUME

ABBREVIATIONS

BIRCH, *Boyle*. Thomas Birch, *The Life and Works of the Honourable Robert Boyle* (two editions: London, 1744, 1772).

BIRCH, *History*. Thomas Birch, *History of the Royal Society*. London, 1756–7.

BREWSTER. Sir David Brewster, *Memoirs of the Life, Writings and Discoveries of Sir Isaac Newton*. Edinburgh, 1855.

CAJORI. *Sir Isaac Newton's Mathematical Principles of Natural Philosophy and his System of the World. Translated into English by Andrew Motte in 1729. The translation revised...by Florian Cajori*. Berkeley, 1946.

EDLESTON. Joseph Edleston, *Correspondence of Sir Isaac Newton and Professor Cotes*. London, 1850.

Opticks. Sir Isaac Newton, *Opticks*. (Fifth edition), London, 1931.

Papers & Letters. I. Bernard Cohen (ed.). *Isaac Newton's Papers & Letters on Natural Philosophy*. Cambridge, Mass., 1958.

Portsmouth Catalogue. *A Catalogue of the Portsmouth Collection of Books and Papers written by or belonging to Sir Isaac Newton*. Cambridge, 1888.

Principia. Is. Newton, *Philosophiae Naturalis Principia Mathematica*. London, 1687. (Facsimile reprint issued by William Dawson & Sons Ltd, London, n.d.) All references, unless otherwise stated, are to this edition.

RIGAUD. S. P. Rigaud, *An Historical Essay on the First Publication of Sir I. Newton's Principia*. Oxford, 1838.

ROUSE BALL. W. W. Rouse Ball, *An Essay on Newton's 'Principia'*. London, 1893.

PLATE II

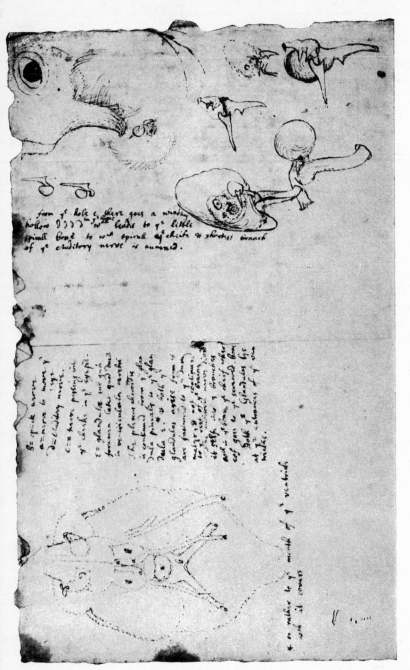

from yᵉ hole c, there goes a winding
hollow ƆƆƆƆ wᶜʰ leads to yᵉ little
spirall bone to wᶜʰ spirall aᵖᶜheite & shortes branch
of yᵉ auditory nerve is annexed.

Anatomical drawings by Newton. *Left*: the brain, probably of a small rodent. *Right*: the eye of a bird, and the bones of the ear. TEXT, *left*: 'a, nerve to move the eye; b, optic nerve; c, a nerve passing into the cheek in the eye-pit; d, auditory nerve; e, glandule beneath which the foramen lies which leads into the ventricles of the brain. The *plexus choroides* is continued from the pineal glandule to the glandule e or rather to the mouth of the ventricle which it covers. Glandules arise from and are fastened to the *dura mater*, and are not continued to the rest of the brain. The auditory nerve divides itself into 3 branches within the bone, the chief whereof goes to the screwed bone. Both the glandules lie at the entrances of the ventricles.' TEXT, *right*: 'from the hole c there goes a winding hollow *dddd* in the bone which leads to the little spirall bone to which spirall the chief and shortest branch of the Auditory nerve is annexed.' On the same sheet is a draft receipt to be signed by Hannah Smith of Woolsthorpe (Newton's half-sister), dated 30 October 1665.

PART I
MATHEMATICS

INTRODUCTION

There are many hundreds of sheets in the Portsmouth Collection relating to pure and applied mathematics, and much material on the controversy over the invention of fluxions and the differential calculus. In pure mathematics there are annotations made by Newton as a student from the authors he read, which led to his own earliest discoveries; fragmentary studies and propositions on various subjects; and drafts of more or less complete works, some of them later incorporated in Newton's published mathematical writings. These latter are more directly derived, however, from the lectures delivered between 1673 and 1683, copies of which are preserved in the Cambridge University Library. This enormous mass of historical material has been largely unexplored hitherto, and existing accounts of Newton's mathematical development have been based almost entirely on published sources.

Although our main interest in making selections for this volume has been in Newton as a theoretical physicist, it seemed appropriate to include one example of his early mathematical work. We have therefore chosen a reasonably complete and coherent tract, which is certainly early in date, and which has the further merit of being in English (No. 1, below). After this, we give an example of a draft in applied mathematics, which shows Newton using fluxions in solving the type of problem treated in the *Principia* (No. 2).

This is not the place to attempt a complete history of Newton's mathematical discoveries and methods based upon the manuscripts, an overdue work of scholarship which we must leave to others; we shall only sketch the development of some of his work. Newton began his more advanced mathematical study by reading Descartes' *Geometry* (1637) and William Oughtred's *Clavis Mathematicae* (1631), in the summer of 1664, and continued with the 'miscellanies' of Frans van Schooten[1] and the publications of John Wallis,[2] of which he made himself master during the winter of 1664–5. These are the authorities he names himself; he certainly read others and

---

[1] Presumably *Exercitationum Mathematicarum Libri quinque*, 1657.
[2] Chiefly *Arithmetica Infinitorum*, 1655.

3

profited as well from the geometrical teaching of Isaac Barrow, then Lucasian Professor. It is clear that Newton's reading as an undergraduate took him to the frontier of knowledge in mathematics, and also that his own original discoveries were a product of continuous development from what had been done before. The first of these discoveries seems to date from the early months of 1665 and resulted from his method of expressing quantities by means of infinite series: he discovered a method of drawing a tangent to a curve at any point, of determining the radius of curvature at any point, and of calculating an area bounded by a curve. By the summer of 1665 he had accomplished the quadrature of the hyperbola.

At first Newton used no algebraic concept or notation unfamiliar to contemporary mathematicians. He was relatively slow to formulate exactly the new concepts latent in his methods, and slower still to devise a new notation to express them, or a system involving them. His problem—one common to other contemporary mathematicians—was to handle continuously varying quantities; for example, in the parabola $y = ax^2$, the height of the ordinate y varies continuously with the value of x; moreover, the rate of change of y (for any given change in x) is dependent on the value of x, and the area enclosed under the curve is in turn dependent upon this rate of change. 'Rate of change' implies that the change takes place with respect to time, and indeed Newton thought of a curve as the line produced by a point moving with an arbitrary velocity, so that the co-ordinates x and y of the point also change with a corresponding and varying velocity, dependent upon the equation to the curve. Similarly, the algebraic expressions denoting the slope of the tangent to the curve at the point, the area enclosed under it and so forth have their appropriate velocities of change. Just as the distance moved in a straight line by a point is calculated as a product of velocities and times, so Newton calculated such algebraic expressions from their velocities of change, themselves derived from the rates of change of the co-ordinates.

His basic ideas about changing quantities may be formulated in his own first published description, from the *Principia*:

Quantities of this kind are products, quotients, roots, rectangles, squares, cubes, square and cubic sides and the like. These quantities

I here consider as variable and indetermined, and increasing or decreasing as it were by a continual motion or flow (*flux*): and I understand their momentary increments or decrements by the name of *moments*: so that the increments may be esteemed as added or positive *moments*, and the decrements as subtracted or negative ones. [Thus, if *Y* varies with *X*, and the *moment* of *X* be *x* and the corresponding *moment* of *Y* be *y*, the *moment* of the rectangle *XY* is $Xy + xY$, the small term xy being negligible.] But take care lest you regard *moments* as little finite quantities (*particulas finitas*). As soon as *moments* become finite quantities they cease to be *moments*. For to be finite is in some way contrary to their perpetual increase or decrease. *Moments* are to be understood as the just nascent beginnings of finite quantities.... It will be the same thing if, instead of *moments*, we use either the velocities of the increments and decrements (which may also be called the motions, mutations, and fluxions of quantities), or any finite quantities proportional to these velocities.[1]

From the above it follows, for instance, that (if the *moment* of *X* be called *x*) the *moment* of aX^n is $anxX^{n-1}$: that is, *x* here corresponds to *dX* in Leibniz's notation for the calculus, and (since $[d(aX^n)]/dx = anX^{n-1}$), the *moment* of aX^n is $d(aX^n)\,dX$. The velocity of the increment or *moment x* therefore, which Newton later called its fluxion and denoted by $\dot{X}$, is dX/dt, but where the velocities of the *moments* are directly comparable, $\dot{v}$ can be taken as *dv*, and $\dot{y}$ as *dy*, etc.

Newton seems to have developed the method of fluxions during the summer of 1665, and to have begun illustrating it by application to particular problems in the autumn. The paper reproduced here (No. 1) was written in October 1666, and appears to be Newton's most complete exposition of his methods up to that time. It was briefly described by Brewster, but has never been printed.[2] It is written on twenty-four sides of eight sheets of paper, folded and stitched down the middle to make thirty-two pages. The handwriting and the peculiarities of spelling are typical of Newton's early manuscripts, but this one is less heavily corrected than many, no doubt because it was not his first attempt to draft these topics.

The tract opens with eight propositions introducing the study of equations: Newton considers the terms of an equation as geometrically formulated, being expressions of the relations

---

[1] Book II, Lemma ii, 250; Cajori, 249.
[2] Brewster, II, 12–14.

between lines described by points moving at different velocities for the same period of time. (The velocity is, in fact, the fluxion of the term.) The seventh proposition explains a method which is virtually differentiation; the eighth, the reverse process of integration.[1] Next (fol. 52v) Newton turns to the demonstration of these propositions, beginning with Proposition 7; Proposition 8 'is the converse of this seventh prop. and may be therefore analytically demonstrated by it'; next, Propositions 1 and 2 are proved. There are no demonstrations of Propositions 3–6. The rest of the paper consists of the application of the propositions to seventeen problems, each illustrated by numerous examples. The problems are:

1. To draw tangents to crooked lines.
2. To find the quantity of crookedness of lines.
3. To find the points distinguishing betwixt the concave and convex points of crooked lines.
4. To find the points at which lines are most or least crooked.
5. To find the nature of the crooked line whose area is expressed by any given equation.
6. The nature of any crooked line being given, to find other lines whose areas may be compared to the area of the given line.
7. The nature of any crooked line being given, to find its area, when it may be [done]. Or more generally, two crooked lines being given, to find the relation of their areas, when it may be [done].
[8. This number is omitted by mistake; Newton left a space, and deliberately altered the next number from 8 to 9.]
9. To find such crooked lines whose lengths may be found, and also to find their lengths.
10. Any curve line being given, to find other lines whose lengths may be compared to its length or its area, and to compare them.
11. To find curve lines whose areas shall be equal (or have any other given relation) to the length of any given curve line drawn into a given right line.[2]
12. To find the length of any given crooked line, when it may be done.
13. To find the nature of a crooked line whose length is expressed by any given equation (when it may be done).

After these purely mathematical problems there are five more concerned with centres of gravity, using the same methods

---

[1] The mathematical use of the words 'differentiate' and 'integrate' is of course post-Newtonian.　　　　[2] Problem 11 is given after Problem 12.

as before.[1] They are preceded by two definitions and two lemmas.

[ ]. To find the centre of gravity in rectilinear plane figures.

[ ]. To find such plane figures which are equiponderate to a given plane figure in respect of an axis of gravity in any given position.

15. To find the gravity of any given plane in respect of any axis, given its position.

16. To find the areas of gravity of any planes.

17. To find the centre of gravity of any plane, when it may be [done].

For mathematical theory the seventh and eighth propositions are the most interesting and important. In the seventh proposition Newton shows how the 'relation of the velocities' (p, q, r) of varying quantities (x, y, z) may be obtained. p, q, and r are in fact the fluxions of x, y, and z, but the term fluxion is never used in this tract and moreover p, q, and r are usually stated as ratios. To derive the ratios containing p, q and r, Newton multiplies each term containing x^n by np/x, each term containing y^n by nq/y and so on, throughout the equation. Or as he otherwise expresses it, he multiplies the terms of the equation by the appropriate term in the series

$$\frac{3p}{x}, \ \frac{2p}{x}, \ \frac{p}{x}, \ 0, \ -\frac{p}{x}, \ -\frac{2p}{x}, \ -\frac{3p}{x} \quad \text{or} \quad \frac{3q}{y}, \ \frac{2q}{y}, \ \frac{q}{y}, \ \dots$$

Then he adds the two or more products thus obtained, gaining a new equation which yields the relation of p, q, and r (or $\dot{x}$, $\dot{y}$ and $\dot{z}$).

For example, if the given equation were

$$x^4 + 2bx^3 + (b^2 + 2)\, x^2 + 2bx + 1 - y^2 = 0,$$

from the x-terms

$$4x^3 p + 6bx^2 p + (2b^2 + 4)\, xp + 2bp,$$

and the y-terms, $\qquad -2yq,$

adding, $\qquad p(4x^3 + 6bx^2 + (2b^2 + 4)\, x + 2b) = 2yq.$

Hence, $\qquad \dfrac{q}{p} = \dfrac{2x^3 + 3bx^2 + (b^2 + 2)\, x + b}{y},$

---

[1] These additional problems were not listed by Brewster.

from which it easily follows that $q/p = 2x + b$, which is the relation between $\dot{y}$ and $\dot{x}$ obtainable from the given equation.

More generally, however, Newton says that p, q, r, etc., are to be obtained by multiplying the terms by the series

$$\frac{ap + 4bp}{x}, \quad \frac{ap + 3bp}{x}, \quad \frac{ap + 2bp}{x}, \quad \text{etc.,}$$

and proceeding as before, a and b signifying two numbers whether rational or irrational.

J. M. Child, who was apparently ignorant of Newton's manuscripts, considered that differentiation by this method (which follows the modern principles) was 'the firstfruits of independent work...the first great step that Newton took, working on his own original lines'.[1] He attributed this step to 1672; in fact, there is evidence throughout this tract that Newton used the method of differentiation by multiplication of the terms of an equation by the terms of a series in 1666. This tract thus shows that historians of mathematics who based their opinion entirely on printed sources have been unduly sceptical in their reaction to the accounts of Newton's mathematical discoveries written by those few nineteenth-century historians who consulted Newton's unpublished papers.

The reverse problem to differentiation, the 'inverse method of fluxions' or integration, is inevitably more complex. Could integration always be done, wrote Newton, all problems whatever might be resolved; but at least by the following rules it might be very often done. As he puts it, if an equation is given expressing the relation between x and the ratio p/q of the velocities of change of x and y (that is, of the fluxions $\dot{x}$ and $\dot{y}$), the problem is to find y. The first result is simple: if $q/p = ax^{m/n}$, then $y = [an/(m+n)]\, x^{(m+n)/n}$, which is the converse of the earlier rule for differentiation. Newton then discusses the integral of x^{-1}, which according to the rule should be $x^0/0$, a meaningless expression. He argues, rather obscurely, that in such a case y is a logarithm (or power) of x, and hence that if x is given, y may be found from a table of (natural) logarithms.

Newton then goes on to discuss the reduction of the fraction q/p, where the denominator of the expression has more than

[1] J. M. Child, 'Newton and the Art of Discovery', in W. J. Greenstreet, *Isaac Newton 1642–1727*, London, 1927, 125. (The equations quoted by Child are misprinted.)

one term, so that it can be integrated. He consolidates his methods in tables of integrals (fol. 50r and v). In explanation, apparently, of these latter arguments, he notes that the velocities

$$\begin{cases} a/(b+cz) \\ \sqrt{(az+bzz)} \\ \sqrt{(a+bzz)} \end{cases} \text{ are to the integrals of } \begin{cases} a/(b+cz) \\ \sqrt{(az+bzz)} \\ \sqrt{(a+bzz)} \end{cases}$$

as the ordinate bc in a conic section is to the area enclosed by it under the curve, z representing the axis ab (Fig. 1). Such quadratures he has supposed to be known, since they are derivable from tables of logarithms and trigonometric functions. Then, finally, Newton says that all equations may be integrated mechanically, by forming them into a

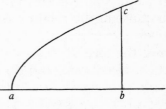

Fig. 1

series (by division or extraction of roots or powers) and then integrating the successive terms. Thus,

$$\frac{a}{b+cx} = \frac{a}{b}\left(1 - \frac{cx}{b} + \frac{c^2x^2}{b^2} - \frac{c^3x^3}{b^3} + \cdots\right),$$

of which the integral is

$$\frac{a}{b}\left(x - \frac{cx^2}{2b} + \frac{c^2x^3}{3b^2} - \frac{c^3x^4}{4b^3} + \cdots\right).$$

At this stage Newton introduces a special symbol for an integral (he has so far used none for the derivative). He gives the integral of $x^2/(ax+b)$ as

$$\frac{x^2}{2a} - \frac{bx}{a^2} + \square\frac{b^2}{a^3x+a^2b} = y,$$

the last term signifying that part of the value of y which is correspondent to the same term in the expansion of $x^2/(ax+b)$ by division; that is, $\square$ is an integral sign, equivalent to $\int$. This sign appears often in Newton's table of integrals: thus,

$$\square\frac{c}{2b}\left(\frac{1}{a+bx^2}\right) = y$$

9

is the integral of $cx/(a+bx^2)$. The terms preceded by the symbol $\square$ are those that have to be derived from tables, but Newton does not explain in detail how this was to be done, or how some of the integrals he lists are derived.

Newton's interpretation of a fluxion (fol. 52v)—though he does not give it that name—in the demonstration to Proposition 7 may be rendered as follows. Let x be the distance moved by a body in a finite time, and y that moved by a second body in the same time, y being a function of x. Let their velocities [that is fluxions] at any moment be p/q; then, though p and q are not constant, the ratio p/q is constant, because of the relation between x and y. Further, if the first body moves the infinitesimal distance po in the brief instant o,[1] and the second the distance qo, $(x+po)$ and $(y+qo)$ may be substituted for x and y in the former equation, $y = f(x)$, the ratio between these combined quantities being the same as the ratio between x and y. Hence, if the equation be the irrational one, $y = 2\sqrt{(a+bx)}/b$, we may write instead of

$$\frac{4a}{b^2}+\frac{4x}{b}-y^2 = 0,$$

$$\frac{4a}{b^2}+\frac{4x}{b}+\frac{4po}{b}-y^2-2yqo-q^2o^2 = 0.$$

Subtracting the known terms of the equation, dividing by o, and omitting any remaining terms in o, we have

$$\frac{4p}{b}-2yq = 0,$$

whence
$$\frac{q}{p}=\frac{2}{by}.[2]$$

This is the basic process for finding the velocity of a changing quantity, giving rise to the rules stated in Proposition 7; we should now write dx for p, and dt for o.

On fol. 55r, Newton introduces another new notation, referring to series of terms arranged in an equation. This

---

[1] o is, of course, not a cipher, but a letter.

[2] Putting dy for q, and dx for p, and substituting for y, gives the differential equation

$$\frac{dy}{dx} = \frac{1}{\sqrt{(a+bx)}}.$$

notation is not that of the later fluxional calculus. In it he writes

$\mathfrak{X}$ for $f(x)$, where each term containing x has been multiplied by the power of x in it;

$\mathfrak{X}$ for $f(x)$, where each term containing y has been multiplied by the power of y in it;

$\mathfrak{X}$ for $f(x)$, where each term containing x has been multiplied by $(2n+1)$, n being the power of x in it;

$\mathfrak{X}$ for $f(x)$, where each term containing y has been multiplied by $(2n+1)$, n being the power of y in it;

$\mathfrak{X}$ for $f(x)$, where each term has been multiplied by $(n+1)$, n being the power of x or y in it.

Evidently he later decided that this was not a convenient notation, and abandoned it. In general, one may say that this tract shows that in 1666 Newton possessed methods of great power, but that these methods involved clumsy and laborious processes, especially in their application to well-known problems of the period, such as tangents and quadratures.

Later examples of Newton's use of fluxions in problems of pure mathematics are fairly common in the Portsmouth Collection; many occur in draft passages later reproduced in his mathematical works as printed during the eighteenth century. While drafts of *Principia* propositions in the synthetic geometrical form are frequently found, examples of the use of fluxions for solving such problems are rather rare. Yet it has long been understood that Newton tackled many problems by fluxional analysis before devising the synthetic proof of the truth of his solution. As early as 1715, in a review of the *Commercium Epistolicum* in the *Philosophical Transactions* inspired or perhaps actually written by Newton himself, it was alleged,

By the help of the New Analysis Mr. Newton found out most of the Propositions in his *Principia Philosophiae* but because the ancients for making things certain admitted nothing into geometry before it was demonstrated synthetically, he demonstrated the propositions synthetically that the system of the heavens might be founded on good geometry. And this makes it now difficult for unskilful men to see the Analysis by which these propositions were found out.[1]

---

[1] *Phil. Trans.* 1715, 206. 'An Account of a Book called *Commercium Epistolicum*.'

An instance of Newton's use of the fluxional calculus in preparing the *Principia* was published in the *Catalogue of the Portsmouth Collection* in 1888,[1] and has been reproduced (in modernized notation) by Cajori.[2] In the Scholium following Proposition XXXV, Book II, Newton defines the surface of revolution *CFGB*, limited by the radii *CO*, *FD* and the axis *OD*, which shall be least resisted by a medium when moving in the direction *OD* (Fig. 2). Apparently he failed to construct a synthetic proof of his definition, for it is given without proof in the *Principia*; the proof by fluxions was given in a letter of 1694 addressed to David Gregory.

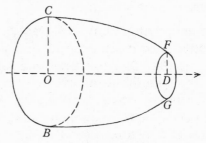

Fig. 2

It would be unwise to generalize excessively from this evidence. The statement in the *Philosophical Transactions* is *parti pris*: it was to Newton's interest, in the dispute with the adherents of Leibniz, to insist on his constant, elaborate and successful use of fluxions before 1687. The letter of 1694 justifies this insistence to a certain extent: it showed that fluxions could be a powerful tool in Newton's hands though it did not prove that he made great use of this tool. Indeed there was no need for him to use a steam-hammer to crack nuts. His mastery of the geometrical form employed in the *Principia* was undoubtedly so great that he could extract most of the results he required directly; and since he had resolved to publish in the geometrical form, double labour could be avoided by using geometrical methods of proof in the first instance. It is practically certain, for example, that the propositions on central forces and elliptical motion were established in just the form in which Newton ultimately communicated and printed

[1] Pp. xxi–xxiii.   [2] Pp. 657–61, and references there cited.

them. There are, to be sure, drafts of calculations on elliptical orbits using fluxions, but they seem to have been abandoned as abortive. It seems reasonable, therefore, in view of the scant evidence of the use of fluxions in mechanical propositions among his papers, to suppose that Newton only resorted to fluxions where geometrical methods failed him, and did not regard them as his first offensive weapon.

The example given below (No. 2) of this relatively rare use of fluxions concerns the determination of the centrifugal force towards any point C within any curved orbit, of which the nature is known. This problem was not treated in the first

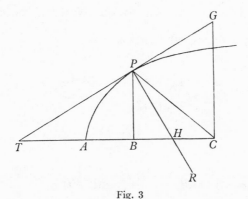

Fig. 3

edition of the *Principia*, even in the more limited case where the orbit is a conic section. In the second edition, the Scholium after Proposition X, Book I, was extended, and a new Scholium after Proposition XVII was added (pp. 47, 58). The addition is the first paragraph of the treatment in No. 2, below, and the new Scholium the second paragraph. Hence, the draft may have been composed at any time between 1687 and 1713.

The draft goes further than the printed version, for here Newton goes on to consider cases where the equation to the orbit does not necessarily involve x^2. He uses two slightly different methods, and two sample equations $ax^n = y^{n+1}$, and $ax = v^n$. His reasoning is as follows.

Let the radius of curvature at any point P of the orbit be PR, so that PR is then normal to the tangent TPG (Fig. 3). The centripetal force towards R in the orbit, at P, is 'the same as that in a circle of the same curvature' (Prop. VII, Cor. 3), that is, as $\dfrac{1}{PR}$.[1] Further, from the same corollary, the centripetal force towards the point C is to the centripetal force towards R as $PC.PR^2$ is to CG^3. Consequently the force towards C is as $\dfrac{PC.PR}{CG^3}$, that is, as $\dfrac{PC}{CG^3}$ multiplied by the radius of curvature of the orbit at C.

To determine the radius of curvature, Newton calculates the fluxion of the fluxion of the equation (indicated by two dots above the algebraic symbol), that is, the second derivative.[2] Taking the second equation, $ax = v^n$, and putting $\dot{x} = 1$, then $\dot{v} = \dfrac{a}{nv^{n-1}}$ and hence $a = n.v^{n-1}\dot{v}$. Newton now finds the fluxion of this expression, treating nv^{n-1} and $\dot{v}$ as distinct functions of v, thus deriving:

$$n(n-1)\,v^{n-2}\dot{v}^2 + \ddot{v}.nv^{n-1} = 0.$$

Hence,
$$(n-1)\,\dot{v}^2 + v.\ddot{v} = 0,$$

and
$$\ddot{v} = -\frac{(n-1)\,\dot{v}^2}{v}.$$

Now, since $\dot{x} = 1$ and $\dfrac{\dot{x}}{\dot{v}} = \dfrac{TB}{PB}$ and $PB = v$, $\dfrac{\dot{v}^2}{v} = \dfrac{PB}{TB^2}$, and therefore the centripetal force is as $\ddot{v}\dfrac{PC}{CG^3}$, or as $(n-1)\dfrac{PB}{TB^2}.\dfrac{PC}{CG^3}$, which Newton concludes may be reduced to $\dfrac{PC}{BH.TC^3}$.

---

[1] Proposition VII is made more general in the second edition; to determine the force towards any point in a circular orbit, not merely a point on the circumference. The corollaries were also added to this edition.

[2] Following Newton's assumptions, and making $\dot{x} = \ddot{x} = 1$, $\dfrac{\dot{v}}{\dot{x}} = \dfrac{1}{n}a^{\frac{1}{n}}x^{\frac{1-n}{n}}$ and $\dfrac{\ddot{v}}{\ddot{x}} = \dfrac{1}{n}\left(\dfrac{1-n}{n}\right)a^{\frac{1}{n}}x^{\frac{1-2n}{n}}$. Substituting for x, $\dfrac{\ddot{v}}{\ddot{x}} = \dfrac{(1-n)\,a^2}{n^2v^{2n-1}}$. But $\dfrac{\dot{v}^2}{v} = \dfrac{a^2}{n^2v^{2n-1}}$, hence $\ddot{v} = -\dfrac{(n-1)\,\dot{v}^2}{v}$.

1

TO RESOLVE PROBLEMS BY MOTION

MS. Add. 3958, fols. 49–63

This tract consists of eight sheets of paper, folded and stitched to make a booklet of 32 pages, of which 24 pages are written upon. It is mentioned in the *Portsmouth Catalogue* (Section I, p. 1, no. 3) along with another version in Latin entitled 'De Solutione Problematum per Motum'. Certain passages have been left incomplete: one example was abandoned because Newton found the computation too tedious. The date 'October 1666' is written at the head of the first page, but in a different ink from the opening lines and may have been added later.

This is probably Newton's first attempt at a finished exposition of new mathematical procedures, though there is earlier evidence of them in a notebook (MS. Add. 4000) where the famous calculation of a hyperbolic area to 52 places of decimals is to be found.

We have attempted to reproduce Newton's freehand sketches in the margin of the manuscript as accurately as possible.

October 1666

To resolve Problems by Motion these following Propositions are sufficient

1. If the body a in the Perimeter of ye cirkle or sphaere *adce* moveth towards its center b, its velocity to each point $(d, c, e,)$ of yt circumference is as ye chords (ad, ac, ae) drawne from that body to those points are [Fig. 4].

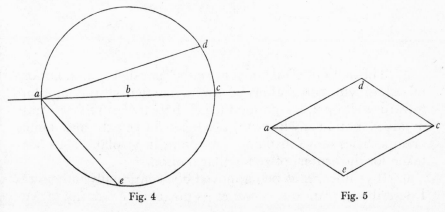

Fig. 4   Fig. 5

2. If ye △s *adc, aec*, are alike viz. *ad = ec* &c (though in divers plaines) & 3 bodys move from the point *a* uniformely & in equall times ye first to *d*, the 2d to *e*, ye 3d to *c* [Fig. 5]; Then is the thirds motion compounded of ye motions of the first & second.

3. All the points of a Body keeping Parallel to it selfe are in equall velocity.

4. If a body move onely $\left(\begin{array}{c}\text{angularly}\\\text{circularly}\end{array}\right)$ about some axis, ye velocity of its points are as their distances from that axis.

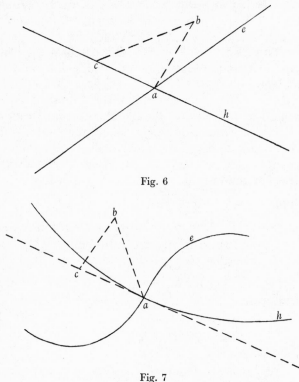

Fig. 6

Fig. 7

5. The motions of all bodys are either parallel or angular, or mixed of ym both, after ye same manner yt the motion towards *c* (Prop 2) is compounded of those towards *d* & *e*. And in mixed motion any line may bee taken for ye axis (or if a line or superficies move in plano, any point in yt plane may bee taken for the center) of ye angular motion.

6. If ye lines *ae, ah* being moved doe continually intersect; I describe ye trapezium *abcd*, & its diagonall *ac*: & say yt, ye

proportion & position of these five lines *ab*, *ad*, *ac*, *cb*, *cd*, being determined by requisite data; shall designe ye proportion & position of these five motions; viz: of ye point *a* fixed in ye line *ae* & moveing towards *b*, of ye point *a* fixed in ye line *ah* & moveing towards *d*; of ye intersection point *a* moveing in ye plaine *abcd* towards *c*, (for those five lines are ever in ye same plaine, though *ae* & *ah* may chanch onely to touch that plaine in their intersection point *a*) [Fig. 6]; of ye intersection point *a* moveing in ye line *ae* parallely to *cb* & according to ye order of ye letters *c*, *b* [Fig. 7]; & of ye intersection point *a* moving in

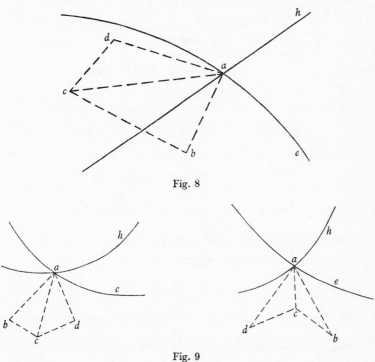

Fig. 8

Fig. 9

ye line *ah* parallely to *cd* & according to ye order of the letters *c*, *d* [Fig. 8].

Note yt one of ye lines as *ah* resting, ye points *d* & *a* are coincident, & ye point *c* shall bee in ye line *ah* if it bee streight, otherwise in its tangent [Fig. 9].

7. Haveing an Equation expressing ye relation twixt two or more lines *x*, *y*, *z* &c: described in ye same time by two or more moveing bodys *A*, *B*, *C*, &c [Fig. 10]: the relation of their

velocitys p, q, r, &c may bee thus found, viz: Set all ye termes on one side of ye Equation that they become equall to nothing. And first multiply each terme by so many times $\dfrac{p}{x}$ as x hath dimensions in yt terme. Secondly multiply each terme by so many times $\dfrac{q}{y}$ as y hath dimensions in it. Thirdly (if there be 3 unknowne quantitys) multiply each terme by so many times $\dfrac{r}{z}$ as z hath dimensions in yt terme.
(& if there bee still more unknowne quantitys doe like to every unknowne quantity). The summe of all these products shall be equall to nothing. Wch Equation gives ye relation of ye velocitys p, q, r, &c.

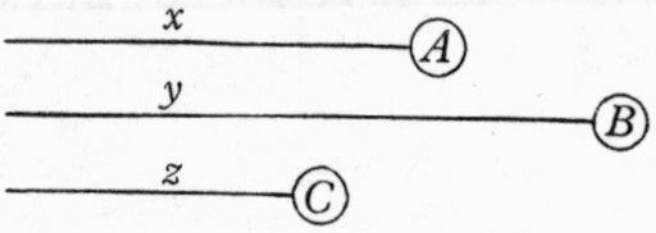

Fig. 10

Or thus. Translate all ye termes to one side of ye Equation, & multiply them being ordered according to x by this progression,

$$\Join \frac{3p}{x} . \frac{2p}{x} . \frac{p}{x} . 0 . \frac{-p}{x} . \frac{-2p}{x} . \frac{-3p}{x} . \frac{-4p}{x} . \&c:$$

& being ordered by ye dimensions of y multiply them by this:

$$\Join . \frac{3q}{y} . \frac{2q}{y} . \frac{q}{y} . 0 . \frac{-q}{y} . \frac{-2q}{y} . \&c.$$

The sume of these products shall bee equall to nothing, which equation gives ye relation of their velocitys p, q, &c.

Or more Generally ye Equation may bee multiplyed by ye terme of these progressions

$$\frac{ap+4bp}{x} . \frac{ap+3bp}{x} . \frac{ap+2bp}{x} . \frac{ap+bp}{x} . \frac{ap}{x} . \frac{ap-bp}{x} . \frac{ap-2bp}{x} . \&c.$$

And

$$\frac{aq+2bq}{y} . \frac{aq+bq}{y} . \frac{aq}{y} . \frac{aq-bq}{y} \&c.$$

(a & b signifying any two numbers whither rationall or irrationall).

8. If two Bodys A & B, by their velocitys p & q describe ye lines x & y. & an Equation bee given expressing ye relation twixt one of ye lines x, & ye ratio $\dfrac{q}{p}$ of their motions q & p; To find ye other line y.

Could this bee ever done all problems whatever might bee resolved. But by ye following rules it may bee very often done. (Note yt $\pm m$ & $\pm n$ are logarithmes or numbers signifying ye dimensions of x).

First get ye valor of $\frac{q}{p}$. Which if it bee rationall & its Denominator consist of but one terme: Multiply yt valor by x & divide each terme of it by ye logarithme of x in yt terme ye quotient shall bee ye valor of y. As if

$$ax^{\frac{m}{n}} = \frac{q}{p}$$

Then is
$$\frac{na}{m+n}x^{\frac{m+n}{n}} = y.$$

Or if
$$ax^{\frac{m}{n}} = \frac{q}{p}$$

Then is
$$\frac{na}{n+m}x^{\frac{n+m}{n}} = y.$$

(Soe if
$$\frac{a}{x} = ax^{\frac{-1}{1}} = \frac{q}{p}$$

Then is
$$\frac{a}{0}x^0 = y.$$

soe yt y is infinite. But note yt in this case x & y increase in ye same proportion yt numbers & their logarithmes doe, y being like a logarithme added to an infinite number $\frac{a}{0}$. But if x bee diminished by c, as if
$$\frac{a}{c+x} = \frac{q}{p};$$

y is also diminished by ye infinite number $\frac{a}{0}c^0$ & becomes finite like a logarithme of ye number x. & so x being given, y may bee mechanically found by a Table of logarithmes, as shall bee hereafter showne.)

Secondly. But if ye denominator of ye valor of $\frac{q}{p}$ consist of more termes yn one, it may bee reduced to such a forme yt ye denominator of each parte of it shall have but one terme, unlesse yt parte bee $\dfrac{a}{c+x}$: Soe yt y may bee yn found by ye precedent rule. Which reduction is thus performed, viz: 1st, If the denominator bee not $a+bx$ nor all its termes multiplyed by x or xx, or x^3 &c: Increase or diminish x untill ye last terme of ye Denominator vanish. 2dly, And when all ye termes in ye Denominator are multiplyed by x, xx, or x^3 &c: Divide ye numerator by ye Denominator (as in Decimall numbers) untill ye Quotient consist of such parts none of whose Denominators are so multiplyed by x, x^2 &c & begin ye Division in those termes in wch x is of its fewest dimensions unlesse ye Denominator be $a+bx$. If yn ye termes in ye valor of $\dfrac{q}{p}$ bee such as was before required ye valor of y may bee found by ye first parte of this Prop: onely it must bee so much diminished or increased as it was before increased or diminished by increasing or diminishing x. But if the denominator of any terme consist of more termes yn one, unlesse yt terme bee $\dfrac{a}{c+x}$: First find those partes of y's valor wch correspond to ye other parts of $\dfrac{q}{p}$ its valor. & yn by ye preceding reductions &c seeke yt parte of y's valor answering to this parte of $\dfrac{q}{p}$ its valor.

Example 1st. If $\dfrac{xx}{ax+b}=\dfrac{q}{p}$.

Then by Division tis

$$\frac{x}{a}-\frac{b}{aa}+\frac{bb}{a^3x+aab}=\frac{xx}{ax+b}=\frac{q}{p}$$

(as may appear by multiplication.) Therefore (by 1st parte of this Prop:) tis

$$\frac{xx}{2a}-\frac{bx}{aa}+\square\,\frac{bb}{a^3x+aab}=y.$$

$\left(\square\,\dfrac{bb}{a^3x+aab}\right.$ signifys yt parte of ye valor of y wch is correspondent to ye terme $\dfrac{bb}{a^3x+aab}$ of ye valor of $\dfrac{q}{p}$, wch well may bee found by a Table of logarithmes as may hereafter appeare.$\Big)$

Example 2d. If $\dfrac{x^3}{aa-xx}=\dfrac{q}{p}$.

I suppose $x=z-a$. Or

$$\frac{z^3-3azz+3aaz-a^3}{2az-zz}=\frac{q}{p}.$$

And by Division $\quad\dfrac{-aa}{2z}-z+a+\dfrac{aa}{4a-2z}=\dfrac{q}{p}$,

(as may appeare by multiplication.) And substituting $x+a$ into ye place of z, tis

$$x-\frac{aa}{2a+2x}+\frac{aa}{2a-2x}=\frac{q}{p}=\frac{x^3}{aa-xx}.$$

And Therefore (by parte 1st of Prop 8)

$$\frac{xx}{2}+\square\frac{-aa}{2x+2a}+\square\frac{aa}{2a-2x}=y.$$

But sometimes the last terme of ye Denominator cannot bee taken away, (as if ye Denominator bee $aa+xx$, or a^4+x^4 or $a^4+bbxx+x^4$ &c). And then it will bee necessary to have in readinesse some examples wth such Denominators to wch all other cases of like denomination may bee by Division reduced. As if

$\dfrac{cx}{a+bxx}=\dfrac{q}{p}$. Make $bxx=z$, Then is $\square\,\dfrac{c}{2ab+2bz}=y$.

$\dfrac{cxx}{a+bx^3}=\dfrac{q}{p}$. Make $bx^3=z$, Then is $\square\,\dfrac{c}{3ba+3bz}=y$.

$\dfrac{cx^3}{a+bx^4}=\dfrac{q}{p}$. Make $bx^4=z$, Then is $\square\,\dfrac{c}{4ba+4bz}=y$ &c.

In Generall if

$\dfrac{cx^{n-1}}{a+bx^n}=\dfrac{q}{p}$. Make $bx^n=z$, & yn is $\square\,\dfrac{c}{nba+nbz}=y$.

Also if $\quad\dfrac{c}{a+bxx}=\dfrac{q}{p}$. Make $\sqrt{\dfrac{c}{a}}-\sqrt{\dfrac{c}{a+bxx}}=z$

& yn is $\quad\dfrac{cx}{a+bxx}+\square\,2\sqrt{\dfrac{2z\sqrt{ac}}{b}-\dfrac{azz}{b}}=y$.

That is, if $$\sqrt{\dfrac{c}{bx}-\dfrac{a}{b}}=\dfrac{q}{p};$$

I make $$x=zz\ \&\ \square\, 2\sqrt{\dfrac{c}{b}-\dfrac{a}{b}}zz=y.$$

Or if $$\dfrac{c}{a+bxx}=\dfrac{q}{p}.$$

Make $$\sqrt{\dfrac{c}{a+bxx}}=z=CB,\quad 2\sqrt{\dfrac{c}{b}-\dfrac{a}{b}}zz=y=BD$$

& $\square=CDV=y$ [Fig. 11].

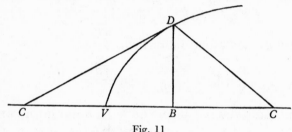

Fig. 11

Thirdly. If ye valor of $\dfrac{q}{p}$ is irrationall being a square roote, The simplest cases may bee reduced to these following examples.

1. If $$\dfrac{cx^n}{x}\sqrt{a+bx^n}=\dfrac{q}{p}.$$

Then $$\dfrac{2ac+2bcx^n}{3nb}\sqrt{a+bx^n}=y.$$

2. If $$\dfrac{cx^{2n}}{x}\sqrt{a+bx^n}=\dfrac{q}{p}.$$

Then $$\dfrac{6bbcx^{2n}+2abcx^n-4aac}{15nbb}\sqrt{a+bx^n}=y.$$

3. If $$\dfrac{cx^{3n}}{x}\sqrt{a+bx^n}=\dfrac{q}{p}.$$

Then

$$\dfrac{30b^3cx^{3n}+6abbcx^{2n}-8aabcx^n+16a^3c}{105nb^3}\sqrt{a+bx^n}=y.$$

4. If
$$\frac{cx^{4n}}{x}\sqrt{a+bx^n}=\frac{q}{p}.$$

Then

$$\frac{210b^4cx^{4n}+30ab^3cx^{3n}-36aabbcx^{2n}+48a^3bcx^n-96a^4c}{945nb^4}\sqrt{a+bx^n}=y.$$

5. If
$$\frac{cx^{5n}}{x}\sqrt{a+bx^n}=\frac{q}{p}.$$

Then

$$\frac{1890b^5cx^{5n}+210ab^4cx^{4n}-240aab^3cx^{3n}+288a^3bbcx^{2n}-384a^4bcx^n+768a^5c}{10395nb^5}\sqrt{a+bx^n}=y.$$

---

1. If
$$\frac{cx^n}{x\sqrt{a+bx^n}}=\frac{q}{p}.$$

Then
$$\frac{2c}{nb}\sqrt{a+bx^n}=y.$$

2. If
$$\frac{cx^{2n}}{x\sqrt{a+bx^n}}=\frac{q}{p}.$$

Then
$$\frac{2bcx^n-4ac}{3nbb}\sqrt{a+bx^n}=y.$$

3. If
$$\frac{cx^{3n}}{x\sqrt{a+bx^n}}=\frac{q}{p}.$$

Then
$$\frac{6bbcx^{2n}-8abcx^n+16aac}{15nb^3}\sqrt{a+bx^n}=y.$$

4. If
$$\frac{cx^{4n}}{x\sqrt{a+bx^n}}=\frac{q}{p}.$$
Then is

$$\frac{30b^3cx^{3n}-36abbcx^{2n}+48aabcx^n-96a^3c}{105nb^4}\sqrt{a+bx^n}=y.$$

5. If
$$\frac{cx^{5n}}{x\sqrt{a+bx^n}}=\frac{q}{p}.$$

Then

$$\frac{210b^4cx^{4n} - 240ab^3cx^{3n} + 288aabbcx^{2n} \atop {}- 384a^3bcx^n + 768a^4c}{945nb^5} \sqrt{a+bx^n} = y.$$

---

1. If
$$\frac{b}{x} - \frac{a}{2x^{n+1}} \sqrt{ax^n + bx^{2n}} = \frac{q}{p}.$$

Then
$$\frac{a+bx^n}{nx^n} \sqrt{ax^n + bx^{2n}} = y.$$

---

1. If
$$\frac{3ax^n + 6bx^{2n}}{x} \sqrt{ax^n + bx^{2n}} = \frac{q}{p}.$$

Then is
$$\frac{2ax^n + 2bx^{2n}}{n} \sqrt{ax^n + bx^{2n}} = y.$$

2. If
$$\frac{-15aax^n + 48bbx^{3n}}{x} \sqrt{ax^n + bx^{2n}} = \frac{q}{p}.$$
Then is
$$\frac{-10a + 12bx^n}{n} \times \overline{ax^n + bx^{2n}} \times \sqrt{ax^n + bx^{2n}} = y$$

3. If
$$\frac{105a^3x^n + 400b^3x^{4n}}{x} \sqrt{ax^n + bx^{2n}} = \frac{q}{p}.$$
Then
$$\frac{70aa - 84abx^n + 96bbx^{2n}}{n} \times \overline{ax^n + bx^{2n}} \times \sqrt{ax^n + bx^{2n}} = y.$$

4. [If]
$$\frac{-945a^4x^n + 5760b^4x^{5n}}{x} \sqrt{ax^n + bx^{2n}} = \frac{q}{p}. \quad -630a^3 + [sic]$$

---

1. If
$$\frac{ax^n + 2bx^{2n}}{x\sqrt{ax^n + bx^{2n}}} = \frac{q}{p}.$$

Then is
$$\frac{2}{n} \sqrt{ax^n + bx^{2n}} = y.$$

24

2. If
$$\frac{3ax^{2n} + 4bx^{3n}}{x\sqrt{ax^n + bx^{2n}}} = \frac{q}{p}.$$

Then is
$$\frac{2x^n}{n}\sqrt{ax^n + bx^{2n}} = y.$$

3. If
$$\frac{15aax^{2n} - 24bbx^{4n}}{x\sqrt{ax^n + bx^{2n}}} = \frac{q}{p}.$$

Then is
$$\frac{10ax^n - 8bx^{2n}}{n}\sqrt{ax^n + bx^{2n}} = y.$$

4. If
$$\frac{105a^3x^{2n} - 192b^3x^{5n}}{x\sqrt{ax^n + bx^{2n}}} = \frac{q}{p}.$$

Then
$$\frac{70aax^n - 56abx^{2n} + 48bbx^{3n}}{n}\sqrt{ax^n + bx^{2n}} = y.$$

---

2. If,
$$\frac{c}{x}\sqrt{ax^n + bx^{2n}} = \frac{q}{p}. \text{ Make } x^n = zz.$$

And yn is
$$\square\,\frac{2c}{n}\sqrt{a + bzz} = y.$$

3. If,
$$\frac{cx^n}{x}\sqrt{ax^n + bx^{2n}} = \frac{q}{p}. \text{ Make } x^n = z.$$

Then is
$$\square\,\frac{c}{n}\sqrt{az + bzz} = y.$$

4. If
$$\frac{cx^{2n}}{x}\sqrt{ax^n + bx^{2n}} = \frac{q}{p}. \text{ Make } x^n = z.$$

Then,
$$\frac{2acx^n + 2bcx^{2n}}{6nb}\sqrt{ax^n + bx^{2n}} - \square\,\frac{ac}{2nb}\sqrt{az + bzz} = y.$$

5. If
$$\frac{cx^{3n}}{x}\sqrt{ax^n + bx^{2n}} = \frac{q}{p}. \text{ Make } x^n = z.$$

Yn is
$$\frac{12bcx^n - 10ac}{48nbb} \times \overline{ax^n + bx^{2n}} \times \sqrt{ax^n + bx^{2n}} + \square\,\frac{5aac}{16nbb}\sqrt{az + bzz} = y.$$

25

1. If $\dfrac{c}{xx^n}\sqrt{ax^n+bx^{2n}}=\dfrac{q}{p}$. Make $x^n=zz$.

Yn is,

$$\dfrac{-2ac-2bcx^n}{nax^n}\sqrt{ax^n+bx^{2n}}+\square\,\dfrac{4bc}{na}\sqrt{a+bzz}=y.$$

---

1. If $\dfrac{cx^n}{x\sqrt{ax^n\pm bx^{2n}}}=\dfrac{q}{p}$. Make $\sqrt{a\pm bx^n}=z$.

Yn is, $\quad\dfrac{-2c}{na}\sqrt{ax^n\pm bx^{2n}}\mp\square\,\dfrac{4c}{na}\sqrt{\dfrac{zz-a}{\pm b}}=y.$

2. If $\dfrac{cx^{2n}}{x\sqrt{ax^n+bx^{2n}}}=\dfrac{q}{p}$. Make $\sqrt{a+bx^n}=z$.

Yn is $\qquad\square\,\dfrac{2c}{nb}\sqrt{\dfrac{zz-a}{b}}=y.$

3. If $\dfrac{cx^{3n}}{x\sqrt{ax^n+bx^{2n}}}=\dfrac{q}{p}$. Make $\sqrt{a+bx^n}=z$.

Yn is $\dfrac{cx^n}{2nb}\sqrt{ax^n+bx^{2n}}-\square\,\dfrac{3ac}{2nbb}\sqrt{\dfrac{zz-a}{b}}=y.$

If $\quad\dfrac{cx^n}{x}\sqrt{a+bx^n+dx^{2n}}=\dfrac{q}{p}$. Make $x^n=z$.

Then is $\qquad\square\,\dfrac{c}{n}\sqrt{a+bz+dz^2}=y.$

If $\quad\dfrac{cx^n\sqrt{d+ex^n}}{x\sqrt{a+bx^n}}=\dfrac{q}{p}$. Make $\sqrt{a+bx^n}=z$.

Then is $\qquad\square\,\dfrac{2c\sqrt{db-ae+ezz}}{nb\sqrt{b}}=y.$

If $\qquad\dfrac{-cx^{2n}\sqrt{a+bx^n}}{x\sqrt{a-3bx^n}}=\dfrac{q}{p}.$

Then is $\dfrac{ac+bcx^n}{6nbb}\sqrt{aa-2abx^n-3bbx^{2n}}=y.$

If $\dfrac{cx^{3n}\sqrt{3a+bx^n}}{x\sqrt{-a+bx^n}}=\dfrac{q}{p}$. Make $\dfrac{2a}{b}+x^n\sqrt{a+bx^n}=z$.

Then is $\qquad \Box\, 2c\dfrac{\sqrt{bbzz-4a^3}}{3nbb}=y$.

1. $\qquad \dfrac{a}{b+cz}:\qquad\qquad \Box\dfrac{a}{b+cz}::$

Note 2. yt $\sqrt{az+bzz}:$ is to $\Box\sqrt{az+bzz}::$

3. $\qquad \sqrt{a+bzz}:\qquad\qquad \Box\sqrt{a+bzz}::$

as ye ordinately applied line bc in some of ye Conick sections: is to its corresponding superficies abc [Fig. 12], ye axis ab being

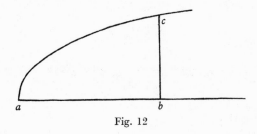

Fig. 12

in like manner related to z. But all those areas (& consequently $\Box\dfrac{a}{b+cz}$, $\Box\sqrt{az+bzz}$, $\Box\sqrt{a+bzz}$) may bee Mechanichally found either by a Table of logarithmes or signes & Tangents. And I have beene therefore hitherto content to suppose ym knowne, as ye basis of most of ye precedent propositions.

Note also yt if ye Valor of $\dfrac{q}{p}$ consists of severall parts each part must bee considered severally, as if:

$$\frac{ax^3-bbxx}{c^4}=\frac{q}{p}.$$

Then is $\qquad \Box\dfrac{ax}{c^4}=\dfrac{ax^4}{4c^4}$ & $\Box\dfrac{bbxx}{c^4}=\dfrac{bbx^3}{3c^4}$

Therefore $\qquad \Box\dfrac{ax^3-bbxx}{c^4}=\dfrac{ax^4}{4c^4}-\dfrac{bbx^3}{3c^4}=y.$

Note also yt if ye denominator of ye valor of $\frac{q}{p}$ consist of both rationall & surde quantitys or of two or more surde quantitys First take those surde quantitys out of ye denominator, & yn seeke (y) by ye precedent theoremes.

But this eighth Proposition may bee ever thus resolved mechanichally. viz: Seeke ye Valor of $\frac{q}{p}$ as if you were resolving ye equation in Decimall numbers either by Division or Extraction of rootes or Vieta's Analyticall resolution of powers; This Operation may bee continued at pleasure, ye farther the better. & from each terme ariseing from this operation may bee deduced a parte of ye valor of y, (by parte ye 1st of this prop).

Example 1. If

$$\frac{a}{b+cx}=\frac{q}{p}.$$

Then by division

$$\frac{q}{p}=\frac{a}{b}-\frac{acx}{bb}+\frac{accxx}{b^3}-\frac{ac^3x^3}{b^4}+\frac{ac^4x^4}{b^5}-\frac{ac^5x^5}{b^6}+\frac{ac^6x^6}{b^7}\&c.$$

And consequently

$$y=\frac{ax}{b}-\frac{acxx}{2bb}+\frac{accx^3}{3b^3}-\frac{ac^3x^4}{4b^4}+\frac{ac^4x^5}{5b^5}\&c.$$

Example 2. If $\qquad \sqrt{aa-xx}=\frac{q}{p}.$

Extract ye roote & 'tis

$$\frac{q}{p}=a-\frac{xx}{2a}-\frac{x^4}{8a^3}-\frac{x^6}{16a^5}-\frac{5x^8}{128a^7}-\frac{7x^{10}}{256a^9}-\frac{21x^{12}}{1024a^{11}}\&c$$

(as may appeare by squaring both parts). Therefore (by 1st parte of Prop 8)

$$y=ax-\frac{x^3}{6a}-\frac{x^5}{40a^3}-\frac{x^7}{112a^5}-\frac{5x^9}{1152a^7}-\frac{7x^{11}}{2816a^9}\&c.$$

Example 3. If $\qquad \frac{q^3}{p^3}*^1-ax\frac{q}{p}-x^3=0$

[1] Newton uses the star symbol to indicate a missing term (here q^2/p^2).

28

☞ Note[1] yt if there happen to bee in any Equation Either a fraction or surde quantity or a Mechanichall one (i:e: wch cannot bee Geometrically computed, but is expressed by ye area or length or gravity or content of some curve line or sollid, &c) To find in what proportion the unknowne quantitys increase or decrease doe thus. 1. Take two letters ye one (as ξ) to signify yt quantity, ye other (as π) its motion of increase or decrease: And making an Equation betwixt ye letter (ξ) & ye quantity signifyed by it, find thereby (by prop 7 if ye quantity bee Geometricall, or by some other meanes if it bee mechanicall) ye valor of ye other letter (π). 2. Then substituting ye letter (ξ) signifying yt quantity, into its place in ye maine Equation esteeme yt letter (ξ) as an unknowne quantity & performe ye worke of seaventh proposition; & into ye resulting Equation instead of those letters ξ & π substitute theire valors. And soe you have ye Equation required.

Example 1. To find p & q ye motions of x & y whose relation is, $yy = x\sqrt{aa-xx}$. first suppose $\xi = \sqrt{aa-xx}$, Or $\xi\xi + xx - aa = 0$. & thereby find π ye motion of ξ, viz: (by prop 7) $2\pi\xi + 2px = 0$.

Or $\dfrac{-px}{\xi} = \pi = \dfrac{-px}{\sqrt{aa-xx}}$. Secondly in ye Equation $yy = x\sqrt{aa-xx}$,

writing ξ in stead of $\sqrt{aa-xx}$, the result is $yy = x\xi$, whereby find ye relation of ye motions p, q, & π: viz (by prop 7) $2qy = p\xi + x\pi$. In wch Equation instead of ξ & π writing theire valors, ye result is,

$$2qy = p\sqrt{aa-xx} - \frac{pxx}{\sqrt{aa-xx}}.$$

wch was required.

[wch equation multiplyed by $\sqrt{aa-xx}$, is

$$2qy\sqrt{aa-xx} = paa - 2pxx.$$

& in stead of $\sqrt{aa-xx}$, writing its valor $\dfrac{yy}{x}$, it is

$$\frac{2qy^3}{x} = paa - 2pxx.$$

Or $2qy^3 = paax - 2pxxx$. Which conclusion will also bee found by taking ye surde quantity out ye given Equation for both

---

[1] These notes are written on fol. 48 v, facing the first page of the tract. As there is nothing to indicate where they belong, they are inserted here, at a natural break.

parts being squared it is $y^4 = aaxx - x^4$. & therefore (by prop 7) $4py^3 = 2qaax - 4x^3$. as before.]

☞ Note also yt it may bee more convenient (setting all ye termes on one side of ye Equation) to put every fractionall, irrationall & mechanicall terme, as also ye summe of ye rationall termes, equall severally to some letter: & then to find ye motions corresponding to each of those letters ye sume of wch motions is ye Equation required.

Example ye 2d. If $x^3 - ayy + \dfrac{by^3}{a+y} - xx\sqrt{ay + xx} = 0$ is ye relation twixt x & y, whose motions p & q are required. I make

$$x^3 - ayy = \tau; \quad \frac{by^3}{a+y} = \phi; \quad \& \quad -xx\sqrt{ay + xx} = \xi.$$

& ye motions of τ, ϕ, & ξ being called β, γ, & δ; ye first Equation $x^3 - ayy = \tau$, gives (by prop 7) $3pxx - 2qay = \beta$. Ye second $by^3 = a\phi + y\phi$, gives $3qbyy = a\gamma + y\gamma + q\phi$; Or

$$\frac{3qbyy - q\phi}{a+y} = \gamma = \frac{3qabyy + 2qby^3}{aa + 2ay + yy}.$$

& ye Third $ayx^4 + x^6 = \xi\xi$ gives

$$qax^4 + 4payx^3 + 6px^5 = 2\delta\xi;$$

Or

$$\frac{-qaxx - 4payx - 6px^3}{2\sqrt{ay + xx}} = \delta.$$

Lastly

$$\beta + \gamma + \delta = 3pxx - 2qay + \frac{3qabyy + 2qby^3}{aa + 2ay + yy}$$

$$-\frac{qaxx - 4payx - 6px^3}{2\sqrt{ay + xx}} = 0,$$

is ye Equation sought.

Example 3d. If $x = ab \perp bc = \sqrt{ax - xx}$. $be = y$. & ye Superficies $abc = z$. Suppose yt $zz + axz - y^4 = 0$, is ye relation twixt x, y & z, whose motions are p, q, & r: & yt p & q are desired [Fig. 13]. The Equation $zz + axz - y^4 = 0$ gives (by prop 7), $2rz + rax + paz - 4qy^3 = 0$. Now drawing $dh \parallel ab \perp ad = 1 = bh$. I consider ye superficies $abhd = ab \times bh = x \times 1 = x$, & $abc = z$ doe increase in ye proportion of bh to bc: yt is, $1 : \sqrt{ax - xx} :: p : r$. Or

$r = p\sqrt{ax-xx}$. Which valor of r being substituted into ye Equation $2rz + rax + paz - 4qy^3 = 0$, gives

$$\overline{2pz + pax} \times \sqrt{ax - xx} + paz - 4qy^3 = 0.$$

wch was required.

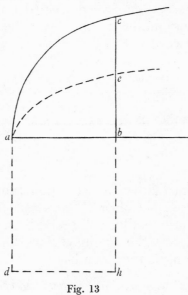

Fig. 13

How to proceede in other cases (as when there are cube rootes, surde denominators, rootes within rootes (as

$$\sqrt{ax + \sqrt{aa - xx}}) \;\&c:$$

in the equation) may bee easily deduced from what hath bee already said.

But ye Demonstrations of wt hath beene said must
not bee wholly omitted.

Proposition 7   Demonstrated

Lemma. If two bodys A, B, move uniformely ye $\begin{smallmatrix}\text{one}\\\text{other}\end{smallmatrix}$ from $\begin{smallmatrix}a\\b\end{smallmatrix}$ to $\begin{smallmatrix}c,d,e,f,\\g,h,k,l,\end{smallmatrix}$ &c: in ye same time [Fig. 14]. Then are ye lines $\begin{smallmatrix}ac,\\bg,\end{smallmatrix}$ & $\begin{smallmatrix}cd,\\gh,\end{smallmatrix}$ & $\begin{smallmatrix}de,\\hk,\end{smallmatrix}$ & $\begin{smallmatrix}ef,\\kl,\end{smallmatrix}$ &c: as their velocitys $\begin{smallmatrix}p.\\q.\end{smallmatrix}$ And though they

31

move not uniformly yet are ye infinitely little lines wch each moment they describe, as their velocitys wch they have while they describe ym. As if ye body A with ye velocity p describe ye infinitely little line $(cd=)\,p\times o$ in one moment, in yt moment ye body B wth ye velocity q, will describe ye line $(gh=)\,q\times o$. For $p:q::po:qo$. Soe yt if ye described lines bee $(ac=)\,x$, & $(bg=)\,y$, in one moment; they will bee $(ad=)\,x+po$, & $(bh=)\,y+qo$ in ye next.

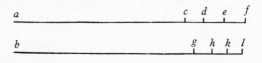

Fig. 14

Demonstration: Now if ye equation expressing ye relation twixt ye lines x & y bee $x^3-abx+a^3-dyy=0$. I may substitute $x+po$ & $y+qo$ into ye place of x & y; because (by ye lemma) they as well as x & y, doe signify ye lines described by ye bodys A & B. By doeing so there results

$$x^3+3poxx+3ppoox+p^3o^3-dyy-2dqoy-dqqoo=0.$$
$$-abx\qquad\ -abpo$$
$$+a^3$$

But $x^3-abx+a^3-dyy=0$ (by supposition). Therefore there remaines onely $3poxx+3ppoox+p^3o^3-2dqoy-dqqoo=0$. Or divi-
$$-abpo$$
ding it by o tis $3px^2+3ppox+p^3oo-2dqy-dqqo=0$. Also those
$$-abp$$
termes are infinitely little in wch o is. Therefore omitting them there rests $3pxx-abp-2dqy=0$. The like may bee done in all other equations.

Hence I observe: First yt those termes ever vanish wch are not multiplyed by o, they being ye propounded equation. Secondly those termes also vanish in wch o is of more yn one dimension, because they are infinitely lesse yn those in wch o is but of one dimension. Thirdly ye still remaining termes being divided by o will have yt form wch, by ye 1st rule in

Prop 7th, they should have (as may partly appeare by ye second termes of Mr Oughtred's latter Analiticall table).

After ye same manner may this 7th Proposition bee demonstrated there being 3 or more unknowne quantitys x, y, z, &c.

Prop 8th is ye Converse of this 7th Prop. & may bee therefore Analytically demonstrated by it.

Prop 1st Demonstrated.

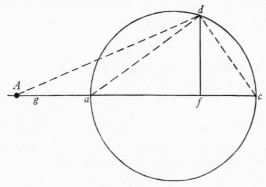

Fig. 15

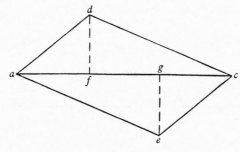

Fig. 16

If some body A move in ye right line $gafc$ from g towards c [Fig. 15]. From any point d draw $df \perp ac$ & call, $df = a$, $fg = x$, $dg = y$. Then is $aa + xx - yy = 0$. Now by Prop 7th, may ye proportion of (p) ye velocity of yt body towards f; to (q) its velocity towards d bee found viz: $2xp - 2yq = 0$. Or $x:y::q:p$. That is $gf:gd::$its velocity to $d:$its velocity towards f or c. & when ye body A is at a, yt is when ye points g & a are coincident then is $ac:ad::ad:af::$velocity to $c:$velocity to d.

Prop. 2d, Demonstrated.

From ye points d & e draw $df \perp ac \perp ge$ [Fig. 16]. And let ye first bodys velocity to d bee called ad, ye seconds to e bee ae, & ye 3ds toward c bee ac. Then shall ye firsts velocity towards c bee af (by Prop 1): & The seconds towards c is ag, (prop 1). but $af=gc$ (for $\triangle adc=\triangle aec$, & $\triangle adf=\triangle gec$ by supposition). Therefore $ac=ag+gc=ag+af$. That is ye velocity of ye third body towards c is equall to ye summ of the velocitys of ye first & second body towards c.

The former Theorems Applyed to Resolving of Problems
Problem 1
To draw Tangents to crooked lines.

Seeke (by prop 7th; or 3d, 4th & 2d, &c) ye motions of those streight lines to wch ye crooked line is cheifely referred, & wth what velocity they increase or decrease: & they shall give (by prop 6t, or 1st or 2d) ye motion of ye point describing ye crooked line; wch motion is in its tangent.

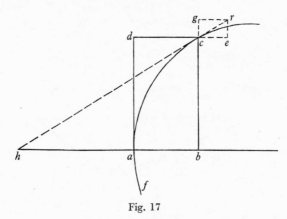

Fig. 17

Tangents to Geometricall lines

Example 1 [Fig. 17]. If ye crooked line fac is described by ye intersection of two lines cb & dc ye one moveing parallely, viz: $cb \parallel ad$, & $dc \parallel ab$; soe yt if $ab=x$, & $bc=y=ad$, Their relation is $x^4 - 3yx^3 + ayxx - 2y^3x + a^4 = 0$. To draw ye tangent hcr; I
$\quad +10a \qquad\qquad -y^4$
consider yt ye point c fixed in ye line cb moves towards e parallely to ab (for so doth ye line cb (by supposition) & consequently all its points): also ye point c fixed in ye line dc

34

moves toward g parallely to ad (by supposition): therefore I draw $ce \,||\, ab$ & $cg \,||\, ad$, & in such proportion as ye motions they designe & so draw $er \,||\, cb$, & $gr \,||\, dc$, & ye diagonall cr, (by Prop 6): yt is, if ye velocity of ye line cb, (yt is ye celerity of ye increasing of ab, or dc; or ye velocity of ye point c from d) bee called p, & ye velocity of ye line cd bee called q; I make

$$ce : gc :: p : q \ (:: ce : er :: hb : cb.)$$

& ye point c shall move in the diagonall line cr (by prop 6) wch is therefore the required tangent. Now ye relation of p & q may bee found by ye foregoing Equation (p signifying ye increase of x, & q of y) to bee

$$4px^3 - 9pyxx + 30paxx + 2payx - 2py^3 - 3qx^3 + qaxx$$
$$- 6qyyx - 4qy^3 = 0.$$

(by Prop 7). And therefore

$$hb = \frac{py}{q} = \frac{3yx^3 - ayxx + 6y^3x + 4y^4}{4x^3 - 9yxx + 30axx + 2ayx - 2y^3}.$$

wch determins ye tangent hc.

☞ Hence may bee observed this Generall Theorem for drawing Tangents to crooked lines thus referred to streight ones; yt is, to such lines in wch $y = bc$ is ordinately applyed to $x = ab$ at any given angle abc. viz: Multiply the termes of ye Equation ordered according to ye dimensions of y, by any Arithmeticall progression wch product shall bee ye Numerator: Againe change ye signes of ye Equation & ordering it according to x, multiply ye termes by any Arithmeticall progression & ye product divided by x shall bee ye Denominator of ye valor of hb, yt is, of x produced from y to ye tangent hc.

As if $rx - \dfrac{rxx}{q} = yy$. Then first $\overset{2.1.}{yy*} + \overset{0.}{\dfrac{r}{q}xx} = 0$, produceth $2yy$,

$$\underset{0. \ -1. \ -2.}{- \ rx}$$

or $2rx - \dfrac{2r}{q}xx$. Secondly, $-\dfrac{r}{q}\overset{2.}{xx} + \overset{1.}{rx} - \overset{0.}{yy}$ produceth $rx - \dfrac{2r}{q}xx$.

Therefore
$$\frac{2yy}{r-\dfrac{2rx}{q}}=bh.$$

Or else
$$\frac{2rx-\dfrac{2r}{q}xx}{r-\dfrac{2r}{q}x}=bh=\frac{2qrx-2rxx}{qr-2rx}.$$

Example 2 [Fig. 18]. If ye crooked line *chm* bee described by ye intersection of two lines *ac*, *bc* circulating about their centers *a* & *b*, soe yt if *ac* = *x*, & *bc* = *y*; their relation is

$$x^3-abx+cyy=0.$$

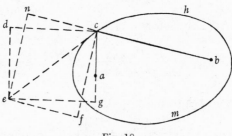

Fig. 18

To draw ye tangent *ec* I consider yt ye point *c* fixed in ye line *bc* moves towards *f* in ye line *cf* ⊥ *bc* (for ye tangent to a circle is perpendicular to its radius). also ye point *c* fixed in ye line *ac* moves towards *d* in ye line *cd* ⊥ *ac* & from those lines *cd* & *cf* I draw two others *de* || *cg* & *ef* || *bc* wch must bee in such proportion one to another as ye motions represented by ym (prop 6), yt is (prop 6) as ye motions of ye intersection point *c* moveing in ye lines *ca* & *cb* to or from ye centers *a* & *b*; yt is, (ye celerity of ye increase of *x* being called *p*, & of *y* being *q*), *de*:*ef*::*p*:*q* [Fig. 19]. Then shall ye diagonall *ce* bee ye required tangent. Or wch is ye same, (for △*ecg* = △*ecd*, & △*ecf* = △*ecn*,) I produce *ac* & *bc* to *g* & *n*, so yt *cg*:*cn*::*p*:*q*. & yn draw *ne* ⊥ *bn*, & *ge* ⊥ *ag*; & ye tangent diagonall *ce* to their intersection point *e*. Now ye relation of *p* & *q* may bee found by ye given Equation to bee, $3pxx-pab+2qcy=0$ (by prop 7). Or $2cy$:*ab* − $3xx$::*p*:*q*::*cg*:*cn*, wch determins ye tangent *ce*.

36

But note yt if *p*, or *q* be negative *cg* or *cn* must bee drawn from *c* towards *a* or *b*; but from *a* or *b* if affirmative.

Hence tis easy to pronounce a Theoreme for Tangents in such like cases & ye like may bee done in all other cases however Geometricall lines bee referred to streight ones.

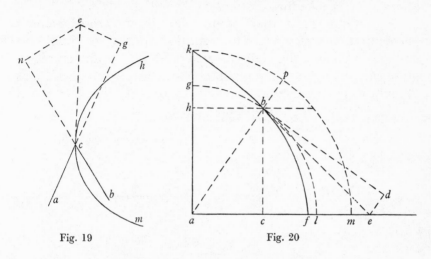

Fig. 19   Fig. 20

Tangents to Mechanichall lines

Example ye 3d [Fig. 20]. If ye Quadratrix *kbf* is described by ye intersection *b* of ye two lines *hb* & *ap*, ye one *hp* ‖ *ma* moving uniformly from *k* to *a*, whilest ye other *ap* circulates from *k* to *m* about ye center *a*. Draw ye circle *gbl* wth ye Rad. *ab*; & make *bl* = *bd* ⊥ *ab* ‖ *de*; & to ye intersection point *e* of ye lines *am* & *ed* draw *eb* wch shall touch ye Quadratrix in *b*. For suppose ye motion of ye point *p* fixed in ye line *ap*, towards *m* to bee *pm*, yn ye motion of ye point *b* fixed in ye line *ab*, towards *d* is *bl* = *bd* (prop 4), & ye motion of ye line *bh* towards *ca*, & therefore of ye point *b* fixed in it towards *c* (prop 3) is *ha* = *bc* (by supposition): Also *ce* ‖ *bh* & *ed* ‖ *ap* (supposition). Therefore (by Prop 6) ye intersection point *b* of these two lines *ap* & *hb*, moves in ye diagonall *eb*, & consequently *eb* toucheth ye Quadratrix in *b*.

Example 4th

Problem 2ᵈ

To find ye quantity of crookednesse of lines [Fig. 21]

Lemma. The crookednesse of equall parts of circles are as

37

their diameters reciprocally. For the crookednesse of a whole circle (*acdea, bfghmb*) amounts to 4 right angles. Therefore there is not more crookednesse in one whole circle *acdea* yn in another *bfghmb*. Suppose ye perimeter *acde*=*bfgh*. Then tis *ar*:*br*::*bfgh*=*acde*:*bfghmb*::crookednesse of *bfgh*:crookednesse of *bfghmb*=crookednesse of *acdea*.

Resolution. Find that point fixed in ye crooked line's perpendicular wch is yn in least motion, for it is ye center of a circle wch passing through ye given point is of equall crookednesse wth ye line at yt given point. Now, since ye crooked line's tangent & perpendicular &c: (at yt moment) circulate about

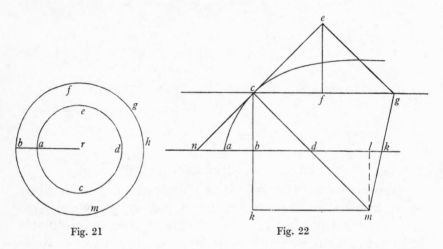

Fig. 21   Fig. 22

yt center; I observe, 1st yt every point fixed in ye Tangent or Perpendicular, or whose position to ym is determined, doth describe a curve line to wch ye right line drawne from yt center is perpendicular; & is also ye radius of a circle of equall crookedness wth it: 2dly yt ye motion of every such point is as its distance from yt center: & so are ye motions of ye intersection points, in wch any radius drawn from yt center intersects two parallel lines.

Example 1 [Fig. 22]. If *cb*=*y* is ordinately applyed to *ab*=*x* at a right angle *abc*, *nc* being tangent & *mc* perpendicular to ye curve line *ac*: I seeke ye motion of two points *c* & *d* fixed in ye perpendicular *cd*; or (wch is better & to ye same purpose) I draw *cg* || *ab*, & seeke ye motions of ye two intersection points *c* & *d* in wch ye perpendicular *cd* intersects those fixed lines *cfg*,

& *abdk*: & yn draw *cg* & *dk* in such proportion as those motions are, & ye line *gkm* drawn by their ends shall intersect ye perpendicular *cd* in ye required center *m*: *mc* being ye radius of a circle of equall crookednesse wth ye curve line *ac* at ye point *c*. Now, Making $\triangle cegfe = $ & like $\triangle ncdbc$, suppose ye motion of ye line *cb* & consequently of ye points *c* & *b* fixed in it & moveing towards *f* & *k*, to bee $p = nb = cf$: Then is $ce = cn$ ye motion of ye point *c* in wch *bc* intersects ye tangent *ne* (by Prop 6), yt is of ye point *c* fixed in ye perpendicular *cd*, & moveing in ye tangent *ne*: & therefore $cg = nd = p + bd$ $(= p + v$ if $bd = v)$, is ye motion of ye intersection point *c* towards *g* in wch point ye perpendicular *cd* intersects *cg* (by Prop 6). If also ye motion of ye intersection point *d* from ye point *b* (yt is ye velocity of ye increase of *v*) bee called *r*, yn is $dk = nb + r = p + r$. soe yt, $cg - dk : cd :: cg : cm$; yt is,

$$v - r : cd = \sqrt{yy + vv} :: p + v : cm = \frac{\overline{p + v} \times \sqrt{yy + vv}}{v - r}.$$

Also $v - r : y :: p + v : ck = \dfrac{py + vy}{v - r}$. Lastly, ye motion of ye point *c* from *b*, (yt is ye velocity wth wch $y = cb$ increaseth) will bee $q = cb = y$.

As if ye nature of ye crooked line bee $x^3 - axy + ayy = 0$. Then is $nb = \dfrac{axy - 2ayy}{3xx - ay} = p$ (by example 1. Prob: 1), &

$$\frac{3xxy - ayy}{ax - 2ay} = v = bd \quad \text{(for } nb : bc :: bc : bd)$$

Soe that $3xxy - ayy - axv + 2ayv = 0$. & therefore (by Prop 7) $6pxy - pav - 2qay + 3qxx + 2qav - rax + 2ray = 0$, & substituting $\dfrac{axy - 2ayy}{3xx - ay}$, & $\dfrac{3xxy - ayy}{ax - 2ay}$, & *y* into ye places of *p*, *v* & *q* in this equation, The product will bee

$$\frac{6axxyy - 12axy^3}{3xx - ay} - 3ayy + 3yxx + \frac{6xxyy - 2ay^3}{x - 2y} - rax + 2ray = 0.$$

Or $\dfrac{9yx^5 - 6ayyx^3 - 12ay^3xx + 24ay^4x + 3aay^3x - 4a^2y^4}{3ax^4 - 12ayx^3 - a^2yxx + 12ayyxx + 4a^2yyx - 4a^2y^3} = r.$

And therefore

$$\frac{-18yyx^4 + 24ay^3xx - 24ay^4x - 2aay^3x + 2aay^4}{3ax^4 - 12ayx^3 - aayxx + 12ayyxx + 4aayyx - 4aay^3} = v - r$$

$$= \frac{2aay^4 - 2aay^3x - 24ay^4x + 24ay^3xx - 18yyx^4}{3x^3 - 6yxx - ayx + 2ayy \text{ in } ax - 2ay}$$

Also $\quad p + v = \dfrac{axy - 2ayy}{3xx - ay} + \dfrac{3xxy - ayy}{ax - 2ay}$

$$= \frac{9x^4y - 6axxyy + 5aay^3 - 4aaxyy + aaxxy}{3xx - ay \times ax - 2ay}.$$

Soe yt

$$ck = \frac{py + vy}{v - r}$$

$$= \frac{\begin{array}{r}9x^4y - 6axxyy + 5aay^3 - 4aaxyy \\ + aaxxy \text{ in } 3x^3 - 6xxy - axy + 2ayy\end{array}}{2aay^3 - 2aaxyy - 24axy^3 + 24axxyy - 18x^4y \text{ in } 3xx - ay}.$$

That is

$$ck = \frac{\begin{array}{r}9x^5 - 6ax^3y + aax^3 - 6aaxxy \\ + 13aaxyy + 12axxyy - 10aay^3 - 18x^4y\end{array}}{2aayy - 2aaxy - 24axyy + 24axxy - 18x^4}.$$

Which Equation gives ye point k & consequently ye point m for $km \parallel abk$.

☞ But in such cases where y is ordinately applyed to x at right angles, From ye consideration of ye Equation $\dfrac{py + vy}{v - r} = ck$; Or rather, $\dfrac{py + vy}{r - v} = ck$: may ye following Theoreme bee pronounced. To wch purpose let $\mathcal{X}$ signify the given Equation, yt is, all ye algebraicall terms (expressing ye nature of ye given line) considered as equall to nothing & not some of ym to others. Let $\mathcal{X}$ signify those termes ordered according to ye dimensions of x & yn multiplyed by any arithmeticall progression. Let $\mathcal{X}$ signify those termes ordered according to ye dimensions of y, & yn multiplyed by any Arithmeticall progression. Let $\mathcal{X}$ signify those termes ordered by x & yn multiplyed by any two arithmeticall progressions one of ym being greater yn ye other by a terme. Let ℭ signify those termes ordered by y & yn multiplied by any two Arithmetical Progressions differing by a terme. Let ℭ signify those termes ordered according to x, & yn multiplyed by ye greater of ye progressions wch multiplyed ℭ; & yn ordered by y & multiplyed by ye greater progression wch multiplyed ℭ. Then (observing yt all these progressions have the same difference & proceede ye same way in respect of ye dimensions of x & y;) will ye 3 Theorems bee

1st.
$$\frac{\text{ℭℭℭ}yy + \text{ℭℭℭℭ}xx}{-\text{ℭℭℭ}y + 2\text{ℭℭℭℭ}y - \text{ℭℭℭℭ}y} = ck$$

$$= \frac{\text{ℭℭ}yy + \text{ℭℭ}xx}{-\dfrac{\text{ℭℭℭ}y}{\text{ℭ}} + 2\text{ℭℭ}y - \text{ℭℭ}y}$$

2d.
$$\frac{\text{ℭℭℭ}yy + \text{ℭℭℭℭ}xx}{-\text{ℭℭℭℭ}x + 2\text{ℭℭℭℭ}x - \text{ℭℭℭℭ}x} = km = bl$$

$$= \frac{\text{ℭℭ}yy + \text{ℭℭ}xx}{-\text{ℭℭ}x + 2\text{ℭℭ}x - \dfrac{\text{ℭℭℭ}x}{\text{ℭ}}}.$$

3d.
$$\frac{\text{ℭℭ}yy + \text{ℭℭ}xx \text{ in } \sqrt{\text{ℭℭ}yy + \text{ℭℭ}xx}}{-\text{ℭℭℭ}yx + 2\text{ℭℭℭℭ}yx - \text{ℭℭℭℭ}yx} = cm$$

$=$ radio circuli aequalis curvitatis cum curva ac in puncto c.

As if ye line bee $x^3 - axy + ayy = 0$. Then is $\text{ℭ} = 3x^3 - axy$.

$$3 \times 2 \; 1 \times 0 \; 0 \times -1$$
$$\text{ℭ} = -axy + 2ayy. \quad \text{ℭ} = 6x^3 = x^3 - axy + ayy.$$

$$0 \times -1 \; 1 \times 0 \; 2 \times 1 \qquad\qquad 3 \times 0 \; 1 \times 1 \; 0 \times 2$$
$$\text{ℭ} = 2ayy = x^3 - axy + ayy. \quad \& \quad \text{ℭ} = -axy = x^3 - axy + ayy.$$

Which valors of ℭ, ℭ &c. being substituted into their places in ye first rule, ye result is

$$\frac{9x^6yy - 6ax^4y^3 + 5aaxxy^4 + aax^4yy - 4aax^3y^3}{\dfrac{-18ax^6y^3 + 12aax^4y^4 - 2a^3xxy^5}{2ayy - axy} + 2aaxxy^3 - 12ax^3y^3} = ck.$$

Which being conveniently reduced is

$$\frac{9x^5 - 6ax^3y + aax^3 - 6aaxxy + 13aaxyy \atop + 12axxyy - 10aay^3 - 18x^4y}{18x^4 - 24axxy + 24axyy + 2aaxy - 2aayy} = ck.$$

As was found before.

$$ck = \frac{1}{2}a + 3y - \frac{3ay}{2x} + \frac{ay - 3xx}{2a - 6x}.$$

Or suppose ye line is a Conick section whose nature $rx + \dfrac{rxx}{q} = yy$. Then is $\mathfrak{X} = rx + \dfrac{2r}{q}x^2 = \overset{2}{\dfrac{r}{q}}\overset{1}{xx} + \overset{0}{rx} - yy.$

$$\mathfrak{X} = -2yy = \overset{0\ 1\ 2}{\dfrac{r}{q}xx + rx * - yy}.$$

$$\mathfrak{X} = 2\frac{r}{q}xx = \dfrac{r}{q}\overset{2\times 1}{xx} + \overset{1\times 0}{rx} - \overset{0\times -1}{yy}.$$

$$\mathfrak{X} = -2yy = \dfrac{r}{q}\overset{0\times -1}{xx} + \overset{0\times -1}{rx} \overset{1\times 0}{*} - \overset{2\times 1}{yy}.$$

$$\mathfrak{X} = 0 = \dfrac{r}{q}\overset{2\times 0}{xx} + \overset{1\times 0}{rx} - \overset{0\times 2}{yy}.$$ Which valors of $\mathfrak{X}$, $\mathfrak{X}$, $\mathfrak{X}$ &c. being substituted into their places in ye first Theoreme, give

$$\frac{rrxxyy + \dfrac{4rr}{q}x^3yy + \dfrac{4rr}{qq}x^4yy + 4xxy^4}{-rrxxy - \dfrac{4rr}{q}x^3y - \dfrac{4rr}{qq}x^4y + \dfrac{4r}{q}xxy^3} = ck$$

$$= \frac{qqrry + 4qrrxy + 4rrxxy + 4qqy^3}{4qryy - qqrr - 4qrrx - 4rrxx}.$$

Or (since $4qryy - 4qrrx - 4rrxx = 0$) tis,

$$-ck = y + \frac{4xy}{q} + \frac{4xxy}{qq} + \frac{4y^3}{rr} = y + \frac{4y^3}{qr} + \frac{4y^3}{rr} = y + \frac{4ry^3 + 4qy^3}{qrr}.$$

So by ye second Theorem tis

$$\frac{rrxxyy + \dfrac{4r}{q} xxy^4 + 4xxy^4}{\dfrac{2rx^2yy + \dfrac{4r}{q} x^3yy - \dfrac{8r}{q} y^4x^3}{rx + \dfrac{2r}{q} xx}} = km = bl$$

$$= \frac{r}{2} + \frac{2yy}{q} + \frac{2yy}{r} + \frac{rx}{q} + \frac{4xyy}{qr} + \frac{4xyy}{qq}.$$

Or $bl = \dfrac{2}{r} + 2x + \dfrac{3rx}{q} + \dfrac{6xx}{q} + \dfrac{6rxx}{qq} + \dfrac{4x^3}{qq} + \dfrac{4rx^3}{q^3}$ And so by ye 3d

Theoreme, tis

$$cm = \frac{rrxxyy + \dfrac{4rr}{q} x^3yy + \dfrac{4rr}{qq} x^4yy + 4xxy^4}{2rrx^3y^3 + 8\dfrac{rr}{q} x^4y^3 + \dfrac{8rr}{qq} x^5y^3 - 8\dfrac{r}{q} x^3y^5}$$

$$\times \sqrt{rrxxyy + \frac{4rr}{q} x^3yy + \frac{4rr}{qq} x^4yy + 4xxy^4}. \quad \text{Or}$$

$$cm = \frac{qqrr + 4qrrx + 4rrxx + 4qqyy \text{ in } \sqrt{qqrr + 4qrrx + rrxx + 4qqyy}}{2q^3rr}.$$

Or

$$cm = \frac{qrr + 4qyy + 4ryy}{2qqrr} \sqrt{qqrr + 4qqyy + 4qryy}$$

Note yt ye curvity of any curve whose ordinates are inclined from right to oblique angles is as the curvity of a circle whose ordinates are in like manner inclined so as to make it becom an Ellipsis.

Problem

To find ye points of curves where they have a given degree of curvity.

Problem 3d

To find ye points distinguishing twixt ye concave & convex portions of crooked lines

Resolution. The lines are not crooked at those points: & therefore ye radius *cm* determining ye crookednesse at yt point must bee infinitely greate. To wch purpose I put ye denominator of its valor (in rule ye 3d) to bee equall to nothing, & so have this Theoreme

$$\mathfrak{X}\mathfrak{X}\mathfrak{C}\mathfrak{X} - 2\mathfrak{C}\mathfrak{X}\mathfrak{C}\mathfrak{X}\mathfrak{X} + \mathfrak{X}\mathfrak{C}\mathfrak{X}\mathfrak{C}\mathfrak{X} = 0$$

Or better perhaps

$$\frac{\mathfrak{X}\mathfrak{C}\mathfrak{X}}{\mathfrak{X}} - 2\mathfrak{X} + \frac{\mathfrak{C}\mathfrak{X}\mathfrak{C}}{\mathfrak{X}} = 0.$$

Example, was this point to be found in ye Concha whose nature is $x^4 + 2bx^3 \begin{smallmatrix} +bb \\ -ccxx \end{smallmatrix} - 2bccx - bbcc = 0$. Then is

$$\mathfrak{X} = 2x^4 + 2bx^3 + 2bccx + 2bbcc. \quad \mathfrak{X} = 2xxyy = \mathfrak{X}.$$

$$\mathfrak{X} = 2x^4 - 4bccx - 6bbcc. \qquad \mathfrak{X} = 0.$$

Which valors subrogated into ye Theoreme, they produce

$$2x^4 + 2bx^3 + 2bccx + 2bbcc + \frac{\overline{2x^4 - 4bccx - 6bbcc} \times 2xxyy}{2x^4 + 2bx^3 + 2bccx + 2bbcc} = 0.$$

And subrogating $-x^4 - 2bx^3 \begin{smallmatrix} -bb \\ +cc \end{smallmatrix} xx + 2bccx + bbcc$ ye valor of $xxyy$ into its stead & twice reducing ye Equation by ye divisor $x + b = 0$, Tis

$$x^3 + bcc + \frac{\overline{x^4 - 2bccx - 3bbcc} \times \overline{cc - xx}}{x^3 + bcc} = 0.$$

Or $\qquad x^4 + 4bx^3 + 3bbxx - 2bccx - 2bbcc = 0.$

Which being againe reduced by $x + b = 0$. Tis

$$x^3 + 3bxx - 2bcc = 0.$$

See Geometr: Chart: [Descartes' Geometry] pag. 259.

Problem 4

To find ye points at wch lines are most or least crooked

Resolution. At those points ye afforesaid radius cm neither increaseth nor decreaseth. So yt ye center m in yt moment doth absolutely rest, & therefore neither ye line bk nor al doth increase or diminish, yt is, ck & bl doe soe much increase or diminish as y & x (cb & ab) doe diminish or increase. Or in a word the point m resteth. Find therefore the motion of al or cm or lm & suppose it nothing.

Thus in the Concha to find the point of least crookednes beyond ye point of reflection, having substituted ye valors of $\mathcal{X}$, $\mathcal{X}$ &c (exprest in the precedent problem) into this The computation is too tedious

Thus to find the point of least crookednesse in ye curve $x^3 = ccy$. By the rule in prob 2, I make $\mathcal{X} = 3x^3$. $\mathcal{X} = -ccy$. $\mathcal{X} = 6x^3$. $\mathcal{X} = 0$ & $\mathcal{X} = 0$. & thence obteine $ck = \dfrac{3}{2}y + \dfrac{cc}{6x}$, or

$$= \frac{3x^3}{2cc} + \frac{cc}{6x}$$ wch is least when $\dfrac{9x^3}{2cc} - \dfrac{cc}{6x} = 0$ or $x = \sqrt[4]{\dfrac{c^4}{27}}$. And then therefore happens the greatest crookednesse.

In like manner if ye curve be $xxy = a^3$ The rule gives $\mathcal{X} = 2a^3$ $\mathcal{X} = a^3$ $\mathcal{X} = -6a^3$ $\mathcal{X} = 0 = \mathcal{X}$ And thence $\dfrac{2}{3}y + \dfrac{xx}{6y} = ck$ which is least when[1] $y = a\sqrt{\dfrac{1}{2}}$.

Soe if ye curve bee $x^3 = byy$. Then is $\mathcal{X} = 3x^3$. $\mathcal{X} = -2byy$. $\mathcal{X} = 6x^3$. $\mathcal{X} = -2yyc$. $\mathcal{X} = 0$. And therefore $ck = 3y + \dfrac{4xx}{3y}$, which hath no least nor the curve any least crookednesse.

Problem 5t

To find ye nature of ye crooked line whose area is expressed by any given equation [Fig. 23].

That is, ye nature of ye area being given to find ye nature of ye crooked line whose area it is.

Resolution. If ye relation of $ab = x$, & $\square\, abc = y$ bee given & ye relation of $ab = x$, & $bc = q$ bee required (bc being ordinately

[1] The value given here is crossed out in the MS.

applyed at right angles to *ab*). Make *de* $||$ *ab* $\perp$ *ad* $||$ *be* $=1$. & yn is $\square$ *abed* $=x$. Now supposing ye line *cbe* by parallel motion from *ad* to describe ye two superficies *ae* $=x$, & *abc* $=y$; The velocity wth wch they increase will bee, as *be*, to *bc*: yt is, ye motion by wch x increaseth being *be* $=p=1$, ye motion by wch y increaseth will bee *bc* $=q$. which therefore may bee found by prop. 7th. viz: $-\dfrac{\mathfrak{X}y}{\mathfrak{X}x}=q=bc$.

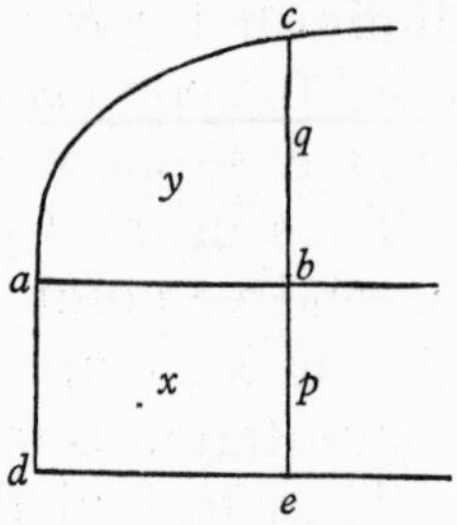

Example 1. If $\dfrac{2x}{3}\sqrt{rx}=y$. Or

$$-4rx^3+9yy=0.$$

Then is $\dfrac{12rxx}{18y}=q=\sqrt{rx}$. Or, $rx=qq$ & there-

fore *abc* is ye Parabola whose area *abc* is

$$\frac{2x}{3}\sqrt{rx}=\frac{2qx}{3}.$$

Fig. 23

Example 2d. If $x^3-ay+xy=0$. Then is $\dfrac{3xx+y}{-x+a}=q$. Or

$$\frac{3axx-2x^3}{aa-2ax+xx}=q=bc.$$

Example 3d. If $\dfrac{na}{n+m}x^{\frac{n+m}{n}}-y=0$. Then is $ax^{\frac{m}{n}}=q$. Or if

$ax^m=bx^n$; yn is $\dfrac{maxx^{m-1}}{nbx^{n-1}}=q$.

Note yt by this probleme may bee gathered a Catalogue of all those lines wch can bee squared. And therefore it will not bee necessary to shew how this Probleme may bee resolved in other cases in wch q is not ordinately applyed to x at right angles.

Problem 6

The nature of any Crooked line being given, to find other lines whose areas may bee compared to ye area of yt given line [Fig. 24]

Resolution. Suppose ye given line to be *ac* & its area *abc* $=s$, ye sought line *df* & its area *def* $=t$; & yt *bc* $=z$ is ordinately applyed to *ab* $=x$, & *ef* $=v$ to *de* $=y$, soe yt $\angle abc=\angle def$; & yt ye velocitys wth wch *ab* & *de* increase (yt is, ye velocity of ye points b & e, of ye lines bc & ef moving from a & d) bee called p & q. Then may ye ordinately applyed lines bc & ef multiplyed by their velocitys p & q, (yt is pz & qv) signify ye velocitys wth wch ye areas $abc = s$ & $def = t$ increase. Now ye relation of ye areas s & t (taken at pleasure) gives ye relation of ye motions pz & qv describing those areas, by Proposition ye 7th; Also ye relation of ye lines $ab = x$ & $de = y$ (taken at pleasure) gives ye relation of p & q, by Prop 7th; wch two equations, together with ye given Equation expressing ye nature of ye line ac, give ye relation of $de = y$ & $ef = x$ yt is ye desired nature of ye line df.

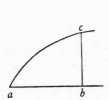 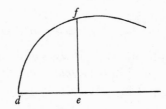

Fig. 24

Example 1. As if $ax + bxx = zz$ is ye nature of ye line ac: & at pleasure I assume $s = t$ to be ye relation of ye areas abc & def; & $x = yy$ to bee ye relation of ye lines ab & de. Then is $pz = qv$ (prop 7), & $p = 2qy$ (by prop 7). Therefore $2yz = v$ (by ye 2 last equations), & $\dfrac{vv}{4yy} = zz = ax + bxx = ayy + by^4$. or $vv = 4ay^4 + 4by^6$. & $v = 2yy\sqrt{a + byy}$: wch is ye nature of ye line def whose area def is equall to ye area abc, supposing $\sqrt{ab} = de = \sqrt{x} = y$.

Example 2. If $ax + bxx = zz$ as before; & I assume, $as + bx = t$, & $x = yy$. Then is $apz + bp = qv$ (by prop 7), & $p = 2qy$.

$$2azy + 2by = v = 2by + 2ay\sqrt{ax + bxx}.$$

Or $v = 2by + 2ayy\sqrt{a + byy}$; The required nature of ye line def.

Example 3d. If $\dfrac{4c}{a}\sqrt{xx - a} = z$: & at pleasure I assume $\dfrac{2c}{a}\sqrt{ay + yy} - s = t$, & $xx - a = y$. Then is

$$4ccay + 4ccyy = aass + 2aast + aatt,$$

47

And (by prop 7)

$$4ccaq + 8ccyq = 2aaspz + 2aatpz + 2aasqv + 2aatqv$$

$$= \overline{2aapz + 2aaqv} \times \frac{2c}{a} \sqrt{ay + yy}. \ \& \ \text{(by prop 7)} \ 2px = q.$$

Therefore $8cax + 16cyx = \overline{4az + 8avx} \times \sqrt{ay + yy}$. But

$$\frac{4c}{a} \sqrt{xx - a} = z = \frac{4c}{a} \sqrt{y}. \ \& \ x = \sqrt{a + y},$$

Therefore

$$\overline{8ca + 16cy} \sqrt{a + y} = 16c \sqrt{y} + 8av \sqrt{a + y} \times \sqrt{ay + yy}.$$

That is $8ca + 2cy = 2cy + av \sqrt{ay + yy}$. Or $v = \dfrac{c}{\sqrt{ay + yy}}$.

Example 4th. If $ax + bxx = zz$ as before, & I assume, $ss = t$, & $x = yy$. Then is $2spz = qv$, & $p = 2qy$ (by prop 7). Therefore $4syz = v = 4sy \sqrt{ax + bxx} = 4syy \sqrt{a + byy}$. Where note yt in this case ye line $v = 4syy \sqrt{a + byy}$ is a Mechanicall one because s ye area of ye line $ax + bxx = zz$ canott bee Geometrically found. The like is to bee observed in other such like cases.

Problem 7

The Nature of any Crooked line being given to find its area, when it may bee. Or more generally, two crooked lines being given to find the relation of their areas, when it may bee.

Resolution. In ye figure of ye fift probleme Let $abc = y$ represent ye area of ye given line acf; $cb = q$ ye motion describing yt area; $abed = x$ another area wch is equall to ye basis $ab = x$ of yt given line acf, (viz: supposing $ab \parallel de \perp be \parallel ad = 1$); & $be = p = 1$ ye motion describing yt other area. Now haveing (by supposition) ye relation twixt $ab = x = abed$, & $bc = q = \dfrac{q}{1} = \dfrac{q}{p}$ given, I seeke ye area $abc = y$ by ye Eight proposition.

Example 1. If ye nature of ye line bee, $\dfrac{ax}{\sqrt{aa - xx}} = bc = \dfrac{q}{p}$. I looke in ye tables of ye Eight proposition for ye Equation corresponding to this Equation wch I find to bee $\dfrac{cx^n}{x \sqrt{a + bx^n}} = \dfrac{q}{p}$,

(For if instead of c, a, b, n I write a, aa, -1, 2, it will bee $\dfrac{ax}{\sqrt{aa-xx}} = \dfrac{q}{p}$.) And against it is ye equation $\dfrac{2c}{nb}\sqrt{a+bx^n} = y$. And substituting a, aa, -1, 2 into ye places of c, a, b, n it will bee, $-a\sqrt{aa-x^n} = y = abc$ ye required area.

Example 2. If $\sqrt{\dfrac{x^3}{a}} - \dfrac{eeb}{x\sqrt{ax-xx}} = bc = \dfrac{q}{p}$. Because there are two termes in ye valor of bc I consider them severally & first I find ye area correspondent to ye terme $\sqrt{\dfrac{x^3}{a}}$, or $\dfrac{1}{a^{\frac12}}x^{\frac32}$; To bee $\dfrac{2}{5a^{\frac12}}x^{\frac52}$, Or $\dfrac{2}{5}\sqrt{\dfrac{x^5}{a}}$ by prop: 8, part 1. Secondly to find ye area corresponding to ye other terme $\dfrac{eeb}{x\sqrt{ax-xx}}$ I looke ye Equation (in prop. 8, part 3) corresponding to it wch is $\dfrac{cx^n}{x\sqrt{a+bx^n}} = \dfrac{q}{p}$ (for if instead of c, a, b, n, I write eeb, -1, a, -1, it will bee

$$\frac{eebx^{-1}}{x\sqrt{-1+ax^{-1}}},$$

Or $\dfrac{eeb}{x\sqrt{-xx+ax}}$): Against which is ye Equation

$$\frac{2c}{nb}\sqrt{a+bx^n} = y.$$

In which writing eeb, -1, a, -1, instead of c, a, b, n, the result will bee $\dfrac{2eeb}{-a}\sqrt{-1+ax^{-1}}$; Or, $\dfrac{-2eeb}{ax}\sqrt{-xx+ax} = y$ wch is the area corresponding to yt other terme. Now to see how these areas stand related one to another, I draw ye annexed scheme [Fig. 25], in which is

$$ab = x. \quad bd = \sqrt{\frac{x^3}{a}}. \quad de = \frac{eeb}{x\sqrt{ax-xx}}. \ \&$$

$$be = \sqrt{\frac{x^3}{a}} - \frac{eeb}{x\sqrt{ax-xx}} = \frac{q}{p} = q.$$

Soe that $abd = \dfrac{2}{5}\sqrt{\dfrac{x^5}{a}}$ is ye superficies corresponding to ye first

49

terme *bd*, wch because it is affirmative must bee extended (or lye) from ye line *bd* towards *a*. Also ye other superficies correspondent to ye 2d terme *de*, being negative must lye on ye other side *bd* from *a*, wch is therefore $gde = \dfrac{-2eeb}{ax}\sqrt{ax-xx}$.

Lastly if $x = ab = r$ Then is $abd = \dfrac{2}{5}\sqrt{\dfrac{r^5}{a}}$. & $gde = \dfrac{-2bee}{ar}\sqrt{ar-rr}$.

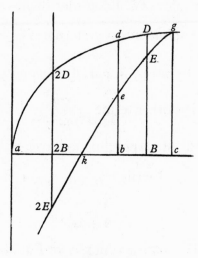

Fig. 25

And if $aB = x = s$, yn is $\dfrac{2}{5}\sqrt{\dfrac{s^5}{a}} = aBD$, And $gDE = \dfrac{-2bee}{as}\sqrt{as-ss}$. Soe yt,

$$\frac{2}{5}\sqrt{\frac{s^5}{a}} - \frac{2}{5}\sqrt{\frac{r^5}{a}} = bBDd.$$

& $\qquad \dfrac{-2bee}{as}\sqrt{as-ss} + \dfrac{2bee}{ar}\sqrt{ar-rr} = DdeE.$

& subtracting *DdeE* from *bBDd* there remaines

$$bBEe = \frac{2\sqrt{s^5}-2\sqrt{r^5}}{5\sqrt{a}} + \frac{2bee}{as}\sqrt{as-ss} - \frac{2bee}{ar}\sqrt{ar-rr} = bBEe.$$

wch is ye required Area of ye given line &c *fkeEg*. Where note yt for ye quantitys $r = ab$ & $s = aB$ taking any numbers you may thereby finde ye area *bBdD* correspond to their difference *bB*.

Note yt sometimes one parte of ye Area may bee Affirmative & ye other negative. as if $a\,2B=r$, & $ab=s$. Then is

$$b\,2B\,2Ee=kbe-k\,2B\,2E$$

$$=\frac{2\sqrt{s^5}-2\sqrt{r^5}}{5\sqrt{a}}+\frac{2bee}{as}\sqrt{as-ss}-\frac{2bee}{ar}\sqrt{ar-rr}$$

$$=b\,2B\,2Ee=kbe-k\,2B\,2E.$$

Problem 9

To find such crooked lines whose lengths may bee found & also to find theire lengths.

Lemma 1 [Fig. 26]. If to any crooked immovable line acg the streight line $cdm\sigma$ moves to & fro perpendicularly every point of ye said line $cdm\sigma$ (as γ, θ, σ, &c) shall describe a curve line (as $\beta\gamma$, $\delta\theta$, $\lambda\sigma$, &c) all which will bee perpendicular to ye said line $cdm\sigma$, & also parallel one to another & to ye line acg.

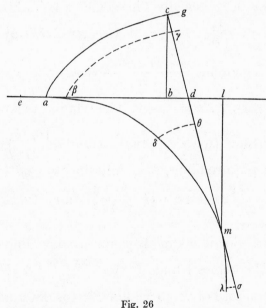

Fig. 26

2. If acg bee not a circle, there may bee drawne some curve line $\beta\delta m\lambda$, wch ye moving line $cdm\sigma$ will always touch in some point or other (as at m) & to wch therefore all ye curve lines $\beta\gamma$, $\delta\theta$, $\lambda\sigma$, &c are perpendiculars.

3. Soe that every point (γ, θ, m, σ, &c) of ye line $cdm\sigma$, when it begineth or ceaseth to touch ye curve line $\beta\delta m\lambda$, doth yn move perpendicularly to or from it: & therefore ye line $\gamma dm\sigma$ doth not at all slide upon ye curve line $\beta\delta m\lambda$, but exactly measure it by applying it selfe to it point by point: & therefore ye correspondent parts of ye said lines are equall (viz: $\beta m = \gamma m$. $\delta m = \theta m$. $\delta\lambda = \theta\sigma$ &c).

Resolution. Take any Equation for ye nature of ye crooked line acg, & by ye 2d probleme find ye center m of its crookednesse at c. That point m is a point of ye required curve line $\beta\delta m\lambda$. For yt point m whereat $cdm\sigma$ ye perpendicular to acg doth touch ye curve line $\beta\delta m\lambda$ is lesse moved yn any other point of ye said perpendicular, being as it were ye hinge & center $\begin{Bmatrix} \text{about} \\ \text{upon} \end{Bmatrix}$ wch ye perpendicular $\begin{Bmatrix} \text{moveth} \\ \text{turneth} \end{Bmatrix}$ at that moment.

Example 1. If acg is a Parabola whose nature (supposing $ab = x \perp bc = y$) is $rx = yy$. By Theoreme ye 1st of Problem 2d I find $cb + lm = y + \dfrac{4y^3}{rr}$. By ye 2d Theoreme, $bl = 2x + \frac{1}{2}r$. And by ye 3d Theoreme $\dfrac{r + 4x \sqrt{r + 4x}}{2r} = cm$. Soe yt supposing $ab = \frac{1}{2}r$. $bl = 3x = z$. $lm = \dfrac{4y^3}{rr} = v$. The relation twixt v & z will bee $27rvv = 16z^3$, $\left(\text{for } r^4vv = 16y^6 = 16r^3x^3 = \dfrac{16z^3r^3}{27} \right)$ wch is ye nature of ye required line $\beta\delta m\lambda$. And since $c\gamma = a\beta = \frac{1}{2}r$. Therefore $\gamma m = \dfrac{r + 4x}{2r} \sqrt{rr + 4xr} - \frac{1}{2}r = \dfrac{3r + 4z}{18r} \sqrt{9rr + 12rz} - \dfrac{r}{2}$ is ye length of its parte $\beta\delta m$.

Example 2: Soe if $aa = xy$, is ye nature of ye line acg: By ye afforesaid Theoremes I find, $cb + lm = \dfrac{xx + yy}{2y}$; &, $bl = \dfrac{xx + yy}{2x}$; whereby ye nature of ye curve line $\beta\delta m\lambda$ is determined. And lastly I find $cm = \dfrac{xx + yy}{2aa} \sqrt{xx + yy}$ which determins its length.

Problem 10

Any curve line being given to find other lines whose lengths may be compared to its length or to its area, & to compare ym.

Resolution [Fig. 27]. Take any Equation for ye relation twixt *ad* & ye perpendicular *cd=y* (whither yt relation bee expressed by an Equation or whither it bee ye same wch some streight line beares to a curved one or to its superficies &c).

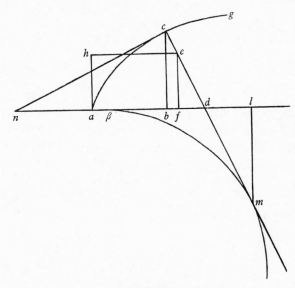

Fig. 27

Then (by prop 7) find ye relation twixt ye increase or decrease (*p* & *q*) of ye lines *ad=x*, & *dc=y*. & say (by prop 6) yt

$$q:p::dc:dn=\frac{py}{q}.$$

And soe is ye triangle *dnc* (rectanguled at *c*) given & consequently ye Nature of ye curve line *acg* to wch *dc* is perpendicular, & *cn* a tangent.

Now ye center (*m*) of ye perpendiculars motion (wch gives ye nature & length of ye required curve line *βm*) may be found as in ye 2d or 9th Probleme, But more conveniently thus.

Draw any fixed line *he* || *ad* ⊥ *ah=a=ef* ⊥ *ad*. Also call *fd=v*. & ye increase or decrease of (*fd*) call *r*. And ye increase or decrease of ye motions *p* & *q*, call *β* & *γ*. Now considering yt ye

53

motion $(p+r)$ of ye intersection point e in ye line he is to ye motion (p) of ye intersection point (d) in ye line (ad), as (em) is to (dm) (see prob 2); That is yt [as] ye difference (r) of those motions is to ye motion (p) of ye point (b) soe is (ed) to (dm): First I find ye valor of v. viz $-\sqrt{pp-qq}:q::cn:cd::ef=a$: $fd=v=\dfrac{-aq}{\sqrt{pp-qq}}$. Or $aaqq+vvqq-vvpp=0$. Secondly by this Equation I find ye valor of r, viz: (by Prob 7),

$$2aaq\gamma+2vvq\gamma+2rvqq-2rvpp-2vvp\beta=0.$$

Thirdly

$$r=\frac{aaq\gamma+vvq\gamma-vvp\beta}{ppv-qqv}:p::ed=\frac{pv}{q}:dm=\frac{p^4zz-ppqqzz}{aaqq\gamma+vvqq\gamma-vvpq\beta}.$$

Lastly supposing ye motion p to bee uniforme yt its increase or decrease β may vanish, & also substituting $ppzz$ ye valor of $aaqq+vvqq$ in its stead in ye Denominator of dm, ye result will bee

$$\frac{pp-qq}{\gamma}=dm=\frac{1-qq}{\gamma}, \text{ if } p=1.$$

And to find ye lines dl & ml, say $-p:q::dm:dl=\dfrac{-q+q^3}{\gamma}$. And

$$p:\sqrt{pp-qq}::dm:ml=\frac{1-qq\sqrt{1-q}}{\gamma}.$$

Soe yt by the equation expressing ye relation twixt x & y first I finde q & yn γ. wch two give me $\dfrac{1-qq}{\gamma}=dm$ &c.

Example 1. If ye relation twixt x & y bee supposed to bee $yy-ax=0$. Then (by Prop 7) I find, first $2qy-ap=0$, Or $2qy-a=0$. & secondly $2\gamma yy+2qq=0$. And substituting these valors of $q=\dfrac{a}{2y}$ & $\gamma=\dfrac{-qq}{y}=\dfrac{-aa}{4y^3}$ in their stead in ye Equation $\dfrac{1-qq}{\gamma}=dm$, &c: The results will bee

$$\frac{aay-4y^3}{aa}=dm. \quad \frac{-aa+4yy}{2a}=dl. \quad \& \quad \frac{aa-4yy}{2aa}\sqrt{4yy-aa}=lm.$$

(And adding $cd=-y$ to dm, & $ad=x=\dfrac{yy}{a}$ to dl, ye result is

$cm = \dfrac{-4y^3}{aa}.$ & $al = \dfrac{3yy}{a} - \dfrac{a}{2}.$) Wch determine ye nature &

crookednesse of ye line βm. (For if $a\beta = \dfrac{a}{4}$. $\beta l = z$. $lm = r$. The

relation twixt z & v will bee $16z^3 = 27rvv$. The length of βm

being $m\gamma = \dfrac{4y^3}{aa} - \dfrac{a}{2} = \dfrac{3a+4z}{18a}\sqrt{9aa + 12az} - \tfrac{1}{2}a$ as before was

found).

Example ye 2d [Fig. 28]. If $x = ad \perp dk = \dfrac{aa}{x}$, is ye nature of

ye crooked line (ye Hyperbola) gkw: And I would find other

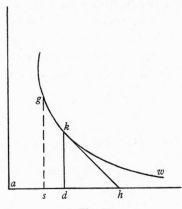

Fig. 28

crooked lines (βm) whose lengths may bee compared wth ye
area $sdkg$ (calling $as = b \perp gs$) of that crooked line gkw. I call
yt area $sdkg = \xi$, & its motion θ. Now since $1 : dk :: p : \theta$, (see
prob 5 & ye Note on prop 7 in Example 3); Therefore
$dk \times p = dk = \theta = \dfrac{aa}{x}$. This being knowne, I take at pleasure any
Equation, in wch ξ is, for ye valor of $cd = y$.

As 1st suppose $ay = \xi$, That (by prop 7) gives $aq = \theta = \dfrac{aa}{x}$, Or
$qx = a$, wch also (by prop 7) gives $\gamma x + qp = 0$. Which valors of
$q = \dfrac{a}{x}$, & $\gamma = \dfrac{-q}{x} = \dfrac{-a}{xx}$; by helpe of ye Theorems $\dfrac{1-qq}{\gamma} = dm$
&c: doe give

$$\dfrac{xx - aa}{-a} = dm. \quad \dfrac{xx - aa}{x} = dl \ \& \ \dfrac{xx - aa}{-ax}\sqrt{xx - aa} = lm.$$

55

wch determines ye nature of ye required curve line βm. The length of yt portion of it wch is intercepted twixt ye point m & the curve line acg being $-cm = cd + dm = \dfrac{\xi + xx - aa}{-a}$, Or

$$mc = \frac{\xi + xx - aa}{a}.$$

☞ [Note, yt in this case although ye area $sdkg = \xi$ cannot bee Geometrically found & therefore ye line acg is a Mechanicall one yet ye desired line βm is a Geometricall one. And ye like will happen in all other such like cases, when in ye Equation taken at pleasure to expresse ye relation twixt x, y, & ξ; neither x, y, nor ξ; doe multiply or divide one another, nor it selfe, nor is in any denominator or roote, except x wch may multiply it selfe & bee in denominators & rootes, when y or ξ are not in those fractions or rootes & herein onely doth this excell the precedent 9th probleme. Such is this Equation

$$a\xi - aby + ax^3 + \frac{axx}{ax - xx} - 5xx\sqrt{ab - xx} = 0. \quad \&c.$$

But not this

$$\xi\xi = a^3 y. \quad \text{nor } \xi = xy. \quad \&c]$$

Secondly, suppose $\xi = xy$. yt (by prop 7) gives $\theta = y + xq = \dfrac{aa}{x}$, Or

$xy + xxq = aa$, & yt (by prop 7) gives $y + qx + 2qx + \gamma xx = 0$.

Which two valors of $\dfrac{aa - xy}{xx} = q$, And $\dfrac{y + 3qx}{-xx} = \gamma$; by meanes of

ye Theoremes $\dfrac{1 - qq}{\gamma} = dm$, $\dfrac{q^3 - q}{\gamma} = dl$, & $\dfrac{1 - qq}{\gamma}\sqrt{1 - qq} = lm$; doe determine ye nature & length of ye desired curve line βm.

Example ye 3d. In like manner to find curve lines whose lengths may bee compared to ye length gk of ye said curve line (Hyperbola) gkw.

Call $gk = \xi$, & its motion θ. Now, drawing kh ye tangent to gkw at k, I consider that $ad = x$ & $gk = \xi$ doe increase in ye proportion of dh to kh; yt is, $dh : kh :: p : \theta$. Now finding (by prob 1) yt $dh = -x$, & $kh = \dfrac{-a}{x}\sqrt{aa + xx}$; therefore is

$$\frac{kh \times p}{dh} = \frac{kh}{dh} = a\frac{\sqrt{aa + xx}}{xx} = \theta.$$

Which being found, I take any equation, in wch ξ is, for the valor of $cd=y$. & yn worke as in ye precedent Example.

Note yt by this or ye Ninth Probleme may bee gathered a Catalogue of whatever lines, whose lengths can bee Geometrically found.

Problem 12

To find ye Length of any given crooked line, when it may bee.

Resolution. The length of any streight line to wch ye curve line is cheifly related being called (x), ye length of ye curve line (y), & theire motions (p & q) first (by prob 1) get an equation expressing ye relation twixt x & $\frac{q}{p}$, & yn seeke ye valor of (y) by ye Eight proposition. [Or find a curve line whose area is equall to ye length of ye given line, by Prob. 11. And then find that area by Prob 7.]

Problem 11

To find curve lines whose Areas shall bee equall (or have any other given relation) to ye length of any given Curve line drawn into a given right line.

Resolution. The length of any streight line, to wch ye given curve line is cheifely referred, being called x, ye length of ye curve line y, & their motions of increase p & q. The valor of $\frac{q}{p}$, (found by ye first probleme) being ordinately applyed at right angles to x, gives ye nature of a curve line whose area is equall to (y) ye length of ye curve line.

And this Line thus found gives (by prob 6) other lines whose areas have any given relation to ye length (y) of ye given curve line.

Problem 13

To find ye nature of a Crooked line whose length is expressed by any given Equation, (when it may bee done).

Resolution [Fig. 29]. Suppose $ab=x$, $bc=y$, $ac=z$. & their motions p, q, r. And let ye relation twixt x & z bee supposed given. Then (by prop 8) finding the relation twixt p & r make $\sqrt{rr-pp}=q$: (For drawing cd tangent to ac at c & $de \perp cb \perp ab$: ye lines de, ec, dc, shall bee as p, q, r. but $\sqrt{dc \times dc - de \times de}=ec$, &

therefore $\sqrt{rr-pp}=q$). Lastly, ye ratio twixt x & $\frac{q}{p}$ being thus knowne, seeke y (by prop 8). Which relation twixt $ab=x$ & $bc=y$ determines ye nature of ye crooked line $ac=z$.

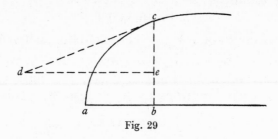

Fig. 29

Of Gravity

Definition 1. I call yt point ye center of Motion in any Body, wch always rests when or howsoever yt Body circulates wthout progressive motion. It would always bee ye same wth ye center of Gravity were ye Rays of Gravity parallel & not converging towards ye center of ye Earth.

Definition 2. And ye right lines passing through yt point I call ye axes of Motion or Gravity.

Lemma 1. The place & distance of Bodys is determined by their centers of Gravity. Which is ye middle point of a right line circle or Parallelogram.

Lemma 2. Those weights doe equiponderate whose quantitys are reciprocally proportionall to their distances from the common axis of Gravity, supposing their centers of Gravity to bee in ye same plaine wth yt common axis of Gravity.

Problem [   ]

To find ye center of Gravity in rectilinear
plaine figures [Fig. 30]

1. In ye Triangle acd make $ab=bc$, & $cf=fd$. & draw db, & af, their intersection point (e) is its center of Gravity.

2. In ye Trapezium $abdc$, draw ad & cb. Joyne ye centers of Gravity e & h, f & g of ye opposite triangles acb & dcb, bad & adc wth ye lines eh, fg [Fig. 31]. Their intersection point n is ye center of Gravity in ye Trapezium. (And so of Pentagons, hexagons &c).

58

Problem [   ]

To find such plaine figures wch are equiponderate to any given plaine figure in respect of an axis of Gravity in any given position [Fig. 32].

Resolution. That ye natures & positions of ye given curvilinear plaine (gbc), & sought plaine (lde) bee such yt they may equiponderate in respect of ye axis (ak) ; I suppose $x = ab \perp bc = z$,

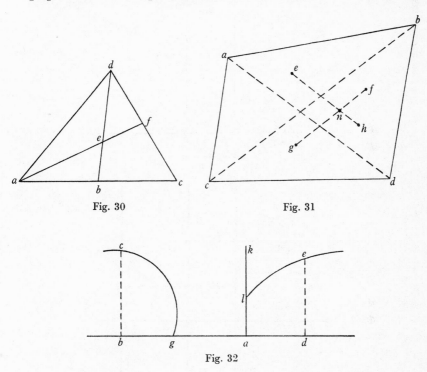

Fig. 30   Fig. 31

Fig. 32

& $y = ad \perp de = v$ to bee either perpendicular or parallel or coincident to ye said axis ak: And ye motions whereby x & y doe increase or decrease (i: e: ye motions of bc, & de to or from ye point a) I call p & q. Now ye ordinatly applyed lines $bc = z$, & $de = v$, multiplyed into their motions p & q (yt is, pz, & qv) may signify ye infinitly little parts of those areas (acb, & lde) wch each moment they describe [Fig. 33]; wch infinitly little parts doe equiponderate (by Lemma 1 & 2), if they multiplied by their distances from ye axis ak doe make equall products [Fig. 34]. (yt is; $pxz = qyv$, in fig [32]: $pxz = \frac{1}{2}qvv$ in fig [33]:

59

$\frac{1}{2}pzz = qv \times fm$, in fig [34]; supposing $dm = me$. &c). And if all ye respective infinitly little parts doe equiponderate ye superficies must do so too.

Now therefore, (ye relation of x & z being given by ye nature of ye curve line cg), I take at pleasure any Equation for ye relation twixt x & y, & thereby (by prop 7) find p & q, & so by ye precedent Theorem find ye relation twixt y & v, for ye nature of ye sought plaine lde.

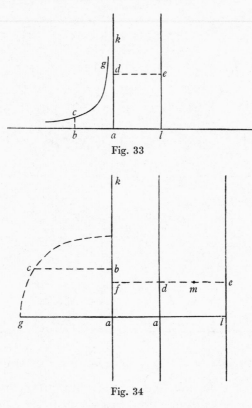

Fig. 33

Fig. 34

Example 1. If cg (fig [33]) is an Hyperbola, soe yt $aa = xz$, & I suppose $2x = y$. yn is $2p = q$ (prop 7). & $paa = pxz = \frac{1}{2}qvv = pvv$. Or $aa = vv$. Or $a = v = de$. Soe yt le is a streight line, & lde a parallelogram. wch equiponderates wth ye Hyperbola $cgkabc$ (infinitly extended towards gk) if $2ab = ad$. $al \times al = ab \times bc$.

Example 2. If cg (fig. [34]) is a circle whose nature is, $\sqrt{aa - xx} = z$, & I suppose at pleasure $3aax - x^3 = 6aay$. Then (by prop 7) I find $3aap - 3xxp = 6aaq$.

And therefore $\frac{1}{2}paa - \frac{1}{2}pxx = \frac{pzz}{2} = qv \times fm = (\text{if } fd = \frac{1}{2}a)$,

$$\frac{qvv + qav}{2} = \frac{vv + av}{2} \times q = \frac{vv + av}{2} \times \frac{aap - xxp}{2aa}.$$

Or $aa \times \overline{paa - pxx} = \frac{vv + av}{2} \times aap - xxp$. Or $2aa = vv + av$.

Or $v = -\frac{1}{2}a \pm \sqrt{\frac{9aa}{4}} = a$. Soe that le is a streight line & $alde$ a parallelogram.

Example 3. If $abcg$ is a parallelogram (fig [35]) whose nature is, $a = z$, & I suppose at pleasure $x = yy - b$. Then (by prop 7) tis $p = 2qy$. Therefore $\frac{1}{2}aap = aaqy = \frac{1}{2}pzz = qvy$. Or $aa = v$. Soe yt aed is a parallelogram.

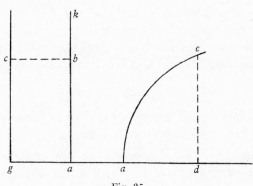

Fig. 35

Or if I suppose at pleasure, $x = y^3 - b$. Then is (prop 7) $p = 3qyy$. & therefore $\frac{1}{2}aap = \frac{3}{2}aaqyy = \frac{1}{2}pzz = qvy$. Or $3aay = 2v$. Soe that aed is a triangle.

Or if I suppose $x = y^4 - b$. yn is, $p = 4qy^3$: & $2aayy = v$. so that aed is a Parabola. [Soe if $xx = y^5$. yn is $2px = 5qy^4$. & $\frac{5a^2qy\sqrt{y}}{4} = \frac{5aaqy^4}{4x} = \frac{1}{2}aap = \frac{1}{2}pzz = qvy$. Or $5aaqy\sqrt{y} = 4qvy$. & $25a^4y = 16vv$. soe yt aed is a Parabola.]

Example 4. If gbc (fig [32]) is an Hyperbola whose nature is $xx - aa = zz$. & I suppose $x = y + b$. Then (by prop 7) is $p = q$.

Therefore $px\sqrt{xx - aa} = pxz = qyv = pyv$.

Or $\overline{y + b} \times \sqrt{yy + 2by + bb - aa} = x\sqrt{xx - aa} = yv$.

61

Or $\overline{yy+2by+bb}$ in $\overline{yy+2by+bb-aa}=yyvv$. &c.

Or if I suppose $xx=2y$. Then is $2px=2q$. Therefore

$$q\sqrt{xx-aa}=pxz=qyv.$$

Or $(xx-aa=)\ 2y-aa=yyvv$.

Or if I suppose $xx=yy+aa$. Then (prop 7) is $2px=2qy$. Therefore $qyz=pxz=qyv$. Or $y=\sqrt{xx-aa}=z=v$. & $y=v$; so yt *aed* is a triangle.

Note yt This Probleme may bee resolved although the lines x, z, y, v, & ak have any other given inclination one to another, but the precedent cases may suffice.

Note also yt if I take a Parallelogram for ye knowne superficies (as in ye 3d Example) I may thereby gather a Catalogue of all such curvilinear superficies whose weight in respect of ye axis, may bee knowne.

Note also I might have shewn how to find lines whose weights in respect of any axis are not onely equall but have also any other given proportion one to another. And yn have made two Problems instead of this, as I did in Problems 5 & 6; 9 & 10.

Problem 15

To find ye Gravity of any given plaine in respect of any axis, given in position, when it may bee done.

Resolution [Fig. 36]. Suppose *ek* to bee ye Axis of Gravity, *acb* the given plaine, $cb=y$, & $db=z$ to bee ordinatly applyed at any angles to $ab=x$. Bisect *cb* at *m* & draw $mn \perp ek$. Now, since $[cb \times mn]$ is ye gravity of ye line $[cb]$, (by Lemmas 1 & 2); if I make $cb \times mn = db = z$, every line *db* shall design ye Gravity of its correspondent line *cb*, yt is, ye superficies *adb* shall designe ye Gravity of ye superficies *acb*. Soe yt finding ye quantity of yt superficies *adb* (by prob 7) I find ye gravity of ye superficies *acb*.

Example 1 [Fig. 37]. If *ac* is a Parabola; soe yt, $rx=yy$, & ye axis *ak* is $\parallel$ to *dcb*. & $nb \perp ak$, &, $ab:nb::d:e$. Then is

$$bc \times nb = y \times \frac{e \times ab}{d} = \frac{ex}{d}\sqrt{rx} = z.$$

Or $eerx^3 = ddzz$, is ye nature of ye curve line *ad* whose area (were *abd* a right angle would be $\frac{2e}{5d}x^{\frac{5}{2}}r^{\frac{1}{2}} = \frac{2e}{5d}\sqrt{rx^5}$ but now it) is

$\frac{2ee}{5dd}\sqrt{rx^5}$, (by prob 7) wch is ye weight of ye area *acb* in respect of ye axis *ak*.

Example 2. If *ac* is a circle

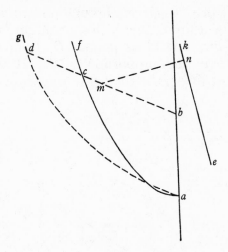

Fig. 36

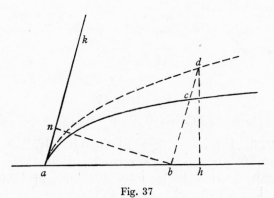

Fig. 37

Problem 16
To find ye Axes of Gravity of any Plaines

Resolution. Find ye quantity of ye Plaine (by Prob 7) wch call *A*, & ye quantity of its gravity in respect of any axis (by prob 15) wch call *B*. & parallell to yt axis draw a line whose distance from it shall bee $\frac{B}{A}$. That line shall bee an Axis of Gravity of ye given plaine.

Or If you cannot find ye quantity of the plane: Then find its gravitys in respect of two divers axes (*AB* & *AC*) [Fig. 38] wch gravitys call *C* & *D*. & through (*A*) ye intersection of those axes draw a line *AD* wth this condition yt ye distances (*DB*; *DC*) of any one of its points (*D*) from the said axes (*AB* & *AC*), bee in such proportion as (*C* to *D*) the gravitys of the plane. That line (*AD*) shall bee an axis of gravity of ye said plane *EF*.

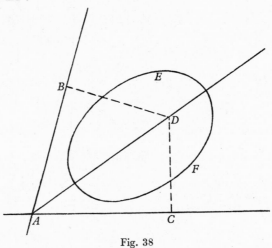

Fig. 38

Problem 17

To find ye Center of Gravity of any Plaine, when it may bee.

Resolution. Find two axes of Gravity, by the precedent Proposition, & their common intersection is ye Center of Gravity desired. If ye figure have any knowne Diameter that may bee taken for one of it axes of Gravity.

PLATE III

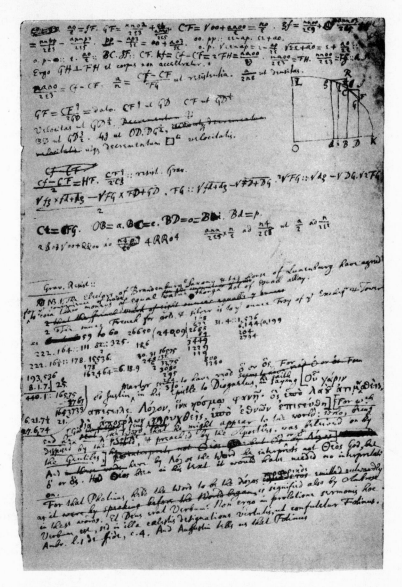

A manuscript leaf illustrating the diversity of Newton's interests in later life. *Top*: calculations on motion in a resisting medium, probably relating to the second edition of the *Principia*. *Centre*: note on currency in Brandenburg and France. *Bottom*: a theological note on the signification of Λόγος.

2

CALCULATIONS OF CENTRIPETAL FORCES

MS. Add. 3965, fols. 4v–5r

The compilers of the *Portsmouth Catalogue* thought this fragment important enough to deserve separate enumeration (Section I, p. 4, viii, no. 2). It consists of little more than notes and calculations on a large folded sheet. The draft is much crossed out and corrected. It begins on the top of fol. 5r and continues (possibly) at the top of 4v. The remainder of fol. 4v is written upside down, beginning with three incomplete sentences followed by the 'Scolium': *Methodus investigandi vires quibus corpora in Orbibus propositis revolventur amplior reddi potest per Propositiones sequentis.* There are three propositions: the first two are in effect corollaries to Proposition X, *Principia*, Book I; and the third is an erroneous expression of the result sought in the following fragment. A fluxional analysis begun at this point was soon abandoned, and so we have not reproduced this part of fol. 4v.

These sketchy drafts of a revision of the *Principia* may belong to the years immediately following its publication.

Et quemadmodum in circulo vel Ellipsi si vires tendunt ad centrum figurae hae vires augendo vel diminuendo ordinatas in ratione quacunque data vel mutando angulum inclinationis Ordinatarum ad Abscissam, semper augentur vel diminuuntur in ratione distantiarum a centro si modo tempora periodica maneant aequalia: sic etiam in figuris universis si Ordinatae augeantur vel diminuuntur in ratione quacunque data vel angulus Ordinationis utcunque mutetur, manente tempore periodico, vires ad centrum quocunque in Abscissa positum tendentes augentur vel diminuuntur in ratione distantiarum a centro.

Scholium

Si corpus moveatur in perimetro datae cujuscunque Sectionis Conicae cujus centrum sit C & requiratur vis centripeta tendens ad punctum quodcunque datum R* [Fig. 39]: vis (per Corol 3 Prop VII) erit ut $\dfrac{SG^{\text{cub}}}{RP^q}$.[1]

---

[1] This paragraph was a false start, and is omitted from the translation, p. 68.

*Radio *RP* parallela ducatur *CG* Orbis tangenti *PG* occurrens in *G* et vis centripeta quaesita (per Prop 10 et cor 3 Prop VII) erit ut $\frac{CG^{\text{cub}}}{RP^{\text{quad}}}$. Haec est Lex Vis centripetae qua corpus in sectione quacunque Conica circa centrum quodcunque movebitur.

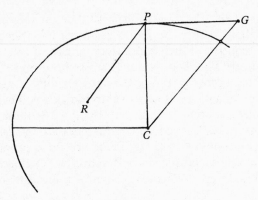

Fig. 39

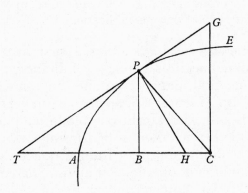

Fig. 40

Si Lex vis centripetae investiganda sit qua corpus in orbe quocunque *APE* circa centrum quodvis *C* movebitur [Fig. 40]: sit $AB=x$ Abscissa & $BP=y$ ordinata ad rectos angulos insistens. Cape *BT* in ea ratione ad *BP* quam habet fluxio *AB* ad fluxionem Ordinatae *BC* et acta *TP* curvam tanget in *P*. Ordinatae *BP* parallela agatur *CG* tangenti occurrens in *G* fluat Abscissa uniformiter & exponatur ejus fluxio per unitatem et si Ordinata BP dicatur v et Vis qua corpus in Orbe APE circa centrum C movebitur erit ut

$$\frac{\ddot{y},\, CP}{\dot{x}\dot{x},\, CG^{\text{cub}}} - \frac{\ddot{v} \times CP}{CG^{\text{cub}}}\, .$$

Exempli gratia dicatur $AB\,x$ et $BC\,y$, & sit $ax = y^n$ aequatio ad Curvam

Exemplum 1. Sit $a^n x = y^{n+1}$ aequatio ad curvam et erit per methodum fluxionum $a^n = \overline{n+1} \times \dot{y} y^n$ et

$$0 = \overline{n+1} \times \ddot{y} y^n + \overline{nn + n} \times \dot{y}^2 y^{n-1}$$

seu $\ddot{y} = \dfrac{-n\dot{y}^2}{y} =$ Et propterea vis centripeta ut $\dfrac{n\dot{y}^2 \times CP}{y \times CG^{\text{cub}}}$, id

est ut $\left(\dfrac{n,\, BP \colon CP}{TB^{\text{q}},\, CG^{\text{cub}}} = \right) \dfrac{n,\, CP}{TB,\, TC,\, CG^{\text{q}}}$

$$= \frac{n,\, CP}{PB,\, TC^{\text{q}},\, CG} \cdot \quad \ddot{y} = \frac{-n\dot{y}^2}{y} = \frac{-n,\, PB}{TB^{\text{q}}}\, .$$

Et propterea vis centripeta ut $\dfrac{n,\, PB,\, CP}{TB^{\text{q}},\, CG^{\text{cub}}}$ id est (ob datum n

et aequalia $TB \times CG$ & $PB \times TC$) ut $\dfrac{CP}{TB,\, TC,\, CG^{\text{q}}}$. vel quod

perinde est ut $\dfrac{CP}{PB \times TC^{\text{q}} \times CG}$.

Si Lex vis centripetae investiganda sit qua corpus P in orbe quocunque APE circa centrum quodcunque datum C movebitur [Fig. 40]: institui potest calculus per methodum sequentem. Sit AB Abscissa Curvae propositae per centrum C transiens sitque BP ejus Ordinata in dato quovis angulo Abscissae insistens. Ducatur TPG Orbem tangens in P & Abscissae occurrens in T et Ordinatae BP parallela agatur CG tangenti occurrens in G. Fluat Abscissa uniformiter & exponatur ejus fluxio per unitatem. Et si Ordinata BC dicatur v Vis centripeta qua corpus P in Orbe APE circa centrum C movebitur erit ut $\dfrac{-CP \times \ddot{v}}{CG^{\text{cub}}}$. Ut si aequatio ad Curvam sit $ax = v^n$, (ubi a quantitatem quamvis datam et x Abscissam denotat, & n index est dignitatis Ordinatae v,) Methodus fluxionum dabit primo $a = n\dot{v}v^{n-1}$ deinde $0 = \overline{nn - n} \times \dot{v}\dot{v}v^{n-2} + n\ddot{v}v^{n-1}$

seu $\ddot{v} = \dfrac{-\overline{n-1} \times \dot{v}\dot{v}}{v}$, id est $\dfrac{-\overline{n-1} \times PB}{TB^q}$. Et propterea vis

centripeta erit ut $\dfrac{\overline{n-1} \times PB \times CP}{TB^q \times CG^{cub}}$ id est (ob datum $n-1$, &

aequalia $TB \times CG$ et $PB \times TC$) ut $\dfrac{CP}{TB \times TC \times CG^q}$, vel quod

perinde est ut $\dfrac{CP}{PB \times TC^q \times CG} = \dfrac{CP}{BH \times TC^{cub}}$

Haec est qualitas omnium in quibus Experimenta instituere licet et propterea per Hypoth III de universis affirmanda est.[1]

And as in the circle or the ellipse, if the forces tend to the centre of the figure, these forces are always increased or diminished (by increasing or decreasing the ordinates in some given proportion, or by changing the angle of inclination of the ordinates to the abscissa) in the ratio of the distances from the centre, if the periodic times remain equal; so also in figures universally: if the ordinates are increased or decreased in some given ratio or the angle of the ordinate is in some way changed, the periodic times being constant, the forces towards a centre, placed anywhere on the abscissa, are increased or diminished in the ratio of the distances from the centre.

Scholium

If a body be moved in the perimeter of a given conic whose centre is C and the centripetal force tending towards any given point R is required [Fig. 39]: let CG be drawn parallel to the radius RP and meeting the tangent to the orbit PG in G, and the centripetal force required (by Prop. 10 and Prop. 7, Cor. 3) will be as CG^3/RP^2. This is the Law of centripetal force by which a body will be moved in any conic section about any centre.

If the Law of centripetal force by which a body will be moved in any orbit whatever APE about any centre C is to be investigated [Fig. 40]: Let $AB = x$ be the abscissa and $BP = y$

---

[1] This last sentence is written sideways along the edge of the sheet.

the ordinate, at right angles to each other. Take *BT* in that ratio to *BP* that the fluxion of *AB* has to the fluxion of the ordinate *BC*, and drawing *TP*, *TP* is a tangent to the curve at *P*. *CG* is drawn parallel to the ordinate *BP* meeting the tangent in *G*: let the abscissa flow uniformly and its fluxion

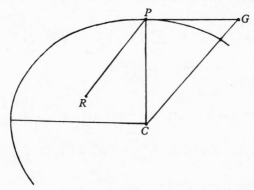

Fig. 39

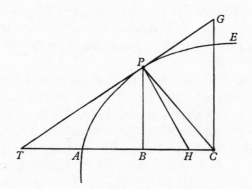

Fig. 40

be expressed by unity, and if the ordinate *BP* is called *v*, the force by which the body will be moved in the orbit *APE* about the centre *C* will be as

$$\frac{\ddot{y}CP}{\dot{x}^2CG^3}\Big] - \frac{\ddot{v}CP}{CG^3}.$$

For example, if *AB* is called *x* and *BC* *y*, and if $ax = y^n$ is the equation expressing the curve.

Example 1. Let $a^n x = y^{n+1}$ be the equation to the curve, and by the method of fluxions

$$a^n = (n+1)\,\dot{y}y^n$$

and $\quad (n+1)\,\ddot{y}y^n + n(n+1)\,\dot{y}^2 y^{n-1} = 0 \quad$ or $\quad \ddot{y} = \dfrac{-n\dot{y}^2}{y}.$

And hence the centripetal force [is] as $(n\dot{y}^2 . CP)/(y . CG^3)$, that is, as

$$\left(\frac{n . BP . CP}{TB^2 . CG^3} = \right)\frac{n . CP}{TB . TC . TG^2} = \frac{n . CP}{PB . TC^2 . CG}.$$

[As before,] $\ddot{y} = \dfrac{-n\dot{y}^2}{y} = \dfrac{-n . PB}{TB^2}.$

And hence the centripetal force [is] as $\dfrac{n . PB . CP}{TB^2 . CG^3}$ that is,

(since n is given, and $TB . GC = PB . TC$), as $\dfrac{CP}{TB . TC . CG^2}$

or what is the same thing, $\dfrac{CP}{PB . TC^2 . CG}.$

If the law of centripetal force by which a body P will be moved in any orbit APE about some given centre C is to be investigated [Fig. 40], the calculation can be effected by the following method.

Let AB be the abscissa to the proposed curve passing through the centre C, and let BP be an ordinate erected upon the abscissa at any given angle. Let TPG be drawn touching the orbit at P and meeting the abscissa at T, and let CG be drawn parallel to the ordinate BP meeting the tangent in G. Let the abscissa flow uniformly and its fluxion be expressed by unity. And if the ordinate BC is called v, the centripetal force by which the body P will be moved about the centre C will be as $-(CP . \ddot{v})/CG^3$. So that if the equation expressing the curve be $ax = v^n$ (where a is a constant and x denotes the abscissa, and n is a power of the ordinate v), the method of fluxions will give: firstly

$$a = n\dot{v}v^{n-1},$$

then $\qquad n(n-1)\,\dot{v}\dot{v}v^{n-2} + n\ddot{v}v^{n-1} = 0$

or $\qquad\qquad \ddot{v} = \dfrac{-(n-1)\,\dot{v}\dot{v}}{v},$

that is,
$$\ddot{v} - \frac{(n-1)\,PB}{TB^2}.$$

And hence the centripetal force will be as
$$\frac{(n-1)\,PB\,.\,CP}{TB^2\,.\,CG^3},$$

that is (since $n-1$ is a constant, and $TB\,.\,CG = PB\,.\,TC$), as
$$\frac{CP}{TB\,.\,TC\,.\,CG^2},$$

or what is the same thing,
$$\frac{CP}{PB\,.\,TC^2\,.\,CG} = \frac{CP}{BH\,.\,TC^3}$$

This is a property of all things that may be investigated experimentally, and should therefore, by Hypothesis III, be affirmed of all things.

PART II

MECHANICS

INTRODUCTION

Although he was a 'mechanical philosopher' in the seven-teenth-century sense, and although the *Principia* is a treatise on mechanics and physics, Newton left few writings on the science of mechanics proper, apart from papers connected with the composition of the *Principia*. His manuscripts are far less rich in such material than those of Huygens, for example, nor did Newton ever lecture on mechanics except when he delivered the material afterwards printed in the *Principia*. We know of his early interest in the calculation of centripetal forces and in the relation of such calculations to the problem of planetary motion,[1] but otherwise, apart from a few scattered references in letters, there is little evidence of his concern even for celestial mechanics before 1679. Of similar studies in other branches of mechanics (such as one would suppose must have preceded the writing of Book II of the *Principia*) there is almost no trace at all.

This lends particular interest to three of the four papers bearing on mechanics generally which we have printed in this section. Although none can be precisely dated, Nos. 1, 2 and 4 clearly antedate the *Principia* by many years; in fact, the first two may well belong to Newton's student days. No. 2, 'The Lawes of Motion', is probably the earliest in the series. It opens with definitions which are similar to the ones treated more elaborately in No. 1, and describes the parallelogram of forces. Next Newton considers the rotation of an irregular body about an internal axis, pointing out that such rotation is unstable unless the forces are equally balanced about the axis of rotation. Finally he discusses the effect of a collision with another body of an irregular body that is both rotating and moving progressively. From the 'Observations' which follow the main treatise it appears that the discussion was designed to have some reference to the collision of physical bodies; it is there pointed out that these do not satisfy the simple condi-tions envisaged earlier in the work by Newton. There is also a remark that the motion of two bodies (defined as the product

[1] A. R. Hall, 'Newton on the Calculation of Central Forces', *Annals of Science*, **13**, 1957, 62–71.

of a body's bulk and its velocity) may be either increased or diminished by their collision. This is of interest for its bearing on Newton's belief that the quantity of motion in the universe is not constant.

Both this short paper and No. 1, 'De Gravitatione et aequipondio fluidorum', appear to spring from Newton's critical reaction to the *Principia Philosophiae* of Descartes. This is indicated by the first few lines of 'The Lawes of Motion', and by its attempt to analyse the motions of bodies after impact, which played so large a role in Cartesian physics. It is obvious throughout No. 1, even without the explicit references to Descartes' work. This, besides being the longest of these papers, is undoubtedly the most important. It is also the most curious. Newton clearly intended to write an elaborate treatise on hydrostatics; but, after completing a long criticism of Descartes, he seems to have lost interest in his original purpose. Only the beginning and the end deal with the equilibrium of fluids; the experiments with which Newton proposed to illustrate his arguments were never furnished; and the enterprise was abandoned when, after many pages of digression, it was at last just begun. As a contribution to hydrostatics, therefore, the document is worthless; but as revealing Newton's thought (early in his life) on many other topics it is of high interest.

The trouble starts when Newton attempts to define the basic concepts, *space*, *place* and *body*—a problem to which he returned, more successfully and more succinctly, in the *Principia*. His two fundamental conceptions, (1) that body is a part of space endowed with certain empirical properties, and (2) that motion is with respect to (absolute) space, oppose those of Descartes. Hence it is necessary to confute Descartes, point by point. Newton's first target is Descartes' definition of rest: that a body is at rest when it is not moved from neighbouring matter that touches it. From this definition, Descartes argued that the Earth is at rest, since it remains always encompassed by the matter of the vortex in which it is borne around the Sun. Newton has no difficulty in showing the absurdity of this position, by which a centrifugal force is attributed to the non-moving Earth, and by which the Sun, fixed stars, planets and Earth are all said to be at rest although they are observed to change their relative positions. From the same definition it must follow that the external particles of bodies move when the whole body moves, but that the internal particles of the same body do not. At the same time, Newton maintains that all relative motions are not true and philosophical motions: for example, while the Earth has a relative motion with respect to Saturn, its true motion is about the Sun, since this gives rise to a force of recession from the Sun, which the other does not. (The argument that centrifugal force is one measure of true motion is repeated in the discussion of absolute motion in the *Principia*.)

Indeed, already at this period, Newton found the Cartesian determination of motion (or rest) in any single particle by measuring its displacement from other particles or bodies one which vitiated the concept of motion. No particles of bodies in the universe are known to be truly at rest, so they cannot serve as milestones; and, what is worse, if the point of origin of any particular motion is designated by the positions of bodies that are themselves moving, the location of that point is lost after any lapse of time and the motion in question rendered indeterminate. 'So it is necessary that the definition of places, and hence of local motion, be referred to some motionless thing such as extension alone or space in so far as it seems to be truly distinct from bodies.' That is, motion can properly be referred only to absolute space, but Newton does not say how space (as distinct from the bodies occupying it) is to be pegged out in feet and inches.

Having gone so far, Newton now feels impelled to complete his destruction of the Cartesian concept of extension, and to define more accurately his own notions of the nature of space, body and extension. And in so doing he encounters the body-mind dualism that he recognizes as the principal foundation of Cartesian philosophy.

Body (or substance) is, for Newton, known from its actions, that is, from its effects on other bodies or on the perceiving mind. Extension has no such effects, and is therefore not to be identified with substance. Since extension (that is, a part of space) can be conceived independently of body, and as existing in space where there are no bodies, it is not an accident or attribute of substance either. Yet extension is not nothingness, for it is defined by length, breadth and depth, whereas nothingness cannot be defined. Extension is, in fact (though these are not Newton's words), the potentiality of figure in space. No figure such as a tetrahedron or a sphere nor any part of the surface of such a figure can actually be perceived in space until a body of that figure occupies that part of space; when that happens, extension, being filled with substance, becomes perceptible. But extension in the form of tetrahedra and spheres and all other figures exists everywhere in potentiality, to become perceptible when, and only when, occupied by substance. 'We firmly believe', says Newton, 'that space was spherical before the sphere occupied it, so that it could contain the sphere; and hence as there are everywhere spaces that can adequately contain any material sphere, it is clear that space is everywhere spherical. And so of other figures.' This is to say that the figure of the material sphere does not, so to speak, imprint itself in shapeless space: rather the material sphere can be anywhere because the potentiality of extension to receive it exists everywhere. The figure of a body is a function of space, rather than of the substance of which the body is composed. To believe otherwise would be to yield to 'that puerile and jejune prejudice according to which extension is inherent in bodies like an accident in a subject without which it cannot possibly exist.'

Newton now goes on to elaborate his conception of space as infinite, eternal, immutable and motionless. This section of the treatise is confused, for he wanders away into further digressions on the distinction between *infinite* and *indefinite*, and *infinite* and *perfect*. Neither space nor duration are completely independent of being, however, for 'when any being is postulated, space is postulated. And the same may be asserted of duration....' How can this be reconciled with the fact that we can conceive of space independently of bodies in it, and of duration without a constantly enduring body, that is, time without a timekeeper? Newton has the answer: God is everywhere, and endures for ever. His being, infinite in time and space, is that which forces us to postulate the infinity of time and space. God did not create space and time when he created the world; indeed it would seem that Newton would argue that God no more created them than he created himself, since they are consequences of his being. As Newton wrote in one draft of the General Scholium, God could not be nowhere (or in no time), for what is nowhere is nothing.

If the existence of God necessitates that of space, does it necessitate that of extension also? Newton answers this question affirmatively. Extension cannot be conceived without God, who eminently contains it within himself. Thus extension is eternal, infinite, uncreated and uniform, whereas body is the opposite in every respect. A body, as defined by Newton, is no more than a volume of space (or 'determinate quantity of extension') which is mobile, impenetrable and able to excite sensation in the human mind. (By *body* he means, of course, an ultimate particle, one of the entities of which gross bodies with their multifarious physical properties are composed.) The creation of substance involved the endowment of extension with these properties, which result directly and continuously from the exercise of the divine will. There is no unintelligible reality commonly called substance residing in bodies, in which all the properties of the bodies are inherent: what we recognize as substance is the image created in our minds by the perceptions which (because of divine will) impenetrable, moving extensions arouse. Substance is an illusion maintained by God alone; if God ceased to keep up the illusion, substance would be annihilated, although extension would not. Nothing is more conducive to atheism than forgetfulness of this truth, and adherence to 'this notion of bodies having as it were a complete, absolute and independent reality in themselves.' And since bodies— that is, the physical universe—cannot exist independently of God, they cannot be understood independently of the idea of God. 'God is no less present in his creatures than they are present in the accidents, so that created substance, whether you consider its degree of dependence or its degree of reality, is of an intermediate nature between God and accident. And hence the idea of it no less involves the concept of God than the idea of accident involves the concept of created substance.' Accidents—the properties of bodies—are, Newton hints, more real than any unintelligible independent substance or matter of which bodies are made, for they are at least intelligible things in themselves; if we should follow this

hint, we might say that there is for Newton in this treatise a
double reality: on the one hand, that of extension which
exists in God, and on the other that of properties like impenetrability and motion with which parts of extension are
continuously endowed by God. Between these two there is
neither room nor reason for an independent reality of matter.

Moreover, it seems clear that if each and every particle of
matter is to be considered as a determinate quantity of
extension, surrounded by empty space, to which God has
given the qualities of impenetrability or hardness and motion,
it must have other properties or accidents in addition to these
two. It must also, for instance, have the power of affecting
human minds (since these can perceive it) as Newton is to
insist later; and, as he has already pointed out, it must be
capable of acting upon other matter too. (Bodies can only
act upon other bodies by causing them to move or rest, or at
least to tend to do so.) In Definitions 5 to 10, towards the end
of the treatise, Newton defines six 'powers' which he presumably attributes to bodies universally: these are force, *conatus*,
impetus, inertia, pressure and gravity. The other five powers
are really all embraced under the first, force, considered in
Definition 5 as 'the causal principle of motion and rest...
either an external one that generates or destroys or otherwise
changes impressed motion in a body, or...an internal
principle by which existing motion or rest is conserved in a
body.' The latter principle (inertia) must, according to
Newton's arguments, be implanted by God in incipient
matter, since extension alone has neither inertia nor mobility.
From inertia follows density and hence mass. By this reasoning mass, like hardness, offers no evidence for the independent
reality of substance. Other forces are, as indicated in
Definition 10 (where centrifugal force is said to be a kind of
gravity) subsumed under the term *gravity*. Newton does not
say how a body acquires gravity, or the power either to exert
or to respond to one or more forces, whether attractive or
repulsive. This is the more unfortunate as the origin of physical
forces is the central problem of Newton's philosophy of
nature. (Cf. below, Introduction to Section III). It seems
probable, however, that at least when writing this treatise,
Newton did not hold the belief that all natural forces are

mechanical in the Cartesian sense, that is, resolvable into the direct impact of particle on particle. On the contrary, he seems to hold that natural forces, whatever their origin in, or association with matter, must result from the divine will. For, if God, extension and duration alone exist in reality, and if the simplest attributes of substance such as hardness and inertia can be conceived as existing only through the continuous exercise of God's will which enforces these and other laws of material existence, then equally the forces of the physical universe revealed in its phenomena must arise in the same way. No cause of material phenomena can be more real and independent than matter itself.

Thus Newton, reasoning that the mechanical philosophy of Descartes promotes atheism by assigning to matter attributes of independence and reality that properly belong only to God, reacts by qualifying matter as an illusion. In doing so, he is, of course, speaking purely as a metaphysician. As a scientist, Newton is content to treat the physical universe as if matter really existed. It is, for him, only necessary to think that it does not when trying to explain how both matter and mind can exist at the same time—which is the point at which Newton found Descartes' arguments totally incomprehensible. Some elements in this discussion of Newton's may recall passages in Bishop Berkeley's *Principles of Human Knowledge*, strange though it may seem that Newton who is commonly regarded as furnishing the scientific substratum of Locke's philosophy should appear on the same side as Locke's critic. Berkeley in his most famous passage wrote that it was perfectly unintelligible that any part of the physical universe should have existence independent of a spirit; that primary qualities have no real existence in an inert, non-sentient substance called matter; and that 'the very notion of what is called matter, or corporeal substance, involves a contradiction in it'.[1] But Berkeley, to whom it was inconceivable that anything should exist unperceived by a sentient being, invoked God who perceives everything in order that matter should exist. Newton, on the other hand, denied matter in order that God should exist. The scientist was more a theologian than the philosopher.

[1] *Principles of Human Knowledge*, Sections VI–IX.

Just as Berkeley justifies belief in the existence of *this* object at *this* moment by awareness of its distinctive qualities in a human mind that perceives them, and the existence of *all* objects at *all* times by the perception of them by God, so Newton draws an analogy between the relation of God to matter and the relation of the human mind to matter. Every man believes that he can move his own body at will, and that other men equally direct their limbs by thought; so God, 'by the sole action of thinking and willing' can make a space impenetrable, and cause it to move, and thus render it body. The human mind in its body is analogous to God in space, 'so that all the difficulties of the [latter] conception may be reduced to that'. Therefore, 'the analogy between the divine faculties and our own is greater than has formerly been perceived by Philosophers,' though human faculties are in all respects feebler. God can both create and move bodies; created minds can only move them, and that in accord with certain divine laws. This is almost tantamount to describing God as the soul of the world and the world as his body, which he created and dominates. Later, in the General Scholium, Newton seems to deny such a conception, writing as he does of God as a being who 'governs all things, not as the soul of the world, but as Lord over all'; but there is no true contradiction. He meant that God was not merely a passive spirit, or an active principle, but (as Newton insists throughout this earlier treatise) the actual creator of the physical world, and its ruler in a far more complete sense than the human mind is ruler of its body.

The analogy would have been more useful if Newton could have thrown fresh light on the relation of mind and body in man: but this he was unable to do. He has to confess that, while we can form no idea of God's attribute of creation, 'nor even of our own power by which we move our bodies, it would be rash to say what may be the substantial basis of mind'. Nevertheless, he is convinced that the dualism of Descartes is incomprehensible: for if mind is not extended in space, it is nowhere and does not exist. There can thus be no complete distinction between thinking entities and extended (that is, material) ones: extension is contained in God, the highest thinking being, and so a body may think, and a thinking being extend. Nor is there a fundamental distinction between 'thinking body' and 'non-thinking body'; the phrase 'living matter' would be a contradiction to Newton. All of what is commonly called substance can constitute, as it were, the machinery of mind. As he rather quaintly puts it, 'From the fact that the parts of the brain, especially the more subtle ones by which the mind is united, are in a continual flux, new ones succeeding to those which fly away, it is manifest that that faculty [of uniting with mind] is in all bodies.' Yet if thought and extension are compatible, they are only made so by God's creative act, for 'there is so great a distinction between the ideas of thinking and extension that it is impossible there should be any basis of connection or relation [between them] except that which is caused by divine power'.

Thus the microcosm-macrocosm analogy between man and God turns out to be disappointing. To say that God in the universe is like the mind in the body is trivial metaphysics, and Newton has after all said little more than this, since the union of body and mind, from which the divine creative process was to be illuminated, proves to be itself a mysterious product of that process. Nor does Newton have anything to substitute for the rejected dualism of Descartes which, thrust out of the front door of his argument, seems to creep in at the back with his avowal that the faculty of thinking is independent of the substance of the thinking organ, the brain. Since Newton fails to establish any real connexion between brain and mind, the existence of a thinking substance must remain a miracle. And on Newton's assumptions, how could it be otherwise? He often considered the problems of perception and cerebration, in relation to vision particularly; here, as in other questions of philosophy, mechanism could account for much, but not for everything. It is true that in Newton's view the human mind is a feeble image of God. But God, who is infinitely extended, who is everywhere the same ('all eye, all ear, all brain, all arm') thinks and perceives without a brain and without organs of sensation:[1] the highest form of thought, accordingly, is unlocalized in space, and requires no material machinery in the form of nerves and brain. Why then should the weak imitation that is human perception and thought

---

[1] Cf. drafts of the General Scholium.

require such localization, or such intimate association with specialized organs as exists in the human body? If we follow Newton's reasoning rightly, created minds *could* exist independently of such trappings, and only God's will can be alleged as the reason if they do not. For to say (as Newton does) that minds cannot exist without extension is, obviously, not the same as saying that minds cannot exist without association with matter, an assertion that perhaps Newton would not have cared to make. Thus, in the end, the union of mind and body remains as inessential and arbitrary, and therefore as inexplicable, as it was at first.

There are, however, possibilities relative to this union inherent in Newton's metaphysics, though he does not explicitly express them. If substance—such as the substance of the brain—is neither real nor independent, but consists of parts of space infused by God with certain properties, then the coincidence of mind and substance in the same space will be, at least, less problematic. The antithesis, 'this entity must be either mind or matter but not both' need not hold good if the 'entity' is (in Newton's manner) considered to be a part of space or a determinate quantity of extension differentiated from other parts or quantities by certain properties infused by God, among which could be the faculties of perception and thought. For it might be argued that, after all sensible qualities have been stripped from a sentient and thinking being, there would remain besides extension the faculties of perceiving and thinking, if we may extrapolate from Newton's own argument that after all the sensible qualities have been stripped from a body what remains is not extension alone but also the faculty or power (essential to the corporeal nature of the body, but not to its extension) by which the body excites perception in thinking beings. But yet—coincidence is not union, nor (as Newton himself was aware) are the disparate ideas of *matter* and *thought* reconciled by attributing both to the same entity. Indeed by conjoining them and thus describing the human brain, for instance, as a 'thinking body', Cartesian dualism is surmounted in only a verbal sense, to reappear as soon as the questions asked about the brain as the seat of reason and emotion are contrasted with other questions about the brain as animal substance. But since Newton does not pose such questions, nor any about the rich range of mental activity, his already shaky arguments are not further embarrassed.

One might suppose that such questions and problems were far removed from the proper content of a treatise on mechanics, and so at last did Newton. After a final attack upon the physical philosophy of Descartes, in which he asserts the vacuity of space between the particles of matter and hence the need for a distinction between space and body, Newton at last returns to the definitions from which he has so long digressed. These definitions are not especially interesting in themselves; they are, in fact, rather old-fashioned. Thus Newton still uses such terms as *conatus* (rather than *vis*) and the scholastic distinction between intension and extension. In dealing with the absolute quantity of 'powers' (not yet 'forces') he is confused: thus he is not clear whether the absolute quantity of motion should be defined as *mv* or *mvt*. Definition 15 is of interest for its remark that density is proportional to inertia [volume being constant], a definition repeated much later in the second edition of the *Principia* (Book III, Prop. VI, Cor. IV); the well-known accusation that Newton's definition of mass as the product of volume and density is circular (since density is usually defined as mass per unit volume) can, in the light of this definition, be seen to be ill-founded.

The treatise ends with two propositions asserting that all the parts of a non-gravitating (and, Newton might have added, uniform) fluid, equally compressed in all directions, compress each other equally without any motion of the parts thereby arising. Such properties in a fluid are, of course, required for the kind of argument that is as old as Archimedes' *On Floating Bodies*. The second proposition would seem to be a consequence of the impossibility of perpetual motion. The propositions are followed by five corollaries and a scholium, and then the treatise ends. It is impossible to tell what further shape it might have taken.

The other two documents are of a different kind. No. 3, 'The Elements of Mechanicks', is a popular exposition, or rather summary, similar to those which we have placed in Section V. It states results which Newton derived in the *Principia* (such as the elliptical form of cometary orbits) in nineteen theorems, without any argument. There is no attempt to correlate the theoretical theorems plainly and directly with those on the celestial phenomena. Yet in eight brief paragraphs we have a clearer and simpler description of the mechanical 'Frame of the World', as Newton saw it, than anywhere else in his writings. It seems possible that Newton drew up this document for someone who desired a succinct statement of his views, in effect an outline of the principal truths of the *Principia*. At least this view is more probable

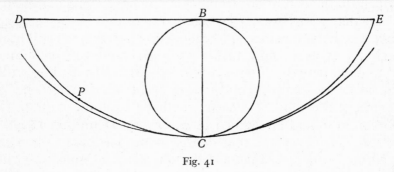

Fig. 41

than the alternative one, that it was a synopsis of a work Newton intended to write. Some statements made here are not found in the *Principia*, for example that about the gravitating power of the stars. More interesting, however, are the remarks on harmonic motion in the first section: this subject, much discussed in the seventeenth century (notably by Marin Mersenne in his *Harmonie Universelle*, 1637), was little treated by Newton elsewhere. However, the point made here, that all vibratory motions—the oscillations of a pendulum, the relaxations and contractions of a spring, or the vibrations of an elastic string—are instances of harmonic motion, is an important one.

The final document, No. 4, 'Gravia in Trochoide descendentia', analyses such an harmonic motion in detail. It consists of mathematical demonstrations of the properties of a body moving in an inverted cycloid, such as *DCE* [Fig. 41], the diameter of the generating circle being *BC*. Newton's results may be summarized as follows:

(1) The acceleration at any point *P* of a body descending along the cycloid to *C* is $g(PC/BC)$.

(2) A body descends to C from any point P in the same time.

(3) Therefore if a body oscillates in a cycloid, oscillations of any amplitude will be of the same duration.

(4) The evolute of a cycloid is an identical cycloid.

(5) Therefore if a bob on a flexible thread oscillates between cycloidal cheeks, its path will be a cycloid and its oscillations isochronous.

(6) An arc of circle of radius $2BC$ differs minimally from the cycloid DCE for a few degrees either side of the centre C.[1]

(7) The time of descent from any point on the cycloid to the centre is $\pi\sqrt{(BC/2g)}$.

(8) During the time of one oscillation of a pendulum of length $2BC$, a body falls freely through the distance $\pi^2 BC$.

The last result is remarked upon by Newton as being of outstanding importance; from it, indeed, it is easy to show that $g = \pi^2 l$, and that $T = \pi\sqrt{(L/g)}$, where l is the length of the pendulum beating seconds, and T is the period of any pendulum of length L.

The geometry of the cycloid had been extensively studied by earlier mathematicians: Roberval, Torricelli, and Descartes had all worked on it in the period after 1630, and in 1658 Blaise Pascal issued his famous challenge to the world to solve certain complex problems concerning it. At about the same time the young Christiaan Huygens, seeking to discover the nature of the curve in which a pendulum would oscillate with perfect isochronism, found that the cycloid satisfied this condition. By the end of 1659 he had demonstrated most of the kinematical properties of the curve enumerated above and had made a clock whose pendulum, by oscillating between cycloidal cheeks, executed a truly isochronous swing. The construction of this clock was soon widely known, but Huygens only revealed the geometrical reasoning on which it was founded in *Horologium Oscillatorium* (1673).

The propositions of this work relevant to Newton's paper are Part II, 25; Part III, 5 and 7; Part IV, 25 and 30. Huygens' proofs are considerably more cumbersome, though also more formal, than those of Newton. He did not begin,

---

[1] For this reason, if the swing of a clock-pendulum is small Huygens' cycloidal cheeks are unnecessary. About 1671 the anchor-escapement, having a pendulum swinging through a small circular arc, supplanted Huygens' cycloidal escapement. Newton's observation of this geometrical point may well indicate a date of writing later than 1671.

as Newton does, by determining the acceleration of the body at any point in the curve, a method which allows Newton to proceed with great facility. It may be noted also that while Newton gives a formula from which the gravitational constant may be calculated, he seems not to have performed accurate pendulum experiments to determine its value. Huygens did make such experiments, finding the length of the seconds' pendulum to be 3·057 feet, Paris measure, from which he computed that $g = 30·16$ Paris feet (980 cm.)/sec^2.[1]

Newton inserted six propositions on cycloidal motion in the *Principia* (Book I, Props. XLVIII—LIII). These are not the same as those of the manuscript, though some of the same results are obtained.

[1] In the *Principia* (Book III, Proposition IV) Newton takes the value of g from Huygens' pendulum experiments. It appears, however, that he became aware of the possibility of measuring g from pendulum experiments about 1666, though he did not then possess a full theory of the pendulum. From some rough experiments he calculated g as about 32 feet/sec^2; in some earlier computations (ignoring the experiments of Riccioli and others) he had followed a value little more than half of this found in Galileo.

1

[DE GRAVITATIONE ET AEQUI-
PONDIO FLUIDORUM]

MS. Add. 4003

This manuscript, accurately described in the *Portsmouth Catalogue* (p. 48, no. 17), is written in a calf-bound notebook on both sides of each leaf from fol. 4 through fol. 23v, where it ceases without any indication that the work was complete. Nothing else is written in the notebook, although there are many blank leaves remaining.

The hand is certainly Newton's. At first sight one might doubt whether Newton was also the author, or whether this was one of the cases where he made long extracts without noting the source, for the matter of the treatise seems un-Newtonian, and some words are found here which are certainly not common in Newton's usual Latin. Any doubt, however, can easily be removed by reading the treatise through, when it appears thoroughly Newtonian. For example, the attack on Descartes' identification of matter with extension occurs elsewhere in Newton's writings; the comparison of the resistance of the aether to a body moving through it with that of air, water or mercury was often made by Newton; and the passages on the relation of God and the created world are extremely similar, even in their language, to parallel passages in the General Scholium which concludes the *Principia* in the later editions. Here as there, for instance, Newton denies that God can be identified with the soul of the world and emphasizes his omnipresence. Moreover, although this manuscript is for the most part neatly written, it contains a sufficient number of alterations and changes of construction, especially in the later portions, to satisfy us that it was composed by the writer.

While it is thus possible to be certain that this piece was written by Newton, it is much more difficult to form any opinion of when and for what reason he wrote it. The form of the hand is suggestive of Newton's youth; by the 1680's his writing was larger and more flowing than it is here, and by then too he normally drafted his thoughts on loose sheets of paper (though he still made entries in his chemical notebook). The structural failure of the essay, which consists for the most part of an enormous digression leading very far from the announced subject of hydrostatics; the immaturity of some of the thought; the pomposity of a good deal of the Latin; the over-elaborate proofs of elementary theorems, all combine to support the

89

same judgement. The mature Newton would never have found such obvious satisfaction in solemnly confuting Descartes by arguments which, if not exactly trivial, are certainly laboured. One might well guess that this was an essay written by a young student who had recently been introduced both to the science of hydrostatics and to Descartes' *Principia Philosophiae*, and who was fired with enthusiasm to show his powers to his master; on this appraisal the manuscript might have been written between, say, 1664 and 1668. It is difficult to believe that it can have been composed later than the published optical paper of 1672 or than *De Aere et Aethere*, although, as the elementary expositions of astronomy written much later reveal, simplicity of content does not entirely guarantee an early date.

Though this treatise does seem to belong to an early stage in Newton's career, it contains, as has been indicated above, many ideas that he developed more fully later. In one draft of the General Scholium, for example (Section IV, no. 8, MS. C), Newton asks, 'Can God be nowhere when the moment of time is everywhere?' This obscure question is illuminated by the discussion in this manuscript of the identity of moments of (absolute) duration in all parts of the universe. It is interesting, too, to find Newton writing on the theory of fluids before the composition of Book II of the *Principia*, for examples of his early interest in this branch of science are rare indeed. Most strikingly, this work emphasizes Newton's protracted interest in the problem of the relation of God to the universe; this has been discussed in the Introduction to this section.

De Gravitatione et aequipondio fluidorum et solidorum in fluidis scientiam duplici methodo tradere convenit. Quatenus ad scientias Mathematicas pertinet, aequum est ut a contemplatione Physica quam maxime abstraham. Et hac itaque ratione singulas ejus propositiones e principijs abstractis et attendenti satis notis, more Geometrarum, stricte demonstrare statui. Deinde cum haec doctrina ad Philosophiam naturalem quodammodo affinis esse censeatur, quatenus ad plurima ejus Phaenomena enucleanda accommodatur, adeoque cum usus ejus exinde praesertim elucescat et principiorum certitudo fortasse confirmetur, non gravabor propositiones ex abundanti experimentis etiam illustrare: ita tamen ut hoc laxius disceptandi genus in Scholia dispositum, cum priori per Lemmata, propositiones et corallaria tradito non confundatur.

Fundamenta ex quibus haec scientia demonstranda est sunt vel definitiones vocum quarundam; vel axiomata et postulata a nemine non concedenda. Et haec a vestigio tradam.

Definitiones

Nomina quantitatis, durationis et spatij notiora sunt quam ut per alias voces definiri possint.

Def: 1. Locus est spatij pars quam res adaequate implet[a].
Def: 2. Corpus est id quod locum implet[a]
Def: 3. Quies est in eodem loco permansio
Def: 4. Motus est loci mutatio[b]

Nota. Dixi corpus implere locum[a], hoc est ita saturare ut res alias ejusdem generis sive alia corpora penitus excludat, tanquam ens impenetrabile. Potuit autem locus dici pars spatij cui res adaequate inest, sed cum hic corpora tantum et non res penetrabiles spectentur, malui definire esse spatij partem quam res implet.

Praeterea cum corpus hic speculandum proponatur non quatenus est Substantia Physica sensibilibus qualitatibus praedita sed tantum quatenus est quid extensum mobile et impenetrabile; itaque non definivi pro more philosophico, sed abstrahendo sensibiles qualitates (quas etiam Philosophi ni fallor abstrahere debent, et menti tanquam varios modos cogitandi a motibus corporum excitatos tribuere,) posui tantum proprietates quae ad motum localem requiruntur. Adeo ut vice Corporis Physici possis figuras abstractas intelligere quemadmodum Geometrae contemplantur cum motum ipsis tribuunt, ut fit in prop 4 & 8, lib 1 Elem Euclid. Et in demonstratione definitionis 10mae, lib 11 debet fieri; siquidem ea inter definitiones vitiose recensetur & potius inter propositiones demonstrari debuit, nisi forte pro axiomate habeatur.

Definivi praeterea motum esse loci mutationem,[b] propterea quod motus, transitio, translatio, migratio &c videntur esse voces synonymae. Sin malueris esto motus transitio vel translatio corporis de loco in locum.

Caeterum in his definitionibus cum supposuerim spatium a corpore distinctum dari, et motum respectu partium spatij istius, non autem respectu positionis corporum contiguorum determinaverim [ne id gratis contra Cartesianos assumatur, Figmenta ejus tollere conabor.

Doctrinam ejus in sequentibus tribus propositionibus complecti possum 1o Quod unicuique corpori unicus tantum motus proprius ex rei veritate competit (Artic 28, 31 & 32 part 2 Princip:) qui definitur esse Translatio unius partis materiae sive unius corporis ex vicinia eorum corporum quae illud immediate contingunt, et tanquam quiescentia spectantur, in viciniam aliorum. (Art 25 part 2, & Artic 28 part 3 Princip). 2do Quod per corpus proprio motu juxta hanc definitionem translatum non tantum intelligitur materiae particula aliqua vel corpus ex partibus inter se quiescentibus compositum, sed id omne quod simul transfertur; etsi rursus hoc ipsum constare possit ex multis partibus quae alios inter se habeant motus. Art 25, part 2, Princip. 3o Quod praeter hunc motum unicuique corpori proprium, innumeri etiam alij motus per pa[r]ticipationem (sive quatenus est pars aliorum corporum alios motus habentium) possunt ipsi revera: in esse (Art 31, part 2 Princip): Qui tamen non sunt motus in sensu philosophico & cum ratione loquendo (Art 29 part 3) & secundum rei veritatem (Art 25, part 2 & Art 28 part 3:) sed improprie tantum et juxta sensum vulgi. (Art 24, 25, 28, & 31 part 2, & Art 29, part 3.) Quod motuum genus videtur (Art 24 part 2, & 28 part 3) describere esse actionem qua corpus aliquod ex uno loco in alium migrat.

Et quemadmodum duplices constituit motus, proprios nempe ac derivativos, sic duplicia loca assignat e quibus isti motus peraguntur, eaque sunt superficies corporum immediate ambientium (Art 15 part 2), et situs inter alia quaecunque corpora (Art 13 part 2 & 29 part 3).

Jam vero quam confusa et rationi absona est haec doctrina non modo absurdae consequentiae convincunt, sed et Cartesius ipse sibi contradicendo videtur agnoscere. Dicit enim Terram caeterosque Planetas proprie et juxta sensum Philosophicum loquendo non moveri, eumque sine ratione et cum vulgo tantum loqui qui dicit ipsam moveri propter translationem respectu fixarum (Art 26, 27, 28, 29 part 3). Sed postea tamen in Terra et Planetis ponit conatum recedendi a Sole tanquam a centro circa quod moventur, quo per consimilem conatum Vorticis gyrantis in suis a Sole distantijs librantur Art 140 part 3. Quid itaque? an hic conatus a quiete Planetarum juxta Cartesium vera et Philosophica, vel potius a motu vulgi et non Philosophico derivandus est? At inquit Cartesius praeterea quod Cometa minus conatur recedere a sole cum primum vorticem ingreditur et positionem inter fixas fere retinens nondum obsequitur Vorticis impetui sed respectu ejus transfertur e vicinia contigui aetheris, adeoque philosophice loquendo circa solem gyrat; quam postea cum Vorticis materia Cometam secum abripuit fecitque ut in ea juxta sensum philosophicum quiesceret. Art 119 et 120 part 3. Haud itaque sibi constat Philosophus jam adhibens motum vulgi pro fundamento Philosophiae, quem paulo ante rejecetrat, et motum illum jam pro nullo rejiciens quem solum antea dixerat esse secundum rei naturam verum et philosophicum. Et cum gyratio Cometae circa Solem in ejus sensu Philosophico non efficit conatum recedendi a centro, quem gyratio in sensu vulgi potest efficere, sane motus in sensu vulgi pro magis philosophico debet agnosci.

Secundo videtur sibi contradicere dum ponit unicum cuique corpori motum juxta rei veritatem competere, et tamen motum istum ab imaginatione nostra pendere statuit, definiendo esse translationem e vicinia corporum non quae quiescunt, sed quae ut quiescentia tantum spectantur etiamsi forte moveant, quemadmodum in Artic 29 et 30, part 2 latius explicatur. Et hinc putat se posse difficultates circa mutuam translationem corporum eludere cur nempe unum potius quam aliud moveri dicatur, et cur navis aqua praeter fluente dicitur quiescere cum positionem inter ripas non mutat. Art 15 part 2. Sed ut pateat contradictio, finge quod materia vorticis a quolibet homine tanquam quiescens spectatur et Terra philosophice loquendo simul quiescet: finge etiam quod alius quisquam eodem tempore spectat eandem Vorticis materiam ut circulariter motam, et terra philosophice loquendo non quiescet. Ad eundem navis in mari simul movebit et non movebit; idque non sumendo motum in laxiori sensu vulgi, quo innumeri sunt motus cujusque corporis, sed in ejus sensu philosophico quo dicit unicum esse in quolibet corpore, et ipsi proprium esse et ex rei (non imaginationis nostrae) natura competere.

Tertio videtur haud sibi constare dum ponit unicum motum cuique corpori secundum rei veritatem competere, et tamen (Art 31 part 2) innumeros motus unicuique corpori revera inesse. Nam motus qui revera insunt alicui corpori, sunt revera motus naturales, adeoque motus in sensu philosophico et secundum rei veritatem, etiamsi contendat esse motus in solo sensu vulgi. Adde quod cum totum aliquod movetur, partes omnes ex quibus una translatis constituitur revera quiescent, nisi concedantur vere moveri participando de motu totius, et proinde motus innumeros juxta rei veritatem habere.

Sed videamus praeterea ex consequentijs quam absurda est haec Cartesij doctrina. Et imprimis quemadmodum acriter contendit Terram non moveri quia non transfertur e vicinia contigui aetheris; sic ex ijsdem principijs consectatur quod corporum durorum internae particulae, dum non transferuntur e vicinia particularum immediate contingentium, non habent motum proprie dictum sed moventur tantum participando de motibus externarum particularum: imo quod externarum partes interiores non moventur motu proprio quia non transferuntur e vicinia partium internarum: Adeoque quod sola superficies externa cujusque corporis movetur motu proprio, et quod tota interna substantia, hoc est totum corpus movetur per participationem motus externae superficiei. Peccat igitur motus definitio fundamentalis quae tribuit id corporibus quod solis superficiebus competit, facitque ut nullus potest esse motus cuivis corpori proprius.

Secundo. Quod si spectemus solam Artic 25 part 2. Unumquodque corpus non unicum tantum sed innumeros sibi proprios motus habebit, dummodo proprie et juxta rei veritatem moveri dicentur quorum totum proprie movetur. Idque quia per corpus cujus motum definit, intelligit id omne quod simul transfertur, etsi hoc ipsum constare potest ex partibus alios motus inter se habentibus; puta Vorticem una cum omnibus planetis vel navem una cum omnibus quae insunt mari innatantem, vel hominem in navi ambulantem una cum rebus quae secum defert, aut horologij rotulam una cum particulis metallum constituentibus. Nam nisi dices quod totius aggregati motus non ponit motum proprie et

94

secundum rei veritatem partibus competentem, fatendum erit quod hi omnes motus rotularum horologij, hominis, navis, & vorticis revera et philosophice loquendo inerunt rotularum particulis.

Ex utraque harum consequentiarum patet insuper quod e motibus nullus prae alijs dici potest verus absolutus et proprius, sed quod omnes, sive respectu contiguorum corporum sive remotorum, sunt similiter philosophici, quo nihil absurdius imaginari possumus. Nisi enim concedatur unicum cujusque corporis motum physicum dari, caeterasque respectuum et positionum inter alia corpora mutationes, esse tantum externas denominationes: sequetur Terram verbi gratia conari recedere a centro Solis propter motum respectu fixarum, et minus conari recedere propter minorem motum respectu Saturni et aetherei orbis in quo vehitur, atque adhuc minus respectu Jovis et aetheris circumducti ex quo orbis ejus conflatur, et iterum minus respectu Martis ejusque orbis aetherei, multoque minus respectu aliorum orbium aethereae materiae qui nullum Planetam deferentes sunt propriores orbi annuo Terrae; respectu vero proprij orbis non omnino conari, quoniam in eo non movetur. Qui omnes conatus et non conatus cum non possunt absolute competere dicendum est potius quod unicus tantum motus naturalis et absolutus Terrae competit, cujus gratia conatur recedere a Sole, et quod translationes ejus respectu corporum externorum sunt externae tantum denominationes.

Tertio. Sequitur e doctrina Cartesiana, motum ubi nulla vis imprimitur generari posse. Verbi gratia si Deus efficeret ut Vortecis nostri gyratio derepente sisteretur, nulla vi in terram impressa quae simul sisteret: diceret Cartesius quod terra propter translationem e vicinia contingentis fluidi, jam in sensu philosophico moveret, quam prius dixit in eodem philosophico sensu quiescere.

Quarto. Ab eadem doctrina sequitur etiam quod Deus ipse in aliquibus motum generare nequit etsi vi maxima urgeat. Verbi gratia si caelum stellatum una cum omni remotissima parte creationis, Deus unquam vi maxima urgebat ut causaretur (puta motu diurno) circa terram convolvi; tamen inde non caelum sed terra tantum juxta Cartesium revera moveri diceretur Art 38 part 3. Quasi perinde esset sive vi ingenti effecerit caelos ab oriente ad occidentem converti, sive vi parva terram in contrarias partes converterit. At quisquam putabit quod partes terrae conantur a centro ejus recedere propter vim caelis solum modo impressam? Vel non est rationi magis consentaneum ut vis caelis indita faciat illos conari recedere a centro gyrationis inde causatae, et ideo solos proprie et absolute moveri; et quod vis terrae impressa faciat ejus partes conari recedere a centro gyrationis inde causatae, et ideo solam proprie et absolute moveri: Etiamsi similis est in utroque casu translatio corporum inter se. Et proinde motus physicus et absolutus aliunde quam ab ista translatione denominandus est, habita translatione ista pro externa tantum denominatione.

Quinto. A ratione videtur alienum ut corpora, absque motu physico, distantias et positiones inter se mutent: sed ait Cartesius quod Terra caeterique planetae et stellae fixae proprie loquendo quiescunt, et tamen mutant positiones inter se.

Sexto. Et contra non minus a ratione videatur alienum esse, ut plura corpora servent easdem positiones inter se quorum alterum physice movetur, et altera quiescunt. Sed si Deus Planetam aliquem sisteret faceretque ut eandem inter stellas fixas positionem continuo servaret annon diceret Cartesius quod stellis non moventibus planeta propter translationem e materia Vorticis jam physice moveretur.

Septimo. Interrogo qua ratione corpus aliquod proprie moveri dicetur quando alia corpora ex quorum vicinia transfertur non spectantur ut quiescentia, vel potius quando non possunt ut quiescentia spectari. Verbi gratia quomodo noster vortex propter translationem materiae juxta circumferentiam, e vicinia consimilis materiae aliorum circumjacentium vorticum potest dici circulariter moveri, siquidem circumjacentium vorticum materia non possit ut quiescens spectari, idque non tantum respectu nostri vorticis, sed etiam quatenus vortices illi non quiescunt inter se. Quod si Philosophus hanc translationem non ad numericas corporeas vorticum particulas refert, sed ad spatium (ut ipse loquitur) genericum in quo vortices illi existunt, convenimus tandem, nam agnoscit motum referri debere ad spatium quatenus a corporibus distinguitur.

Denique ut hujus positionis absurditas quam maxima pateat, dico quod exinde sequitur nullam esse mobilis alicujus determinatam velocitatem nullamque definitam lineam in qua movetur. Et multo magis quod corporis sine impedimentis moti velocitas non dici potest uniformis, neque linea recta in qua motus perficitur. Imo quod nullus potest esse motus siquidem nullus potest esse sine aliqua velocitate ac determinatione.

Sed ut haec pateant, imprimis ostendendum est quod post motum aliquem peractum nullus potest assignari locus juxta Cartesium in quo corpus erat sub initio motus peracti, sive dici non potest unde corpus movebat. Et ratio est quod juxta Cartesium locus non definiri et assignari potest nisi ex positione circumjacentium corporum, et quod post motum aliquem peractum positio corporum circumjacentium non amplius manet eadem quae fuit ante. Verbi gratia si Jovis Planetae locus ubi erat ante annum jam peractum quaeratur; qua ratione, quaeso, Philosophus Cartesianus describet? Non per positiones particularum fluidae materiae, siquidem istae particulae positiones quas ante annum habuere, quam maxime mutaverint. Neque describet per positiones solis et fixarum stellarum, quoniam inaequalis influxus materiae subtilis per polos vorticum in sidera centralia, (Part 3 Art 104), Vorticum undulatio (Art 114,) inflatio (Art 111,) et absorptio, aliaeque veriores causae, ut solis et astrorum circa propria centra gyratio, generatio macularum, et cometarum per caelos trajectio, satis mutant et magnitudines siderum, et positiones, ut forte non sufficiant ad locum quaesitum sine aliquot miliarium errore designandum, et multo minus ut ipsarum ope locus accurate describi ac determinari possit, quemadmodum Geometra describi postularet. Nulla equidem in mundo reperiuntur corpora quorum positiones inter se diuturnitate temporis non mutantur et multo minus, quae non moventur in sensu Cartesij hoc est vel quatenus transferuntur e vicinia contiguorum corporum vel quatenus sunt partes aliorum corporum sic translatorum: Et proinde nullum datur fundamentum quo locus qui fuit in tempore praeterito, jam in praesentia designari possit, vel unde possumus dicere talem locum jam amplius in rerum natura reperiri. Nam cum locus juxta Cartesium nihil aliud sit quam superficies corporum ambientium vel positio inter alia quaelibet remotiora corpora: impossibile est ex ejus doctrina ut in rerum natura diutius existat quam manent eaedem illae corporum positiones ex quibus individuam denominationem sumpsit. Et proinde de loco Jovis quem ante annum habuit, parique ratione de praeterito loco cujuslibet mobilis manifestum est juxta Cartesij doctrinam, quod ne quidem Deus ipse (stante rerum novato statu) possit accurate et in sensu Geometrico describere, quippe cum propter mutatas corporum positiones, non amplius in rerum natura existit.

Jam itaque post motum aliquem completum cum locus in quo inchoabatur hoc est initium trajecti spatij non assignari potest nec amplius esse: illius trajecti spatij non habentis initium nulla potest esse longitudo; et proinde cum velocitas pendet ex longitudine spatij in dato tempore transacti, sequitur quod moventis nulla potest esse velocitas; quemadmodum volui primo ostenderere. Praeterea quod de initio spatij transacti dicitur de omnibus intermedijs locis similiter debet intelligi; adeoque cum spatium nec habet initium nec partes intermedias sequitur nullum fuisse spatium transactum et proinde motus nullam determinationem, quod volui secundo indicare. Quin imo sequitur motum Cartesianum non esse motum, utpote cujus nulla est velocitas, nulla determinatio et quo nullum spatium, distantia nulla trajicitur. Necesse est itaque ut locorum determinatio adeoque motus localis ad ens aliquod immobile referatur quale est sola extensio vel spatium quatenus ut quid a corporibus revera distinctum spectatur. Et hoc lubentius agnoscet Cartesianus Philosophus si modo advertat quod Cartesius ipse extensionis hujus quatenus a corporibus distinctae ideam habuit, quam voluit ab extensione corporea discriminare vocando genericam. Art 10, 12, & 18, part 2 Princip. Et quod vorticum gyrationes, a quibus vim aetheris recedendi a centris, adeoque totam ejus mechanicam Philosophiam deduxit, ad extensionem hance genericam tacite referuntur.

Caeterum cum Cartesius in Art 4 & 11 Part 2 Princip demonstrasse videtur quod corpus nil differt ab extensione; abstrahendo scilicet duritiem, colorem, gravitatem, frigus, calorem caeterasque qualitates quibus corpus carere possit ut tandem unica maneat ejus extensio in longum latum et profundum quae proinde sola ad essentiam ejus pertinebit. Et cum haec apud plurimos pro demonstratione habetur, estque sola ut opinor causa propter quam fides huic opinioni constringi potest: ideo ne ulla circa naturam motus supersit dubitatio, respondebo huic argumento dicendo quid sit Extensio, quid corpus et quomodo ab invicem differunt. Cum enim distinctio substantiarum in cogitantes et extensas vel potius in cogitationes et extensiones sit praecipuum Philosophiae Cartesianae fundamentum, quod contendit esse vel mathematicis demonstrationibus notius: eversionem ejus ex parte extensionis, ut veriora Mechanicarum scientiarum fundamenta substruantur, haud parvi facio.

De extensione jam forte expectatio est ut definiam esse vel substantiam vel accidens aut omnino nihil. At neutiquam sane, nam habet quendam sibi proprium existendi modum qui neque substantijs neque accidentibus competit. Non est substantia tum quia non absolute per se, sed tanquam Dei effectus emanativus, et omnis entis affectio quaedam subsistit; tum quia non substat ejusmodi proprijs affectionibus quae substantiam denominant, hoc est actionibus, quales sunt cogitationes in mente et motus in corpore. Nam etsi Philosophi non definiunt substantiam esse ens quod potest aliquid agere, tamen omnes hoc tacite de substantijs intelligunt, quemadmodum ex eo pateat quod facile concederent extensionem esse substantiam ad instar corporis si modo moveri posset et corporis actionibus frui. Et contra haud concederent corpus esse substantiam si nec moveri posset nec sensationem aut perceptionem aliquam in mente qualibet excitare. Praeterea cum extensionem tanquam sine aliquo subjecto existentem possumus clare concipere, ut cum imaginamur extramundana spatia aut loca quaelibet corporibus vacua; et credimus existere ubicunque imaginamur nulla esse corpora, nec possumus credere periturum esse cum corpore si modo Deus aliquod annihilaret, sequitur eam non per modum accidentis inhaerendo alicui subjecto existere. Et proinde non est accidens. Et multo minus dicetur nihil, quippe quae magis est aliquid quam accidens et ad naturam substantiae magis accedit. Nihili nulla datur Idea neque ullae sunt proprietates sed extensionis Ideam habemus omnium clarissimam abstrahendo scilicet affectiones et proprietates corporis ut sola maneat spatij in longum latum et profundum uniformis et non limitata distensio. Et praeterea sunt ejus plures proprietates concomitantes hanc Ideam, quas jam enumerabo non tantum ut aliquid esse sed simul ut quid sit ostendam.

1. Spatium omnifariam distingui potest in partes quarum terminos communes solemus dicere superficies; et istae superficies omnifariam distingui possunt in partes, quarum terminos communes nominamus lineas; et rursus istae lineae omnifariam distingi possunt in partes quas dicimus puncta. Et hinc superficies non habet profunditatem, nec linea latitudinem, neque punctum quamlibet dimensionem; nisi dicas quod spatia contermina se mutuo ad usque profunditatem interjectae superficiei penetrant, utpote quam dixi esse utriusque terminum sive extremitatem communem: & sic de lineis et puncti. Praeterea spatia sunt ubique spatijs contigua, et extensio juxta extensionem ubique posita, adeoque partium contingentium ubique sunt termini communes, hoc est, ubique superficies disterminantes solida hinc inde, et ubique lineae in quibus partes superficierum se contingunt, et ubique puncta in quibus linearum partes continuae nectuntur. Et hinc ubique sunt omnia figurarum genera, ubique sphaerae, ubique cubi, ubique triangula, ubique lineae rectae, ubique circulares, Ellipticae, Parabolicae, caeteraeque omnes, idque omnium formarum et magnitudinum, etiamsi non ad visum delineatae. Nam materialis delineatio figurae alicujus non est istius figurae quoad spatium nova productio, sed tantum corporea representatio ejus ut jam sensibus appareat esse quae prius fuit insensibilis in spatio. Sic enim credimus ea omnia spatia esse spaerica [sic] per quae sphaera aliqua progressive mota in singulis momentis transijt unquam, etiamsi sphaerae istius inibi sensibilia vestigia non amplius manent. Imo spatium credimus prius fuisse sphaericum quam sphaera occupabat, ut ipsam posset capere; et proinde cum ubique sunt spatia quae possunt sphaeram quamlibet materialem adaequate capere, patet ubique esse spatia sphaerica. Et sic de alijs figuris. Ad eundem modum intra aquam claram etsi nullas videmus materiales figuras, tamen insunt plurimae quas aliquis tantum color varijs ejus partibus inditus multimodo faceret apparere. Color autem si inditus esset, non constitueret materiales figuras sed tantum efficeret visibiles.

2. Spatium in infinitum usque omnifariam extenditur. Non possumus enim ullibi limitem imaginari quin simul intelligamus quod ultra datur spatium. Et hinc omnes lineae rectae paraboliformes, hyperboliformes, et omnes coni et cilindri, et ejusmodi caeterae figurae in infinitum usque progrediuntur, Et nullibi limitantur etsi passim a lineis et superficiebus omnigenis transversim pergentibus intercipiuntur, et figurarum segmenta cum ipsis omnifariam constituunt. Verum ut infiniti specimen aliquod habeatis; fingite triangulum aliquod cujus basis cum uno crure quiescat et crus alterum circa terminum ejus basi contiguum ita gyret in plano trianguli ut triangulum in vertice gradatim apperiatur: et interea advertite puncta animis vestris, ubi crura duo concurrerent si modo eo usque producerentur, et manifestum est quod ista omnia puncta reperiuntur in linea recta in qua crus quiescens jacet, et quod eo longius perpetim distant quo crus mobile diutius convolvitur eo usque dum alteri cruri parallelum evadat et non potest amplius cum eo alicubi concurrere. Rogo jam quanta fuit distantia puncti ultimi in quo crura concurrebant? Certe major fuit quam ulla potest assignari, vel potius nullum e punctis fuit ultimum, et proinde recta linea in qua omnia illa concursuum puncta reperiuntur est actu plusquam finita. Neque est quod aliquis dicat hanc imaginatione tantum et non actu infinitam esse; nam si triangulum sit actu adhibitum, ejus crura semper actu dirigentur versus aliquod commune punctum, ubi concurrerent ambo si modo producerentur, et proinde tale punctum ubi productae concurrerent semper erit actu, etiamsi fingatur esse extra mundi corporei limites; atque adeo linea quam ea omnia puncta designant erit actualis, quamvis ultra omnem distantiam progrediatur.

Siquis jam objiciat quod extensionem infinitam esse non possumus imaginari; concedo: Sed interea contendo quod possumus intelligere. Possumus imaginari majorem extensionem ac majorem deinde, sed intelligimus majorem extensionem existere quam unquam possumus imaginari. Et hinc obiter facultas intelligendi ab imaginatione clare distinguitur.

Sin dicat praeterea quod non intelligimus quid sit ens infinitum nisi per negationem limitum finiti, et quod haec est negativa adeoque vitiosa conceptio: Renuo. Nam limes vel

terminus est restrictio sive negatio pluris realitatis aut existentiae in ente limitato, et quo minus concipimus ens aliquod
limitibus constringi, eo magis ,liquid sibi poni deprehendi-
mus, hoc est eo magis positive concipimus. Et proinde
negando omnes limites conceptio evadet maxime positiva.
Finis est vox quoad sensum negativa, adeoque infinitas cum
sit negatio negationis (id est finium) erit vox quoad sensum et
conceptum nostrum maxime positiva, etsi grammatice negativa videatur. Adde quod plurimarum superficierum longitudine infinitarum positivae et finitae quantitates a Geo-
metris accurate noscuntur. Et sic plurimorum solidorum tum
longitudine tum latitudine infinitorum quantitates solidas
positive et exacte determinare possum, et ad data finita
soli[d]a aequiparare. Sed id non est hujus loci.

Quod si Cartesius jam dicat extensionem non infinitam fore
sed tantum indefinitam, a Grammaticis corrigendus est. Nam
vox indefinita nunquam dicitur de eo quod actu est sed
semper respicit futuri possibilitatem, tantum denotans aliquid
esse nondum determinatum ac definitum. Sic antequam Deus
aliquid de Mundo creando statuerat, (siquando non statuerat,) materiae quantitas stellarum numerus caeteraque
omnia fuerunt indefinita, quae jam mundo creato definiuntur.
Sic materia est indefinite divisibilis, sed semper vel finite vel
infinite divisa. Art 26, P:1. & 34. P:2. Sic linea indefinita est
quae nondum determinatur cujusnam sit futurae longitudinis.
Et sic spatium indefinitum est quod nondum determinatur
cujusnam sit futurae magnitudinis, quod vero jam actu est
non est definiendum sed vel habet terminos vel non habet,
adeoque vel finitum est vel infinitum. Nec obstat quod dicit
esse indefinitum quoad nos, hoc est nos tantum ignorare fines
ejus et non positive scire nullos esse, (Art 27 Part 1:) tum quia
nobis nescientibus, Deus saltem non indefinite tantum sed
certe et positive intelligit nullos esse, tum quia nos etiam
quamvis negative imaginamur, tamen positive et certissime
intelligimus id limites omnes transcendere. Sed video quid
metuit Cartesius, nempe si spatium poneret infinitum,
Deum forte constitueret propter infinitatis perfectionem.
At nullo modo, nam infinitas non est perfectio nisi quatenus
perfectionibus tribuitur. Infinitas intellectus, potentiae,
faelicitatis &c est summa perfectio; infinitas ignorantiae,

impotentiae, miseriae &c summa imperfectio; et infinitas extensionis talis est perfectio qualis est extendi.

3. Partes spatij sunt immobiles. Si moveantur, vel dicendum est quod cujusque motus sit translatio e vicinia aliarum contingentium partium, quemadmodum Cartesius definivit motus corporum, et hoc absurdum esse satis ostendi; vel dicendum est quod sit translatio de spatio in spatium, hoc est de seipsis, nisi forte dicatur quod duo ubique spatia coincidunt, mobile et immobile. Caeterum spatij immobilitas optime per durationem illustrabitur. Quemadmodum enim durationis partes per ordinem individuantur, ita ut (instantiae gratia) dies hesternus si ordinem cum hodierno die commutare posset et evadere posterior, individuationem amitteret et non amplius esset hesternus dies sed hodiernus: Sic spatij partes per earum positiones individuantur ita ut si duae quaevis possent positiones commutare, individuationem simul commutarent, et utraque in alteram numerice converteretur. Propter solum ordinem et positiones inter se partes durationis et spatij intelliguntur esse eaedem ipsae quae revera sunt; nec habent aliud individuationis principium praeter ordinem et positiones istas, quas proinde mutare nequeunt.

4. Spatium est entis quatenus ens affectio. Nullum ens existit vel potest existere quod non aliquo modo ad spatium refertur. Deus est ubique, mentes creatae sunt alicubi, et corpus in spatio quod implet, et quicquid nec ubique nec ullibi est id non est. Et hinc sequitur quod spatium sit entis primario existentis effectus emanativus, quia posito quolibet ente ponitur spatium. Deque Duratione similia possunt affirmari: scilicet ambae sunt entis affectiones sive attributa secundum quae quantitas existentiae cujuslibet individui quoad amplitudinem praesentiae et perseverationem in suo esse denominatur. Sic quantitas existentiae Dei secundum durationem aeterna fuit, et secundum spatium cui adest, infinita; et quantitas existentiae rei creatae secundum durationem tanta fuit quanta duratio ab inita existentia, et secundum amplitudinem praesentiae tanta ac spatium cui adest.

Caeterum nequis hinc imaginetur Deum ad instar corporis extendi et partibus divisibilibus constare: sciendum est ipsissima spatia non esse actu divisibilia, et insuper ens quodlibet habere modum sibi proprium quo spatijs adest. Sic enim durationis longe alia est ad spatium relatio quam corporis. Nam diversis partibus spatij non ascribimus diversam durationem, sed dicimus omnes simul durare. Idem est durationis momentum Romae et Londini, idem Terrae et astris caelisque universis. Et quemadmodum unumquodque durationis momentum sic per universa spatia, suo more, sine aliquo partium ejus conceptu diffundi intelligimus: ita non magis contradicit ut Mens etiam suo more sine aliquo partium conceptu per spatium diffundi possit.

5. Corporum positiones, distantiae, et motus locales ad spatij partes referendae sunt. Et hoc patet e prima et quarta recensita proprietate spatij, et manifestius erit si inter corpuscula concipias vacuitates esse disseminatas, vel attendas ad ea quae de motu prius dixi. His praeterea subnecti potest quod spatio non inest vis aliqua impediendi aut promovendi vel qualibet ratione mutandi motus corporum. Et hinc corpora projectilia lineas rectas uniformi motu describunt si non aliunde occurrant impedimenta. Sed de his plura posthac.

6. Denique spatium est aeternae durationis et immutabilis naturae, idque quod sit aeternis et immutabilis entis effectus emanativus. Siquando non fuerit spatium, Deus tunc nullibi adfuerit, et proinde spatium creabat postea ubi ipse non aderat, vel quod non minus rationi absonum est, creabat suam ubiquitatem. Porro quamvis fortasse possumus imaginari nihil esse in spatio tamen non possumus cogitare non esse spatium; quemadmodum non possumus cogitare durationem non esse, etsi possibile esset fingere nihil omnino durare. Et hoc per extramundana spatia manifestum est, quae (cum imaginamur mundum esse finitum) non possumus cogitare non esse, quamvis nec a Deo nobis revelata sunt, nec per sensus innotescunt nec a spatijs intramundanis quoad existentiam dependent. Sed de spatijs istis credi solet quod sunt nihil. Imo vero sunt spatia. Spatium etsi sit corpore vacuum tamen non est seipso vacuum. Et est aliquid quod sunt spatia quamvis praeterea nihil. Quinimo fatendum est quod spatia non sunt magis spatia ubi mundus existit quam ubi nullus est, nisi forte dices quod Deus cum mundum in hoc spatio creabat, spatium simul creabat in seipso vel quod Deus si

mundum in his spatijs posthac annihilaret, etiam spatia annihilaret in seipsis. Quicquid itaque est pluris realitatis in uno spatio quam in altero, illud corporis est et non spatij; quemadmodum clarius patebit si modo puerile illud et ab infantia derivatum praejudicium deponatur quod extensio inhaeret corpori tanquam accidens in subjecto sine quo revera nequit existere.

Descripta extensione natura corporea ex altera parte restat explicanda. Hujus autem, cum non necessario sed voluntate divina existit, explicatio erit incertior propterea quod divinae potestatis limites haud scire concessum est, scilicet an unico tantum modo materia creari potuit, vel an plures sunt quibus alia atque alia entia corporibus simillima producere licuit. Et quamvis haud credibile videtur Deum posse entia corporibus simillima creare quae omnes eorum actiones edant et exhibeant phaenomena et tamen in essentiali et metaphysica constitutione non sint corpora: cum tamen ejus rei nondum habeo claram ac distinctam perceptionem, non ausim contrarium affirmare, et proinde nolo positive dicere quaenam sit corporea natura sed potius describam quoddam genus entium corporibus per omnia similium quorum creationem esse penes Deum non possumus non agnoscere, et proinde quae haud possumus certo dicere non esse corpora.

Cum quisque hominum sit sibi conscius quod pro arbitrio possit corpus suum movere et credit etiam quod alijs hominibus eadem inest potestas qua per solas cogitationes sua corpora similiter movent: potestas movendi quaelibet pro arbitratu corpora Deo neutiquam deneganda est, cujus cogitationum infinite potentior est et promptior facultas. Et pari ratione concedendum est quod Deus sola cogitandi aut volendi actione impedire posset ne corpora aliqua spatium quodlibet certis limitibus definitum ingrediantur.

Quod si potestatem hancce exerceret, efficeretque ut spatium aliquod super terra ad instar montis vel corporis cujuslibet terminatum evaderet corporibus impervium, adeoque lucem omniaque impingentia sisteret aut reflecteret; impossibile videtur ut ope sensuum nostrorum (qui soli in hac re judices constituerentur) hoc spatium non revera corpus esse detegeremus; foret enim tangibile propter impenetrabilitatem, et visibile opacum et coloratum propter reflectionem lucis, et percussum resonaret propterea quod aer vicinus percussione moveretur.

Fingamus itaque spatia vacua per mundum disseminari quorum aliquod certis limitibus definitum, divina potestate evadat corporibus impervium, et ex hypothesi manifestum est quod hoc obsisteret motibus corporum et fortasse reflecteret, et particulae corporeae proprietates omnes indueret, nisi quod foret immobile. Sed si fingamus praeterea illam impenetrabilitatem non in eadem spatij parte semper conservari sed posse huc illuc juxta certas leges transferri ita tamen ut illius spatij impenetrabilis quantitas et figura non mutetur, nulla foret corporis proprietas quae huic non competeret. Esset figuratum, tangibile, et mobile, reflecti posset, et reflectere, et in aliqua rerum compagine partem non minus constituere quam aliud quodvis corpusculum, et non video cur non aeque posset agere in mentes nostras et vicissim pati, cum sit nihil aliud quam effectus mentis divinae intra definitam spatij quantitatem elicitus. Nam certum est Deum voluntate sua posse nostras perceptiones movere, et proinde talem potestatem effectibus suae voluntatis adnectere.

Ad eundem modum si plura hujusmodi spatia et corporibus et seipsis impervia fierent, ea omnia vices corpusculorum gererent, eademque exhiberent phaenomena. Atque ita si hic mundus ex hujusmodi entibus totus constitueretur, vix aliter se habiturum esse videtur. Et proinde haec entia vel corpora forent vel corporibus simillima. Quod si forent corpora, tum corpora definire possemus esse *Extensionis quantitates determinatas quas Deus ubique praesens conditionibus quibusdam afficit*: quales sunt (1) ut sint mobiles, et ideo non dixi esse spatij partes numericas quae sunt prorsus immobiles, sed tantum definitas quantitates quae de spatio in spatium transferri queant. (2) Ut ejusmodi duo non possint qualibet ex parte coincidere, sive ut sint impenetrabiles et proinde ut occurrentes mutuis motibus obstent certisque legibus reflectantur. (3) Ut in mentibus creatis possint excitare varias sensuum et phantasiae perceptiones, et ab ipsis vicissim moveri, nec mirum cum originis descriptio in hoc fundatur.

Caeterum de jam explicatis juvabit annotare sequentia. (1) Quod ad horum entium existentiam non opus est ut effingamus aliquam substantiam non intelligibilem dari cui tanquam subjecto forma substantialis, inhaereat: sufficiunt extensio et actus divinae voluntatis. Extensio vicem substantialis subjecti gerit in qua forma corporis per divinam voluntatem conservatur; et effectus iste divinae voluntatis est forma sive ratio formalis corporis denominans omnem spatij dimensionem in qua producitur esse corpus.

(2) Haec entia non minus forent realia quam corpora, nec minus dici possent substantiae. Quicquid enim realitatis corporibus inesse credimus, hoc fit propter eorum Phaenomena et sensibiles qualitates. Et proinde haec Entia, cum forent omnium istius modi qualitatum capacia, et possent ea omnia phaenomena similiter exhibere, non minus realia esse judicaremus, si modo existerent. Nec minus forent substantiae, siquidem per solum Deum pariter subsisterent et substarent accidentibus.

(3) Inter extensionem et ei inditam formam talis fere est Analogia qualem Aristotelici inter materiam primam et formas substantiales ponunt; quatenus nempe dicunt eandem materiam esse omnium formarum capacem, et denominationem numerici corporis a forma mutuari. Sic enim pono quamvis formam per quaelibet spatia transferri posse, et idem corpus ubique denominare.

(4) Differunt autem quod extensio (cum sit et quid, et quale, et quantum) habet plus realitatis quam materia prima, atque etiam quod intelligi potest, quemadmodum et forma quam corporibus assignavi. Siqua enim est in conceptione difficultas id non est formae quam Deus spatio indaret, sed modi quo indaret. Sed ea pro difficultate non habenda est siquidem eadem in modo quo membra nostra movemus occurrit, et nil minus tamen credimus nos posse movere. Si modus iste nobis innotesceret, pari ratione sciremus quo pacto Deus etiam corpora posset movere et a loco aliquo data figura terminato expellere et impedire ne expulsa vel alia quaevis possent rursus ingredi, hoc est, efficere ut spatium istud foret impenitrabile et formam corporis indueret.

(5) Hujus itaque naturae corporeae descriptionem a facultate movendi corpora nostra deduxi ut omnes in conceptu difficultates eo tandem redirent; et praeterea ut (nobis intime conscijs) pateret Deum nulla alia quam volendi actione creasse mundum, quemadmodum et nos sola volendi

actione movemus corpora nostra; et in super ut Analogiam inter nostras ac Divinas facultates majorem esse ostenderem quam hactenus animadvertere Philosophi. Nos ad imaginem Dei creatos esse testatur sacra pagina. Et imago ejus in nobis magis elucescet si modo creandi potestatem aeque ac caetera ejus attributa in facultatibus nobis concessis adumbravit neque obest quod nosmet ipsi sumus creaturae adeoque specimen hujus attributi nobis non pariter concedi potuisse. Nam etsi ob hanc rationem potestas creandi mentes non delineetur [sic] in aliqua facultate mentis creatae, tamen mens creata (cum sit imago Dei) est naturae longe nobilioris quam corpus ut forsan eminenter in se contineat. Sed praeterea movendo corpora non creamus aliquid nec possumus creare sed potestatem creandi tantum adumbramus. Non possumus enim efficere ut spatia aliqua sint corporibus impervia, sed corpora tantum movemus, eaque non quaelibet sed propria tantum, quibus divina constitutione et non nostra voluntate unimur, neque quolibet modo sed secundum quasdem leges quas Deus nobis imposuit. Siquis autem maluit hanc nostram potestatem dici finitum et infimum gradum potestatis quae Deum Creatorem constituit, hoc non magis derogaret de divina potestate quam de ejus intellectu derogat quod nobis etiam finito gradu competit intellectus; praesertim cum non est propriae et independentis potestatis sed legis a Deo nobis impositae quod corpora nostra movemus. Quinetiam si quis opinatur possibile esse ut Deus intellectualem aliquam creaturam tam perfectam producat quae ope divini concursus possit inferioris ordinis creaturas rursus producere, hoc adeo non derogaret divinae potestati, ut longe, ne dicam infinite majorem poneret, a qua scilicet creaturae non tantum immediate sed mediantibus alijs creaturis elicerentur. Et sic aliqui fortasse maluerint ponere animam mundi a Deo creatam esse cui hanc legem imponit ut spatia definita corporeis proprietatibus afficiat quam credere hoc officium a Deo immediate praestari. Neque ideo mundus diceretur animae illius creatura sed Dei solius qui crearet constituendo animam talis naturae ut mundus necessario emanaret. Sed non video cur Deus ipse non immediate spatium corporibus informet; dummodo corporum ratio formalis ab actu divinae voluntatis distinguamus. Contradicit enim ut sit ipse actus volendi, vel aliud quid quam effectus tantum quem actus ille in spatio producit. Qui quidem effectus non minus differt ab actu illo quam spatium Cartesianum, aut substantia corporis juxta vulgi conceptum; si modo ista creari, hoc est existentiam a voluntate mutuari sive esse entia rationis divinae supponimus.

Denique descriptae corporum Ideae usus maxime elucescit quod praecipuas Metaphysicae veritates clare involvit optimeque confirmat et explicat. Non possumus enim hujusmodi corpora ponere quin simul ponamus Deum existere, et corpora in inani spatio ex nihilo creasse, eaque esse entia a mentibus creatis distincta, sed posse tamen mentibus uniri. Dic sodes quaenam opinionum jam vulgatarum quampiam harum veritatum elucidat aut potius non adversatur omnibus et perplexas reddit. Si cum Cartesio dicamus extensionem esse corpus, an non Atheiae viam manifeste sternimus, tum quod extensio non est creatura sed ab aeterno fuit, tum quod Ideam ejus sine aliqua ad Deum relatione habemus absolutam, adeoque possumus ut existentem interea concipere dum Deum non esse fingimus. Neque distinctio mentis a corpore juxta hanc Philosophiam intelligibilis est, nisi simul dicamus mentem esse nullo modo extensam, adeoque nulli extensioni substantialiter praesentem esse, sive nullibi esse; quod perinde videtur ac si diceremus non esse; aut minimum reddit unionem ejus cum corpore plane intelligibilem, ne dicam impossibilem. Praeterea si legitima et perfecta est distinctio substantiarum in cogitantes et extensas; tum Deus extensionem in se non continet eminenter et proinde creare nequit; sed Deus et extensio duae erunt substantiae seorsim completae absolutae et univoce dictae. Aut contra si extensio in Deo sive summo ente cogitante eminenter continetur, certe Idea extensionis in Idea Cogitationis eminenter continebitur, et proinde distinctio Idearum non tanta erit quin ut ambae possint eidem creatae substantiae competere, hoc est corpora cogitare vel res cogitantes extendi. Quod si vulgarem corporis Ideam aut potius non Ideam amplectimur, scilicet quod in corporibus latet aliqua non intelligibilis realitas quam dicunt substantiam esse in qua qualitates eorum inhaerent: Hoc (super quam quod non est intelligibile,) ijsdem incommodis ac sententia Cartesiana comitatur. Nam cum nequit intelligi, impossibile est ut distinctio ejus a substantia mentis intelligatur. Non enim sufficit discrimen a forma substantiali vel substantiarum attributis desumptum; Nam si denudatae substantiae non habent essentialem differentiam, eaedem formae substantiales vel attributa possunt alterutri competere et efficere ut vicibus saltem si non simul sit mens et corpus. Adeoque si illam substantiarum attributis denudatarum differentiam non intelligimus, non possumus scientes affirmare quod mens et corpus substantialiter differunt. Vel si differunt, non possumus aliquod unionis fundamentum deprehendere. Praeterea huic corporum substantiae sine qualitatibus et formis spectatae realitatem in verbis quidem minorem sed in conceptu non minorem tribuunt, quam substantiae Dei, abstractae ab ejus attributis. Ambas nude spectatas similiter concipiunt, vel potius non concipiunt sed in communi quadam realitatis non intelligibilis apprehensione confundunt. Et hinc non mirum est quod Athei nascuntur ascribentes id substantijs corporeis quod soli divinae debetur. Quinimo circumspicienti nulla alia fere occurrit Atheorum causa quam haec notio corporum quasi habentium realitatem in se completam absolutam et independentem, qualem plerique omnes a pueris ni fallor per incuriam solemus mente concipere, ut ut verbis dicamus esse creatam ac dependentem. Et hoc praejudicium in causa fuisse credo quod in Scholis nomen substantiae Deo et creaturis univoce tribuitur, et quod in Idea corporis efformanda haerent Philosophi et hallucinantur, utpote dum rei a Deo dependentis Ideam independentem efformare conantur. Nam certe quicquid non potest esse independenter a Deo, non potest vere intelligi independenter ab Idea Dei. Deus non minus substat creaturis quam ipsae substant accidentibus, adeo ut substantia creata, sive graduatam dependentiam spectes sive gradum realitatis, est intermediae naturae inter Deum et accidens. Et proinde Idea ejus non minus involvit conceptum Dei quam accidentis Idea conceptum substantiae creatae. Adeoque non aliam in se realitatem quam derivativam et incompletam complecti debet. Deponendum est itaque praedictum praejudicium et substantialis realitas ejusmodi Attributis potius ascribenda est quae per se realia sunt et intelligibilia et non egeant subjecto cui inhaereant, quam subjecto cuidam quod non possumus ut dependens concipere nedum ullam ejus Ideam efformare. Et hoc vix gravate faciemus si (praeter Ideam corporis supra expositam) animis nostris advertimus nos posse spatium sine aliquo subjecto existens concipere, dum vacuum cogitamus. Et proinde huic aliquid substantialis realitatis competit. Sed si praeterea mobilitas partium (ut finxit Cartesius) in Idea ejus involveretur, nemo sane non facile concederet esse substantiam corpoream. Ad eundem modum si Ideam Attributi sive potestatis istius haberemus quo Deus sola voluntatis actione potest entia creare: forte conciperemus Attributum istud tanquam per se sine aliqua subjecta substantia subsistens et involvens caetera ejus attributa. Sed interea dum non hujus tantum Attributi, sed et potestatis propriae qua nostra corpora movemus, non possumus Ideam efformare: temeritatis esset dicere quodnam sit mentium substantiale fundamentum.

Hactenus de natura corporea: in qua explicanda me satis praestitisse arbitror quod talem exposui cujus creationem esse penes Deum clarissime constet, et ex qua creata si mundus hicce non constituitur, saltem alius huic simillimus constitui potest. Et cum materiarum quoad proprietates et naturam nulla esset differentia, sed tantum in methodo qua Deus aliam atque aliam crearet: sane corporis ab extensione distinctio ex hisce satis elucescet. Quod nempe extensio sit aeterna, infinita, increata, passim uniformis, nullatenus mobilis, nec motuum in corporibus vel cogitationum in mentibus mutationem aliquam inducere potens: corpus vero in his omnibus contrario modo se habet, saltem si Deo non placuit semper et ubique creasse. Nam Deo potestatem hancce non ausim denegare. Et siquis aliter sentit, dicat ubi primum materia creari potuit et unde potestas creandi tunc Deo concessa est. Aut si potestatis illius non fuit initium, sed eandem ab aeterno habuit quam jam habet, tunc ab aeterno potuit creasse. Nam idem est dicere quod in Deo nunquam fuit impotentia ad creandum, vel quod semper habuit potestatem creandi potuitque creasse, et quod materia semper potuit creari. Ad eundem modum assignetur spatium in quo materia non potuit subinitio creari, aut concedatur Deum potuisse tunc ubique creasse.[1]

---

[1] The last four sentences of this paragraph were added in a marginal note.

Caeterum ut Cartesij argumento jam strictius respondeam: tollamus e corpore (sicut ille jubet) gravitatem duritiem et omnes sensibiles qualitates, ut nihil tandem maneat nisi quod pertinet ad essentiam ejus. An itaque jam sola restabit extensio? neutiquam. Nam rejiciamus praeterea facultatem sive potestatem illam qua rerum cogitantium perceptiones movent. Nam cum tanta est inter Ideas cogitationis et extensionis distinctio ut non pateat aliquod esse connectionis aut relationis fundamentum nisi quod divina potestate causetur: illa corporum facultas salva extensione potest rejici, sed non rejicietur salva natura corporea. Scilicet mutationes quae corporibus a causis naturalibus induci possunt sunt tantum accidentales et non denominant substantiam revera mutari. Sed siqua inducitur mutatio quae causas naturales transcendit, ea plusquam accidentalis est et substantiam radicitus attingit. Juxtaque sensum Demonstrationis ea sola rejicienda sunt quibus corpora vi naturae carere et privari possunt. Sed nequis objiciat quod corpora quae mentibus non uniuntur, perceptiones earum immediate movere nequeunt. Et proinde cum corpora dantur nullis mentibus unita, sequitur hanc potestatem non esse de illorum essentia. Animadvertendum est quod de actuali unione hic non agitur sed tantum de facultate corporum qua viribus naturae sunt istius unionis capacia. Quam quidem facultatem inesse omnibus corporibus ex eo manifestum est quod partes cerebri, praesertim subtiliores quibus mens unitur sunt in continuo fluxu, at avolantibus novae succedunt. Et hanc tollere sive spectes actum divinum sive naturam corpoream non minoris est quam tollere facultatem alteram qua corpora mutuas actiones in se invicem transferre valeant, hoc est quam corpus in inane spatium redigere.

Cum autem aqua minus obstat motibus trajectorum solidorum quam argentum vivum et aer longe minus quam aqua, et spatia aetherea adhuc minus quam aërea rejiciamus praeterea vim omnem impediendi motus trajectorum et sane naturam corpoream penitus rejiciemus. Quemadmodum si materia subtilis vi omni privaretur impediendi motus globulorum, non amplius crederem esse materiam subtilem sed vacuum disseminatum. Atque ita si spatium aëreum vel aethereum ejusmodi esset ut Cometarum vel corporum quorumlibet projectilium motibus sine aliqua resistentia cederet crederem esse penitus inane. Nam impossibile est ut fluidum corporeum non obstet motibus trajectorum, puta si non disponitur ad motum juxta cum eorum motu velocem (Part 2 Epist 96 ad Mersennum),[1] quemadmodum suppono.

Hanc autem vim omnem a spatio posse tolli manifestum est si modo spatium et corpus ab invicem differunt; et proinde tolli posse non est denegandum antequam probantur non differe, ne paralogismus, petendo principium admittatur.

Sed nequa supersit dubitatio, ex praedictis observandum venit quod inania spatia in rerum natura dantur. Nam si aether esset fluidum sine poris aliquibus vacuis penitus corporeum, illud, utcunque per divisionem partium sub-tiliatum, foret aeque densum atque aliud quodvis fluidum, et non minori inertia motibus trajectorum cederet, imo longe majori, si modo projectile foret porosum; propterea quod intimos ejus poros ingrederetur, et non modo totius externae superficiei sed et omnium internarum partium superficiebus occurreret et impedimento esset. Sed cum aetheris e contra tam parva est resistentia ut ad resistentiam argenti vivi collata videatur esse plusquam decies vel centies mille vicibus minor: sane spatij aetherei pars longe maxima pro vacuo inter aetherea corpuscula disseminato haberi debet. Quod idem praeterea ex diversa gravitate horum fluidorum conjicere liceat, quam esse ut eorum densitates sive ut quantitates materiae in aequalibus spatijs contentae monstrant tum gravium descensus tum undulationes pendulorum. Sed his enucleandis jam non est locus.

Videtis itaque quam fallax et infida est haecce Cartesij argumentatio, siquidem rejectis corporum accidentibus, non sola remansit extensio ut ille finxerat, sed et facultates quibus tum perceptiones mentium tum alia corpora movere valeant. Quod si praeterea facultates hasce omnemque movendi potestatem rejiciamus ut sola maneat spatij uniformis praecisa conceptio: ecquos Vortices, ecquem mundum Cartesius ex hac extensione fabricabit? sane nullos nisi Deum prius invo-cet qui solus corpora de novo (restituendo facultates istas sive Naturam corpoream prout ante explicui) in spatijs istis

---

[1] The note in the bracket is written in the margin of the MS, with an asterisk to show its place in the text.

procreare possit. Adeoque in superioribus recte assignavi naturam corpoream in facultatibus jam recensitis consistere.

Atque ita tandem cum spatia non sunt ipsissima corpora sed loca tantum in quibus insunt et moventur quae de motu locali definivi satis firmata esse puto. Nec video quid amplius in hac re desiderari queat nisi forte ut quibus haec non satisfaciunt, ipsos moneam ut per spatium cujus partes esse corporum implentium loca definivi intelligant spatium generi-cum Cartesianum in quo spatia singulariter spectata, sive corpora Cartesiana moventur, et sane vix habebunt quod in definitionibus nostris reprehendant.

Jam satis digressi redeamus ad propositum.

Def 5. Vis est motus et quietis causale principium. Estque vel externum quod in aliquod corpus impressum motum ejus vel generat vel destruit, vel aliquo saltem modo mutat, vel est internum principium quo motus vel quies corpori indita conservatur, et quodlibet ens in suo statu perseverare conatur & impeditum reluctatur.

Def 6. Conatus est vis impedita sive vis quatenus resistitur.

Def 7. Impetus est vis quatenus in aliud imprimitur.

Def 8. Inertia est vis interna corporis ne status ejus externa vi illata facile mutetur.

Def 9. Pressio est partium contiguarum conatus ad ipsarum dimensiones mutuo penetrandum. Nam si possent penetrare cessaret pressio. Estque partium contiguarum tantum, quae rursus premunt alias sibi contiguas donec pressio in remotis-simas cujuslibet corporis duri mollis vel fluidi partes trans-feratur. Et in hac actione communicatio motus mediante puncto vel superficie contactus fundatur.

Def 10. Gravitas est vis corpori indita ad descendendum incitans. Hic autem per descensum non tantum intellige motum versus centrum terrae sed et versus aliud quodvis punctum plagamve, aut etiam a puncto aliquo peractum. Quemadmodum si aetheris circa Solem gyrantis conatus recedendi a centro ejus pro gravitate habeatur, descendere dicetur aether qui a Sole recedit. Et sic analogiam observando, planum dicetur horizontale quod gravitatis sive conatus determinationi directe opponitur.

Caeterum harum potestatum, nempe motus, vis, conatus, impetus, inertiae, pressionis, et gravitatis quantitas duplici ratione aestimatur; utpote vel secundum intensionem earum vel extensionem.

Def. 11. Intensio potestatis alicujus praedictae est ejus qualitatis gradus.

Def 12. Extensio ejus est spatij vel temporis quantitas in quo exercetur.

Def 13. Ejusque quantitas absoluta est quae ab ejus intensione et extensione componitur. Quemadmodum si quantitas intensionis sit 2, et quantitas extensionis 3, duc in seinvicem et habebitur quantitas absoluta 6.

Caeterum hasce definitiones in singulis potestatibus illustrare juvabit. Sic itaque Motus intensior est vel remissior quo spatium majus vel minus in eodem tempore transigitur, qua quidem ratione corpus dici solet velocius vel tardius moveri. Motus vero magis vel minus extensus est quocum corpus majus vel minus movetur, sive qui per majus vel minus corpus diffunditur. Et motus absoluta quantitas est quae componitur ex utrisque velocitate et magnitudine corporis moti. Sic vis, conatus, impetus, et inertia intensior est quae est in eodem vel aequali corpore major; extensior est quae est in majori corpore; et ejus quantitas absoluta quae ab utrisque oritur. Sic pressionis intensio est ut eadem superficiei quantitas magis prematur, extensio ut major superficies prematur, et absoluta quantitas quae resultat ab intensione pressionis et quantitate superficiei pressae. Sic denique gravitatis intensio est ut corpus habeat majorem gravitatem specificam, extensio est ut corpus grave sit majus, et absolute loquendo gravitatis quantitas est quae resultat ex gravitate specifica et mole corporis gravitantis. Et haec quisquis non clare distinguit, ut in plurimos errores circa scientias mechanicas incidat necesse est.

Potest insuper quantitas harum potestatum secundum durationis intervallum nonnunquam aestimari: qua quidem ratione quantitas absoluta erit quae ex omnibus intensione extensione ac duratione componitur. Quemadmodum si corpus 2 velocitate 3 per tempus 4 movetur: totus motus erit $2 \times 3 \times 4$, sive 12.

Def. 14. Velocitas est motus intensio, ac tarditas remissio ejus.

Def. 15. Corpora densiora sunt quorum intensior est inertia, et rariora quorum est remissior.

Caeteris praefatarum potestatum speciebus desunt nomina.

Est autem notandum, si cum Cartesio vel Epicuro rarifactionem et condensationem per modum spongiae relaxatae vel compressae, hoc est per pororum sive materia aliqua subtilissima plenorum sive vacuorum dilatationem et contractionem fieri supponimus, quod totius corporis magnitudinem ex quantitate tum partium ejus tum pororum in hac def 15 aestimare debemus; ut inertia per augmentationem pororum remitti concipiatur et per diminutionem intendi; tanquam si pori rationem partium haberent quibus nulla inest ad mutationes subeundas inertia, et quarum mistura cum partibus vere corporeis oriuntur totius varij gradus inertiae.

Ast quo compositum hocce tanquam corpus uniforme concipias, finge partes ejus infinite divisas esse, et per poros passim dispersas, ut in toto Composito ne minima quidem sit extensionis particula in qua non sit partium et pororum sic infinite divisorum mistura perfectissima. Scilicet hac ratione convenit Mathematicos contemplari; aut si malueris, pro more Peripateticorum: etsi in Physica res aliter se habere videtur.

Def 16. Corpus elasticum est quod vi pressionis condensari sive intra spatij angustioris limites cohiberi potest: et non elasticum quod vi ista condensari nequit.[1]

Def 17. Corpus durum est cujus partes inter se nulli pressioni cedunt.

Def 18. Fluidum est cujus partes inter se praepollenti pressioni cedunt.[2] Caeterum pressiones quibus fluidum versus omnes quaquaversum plagas urgetur, (sive in externam tantum superficiem exerceantur, sive in internas partes medianti gravitate vel alia quavis causa) dicuntur aequipollere cum efficiunt ut stet in aequilibrio. Quo posito, si pressio versus aliquam plagam et non versus alias omnes simul intendatur.

Def 19. Vas fluidi est limes quo continetur sive superficies corporis ambientis, ut ligni, vitri, vel partis exterioris ejusdem fluidi continentis partem aliquam interiorem.

[1] This was inserted, and the remaining definitions renumbered, after the completion of this section of the text.

[2] Originally this definition was complete in one sentence; the rest was added as a marginal note after the inclusion of the new definition 16. Definitions 18 and 19 are much corrected.

In hisce autem definitionibus ad corpora absolute dura fluidave solummodo respecto, nam circa mediocria propter innumeras in minutissimarum particularum figuris motibus et contextura circumstantias non licet mathematice ratiocinari. Fingo itaque fluidum non ex duris particulis constare sed ejusmodi esse ut nullam habeat portiunculam particulamve quae non sit similiter fluida. Et praeterea cum causa physica fluiditatis hic non spectatur, partes inter se non motas esse sed mobiles tantum definio, hoc est ita ab invicem ubique divisas esse, ut quamvis se mutuo contingere et inter se quiescere fingantur tamen non cohaerent quasi conglutinatae, sed a vi qualibet impressa seorsim moveri possunt[1] et statum quietis non difficilius mutare quam statum motus si inter se moverent. Durorum vero partes non tantum sese contingere et inter se quiescere sed insuper tam arcte et firmiter cohaerere et quasi glutino aligari suppono ut nulla moveri potest quin caeteras omnes secum rapiat: vel potius durum corpus non ex partibus conglomeratis conflari, sed esse unicum indivisum et uniforme corpus quod figuram firmissime conservat, fluidum vero in omni puncto uniformiter divisum esse.

Atque ita definitiones hasce non ad res physicas sed mathematica ratiocinia accommodavi, sicut Geometrae definitiones figurarum non accommodant ad irregularitates physicorum corporum. Et quemadmodum dimensiones corporum physicorum ab illorum Geometria optime determinantur (ut agri dimensio a Geometria plana etsi ager non sit revera planus, vel Globi terrestris dimensio a doctrina de globo etsi Terra non sit praecise globus:) sic fluidorum, solidorumve physicorum proprietates optime a doctrina hacce Mathematica noscentur, etsi forte nec sint absolute nec uniformiter fluida solidave prout hic definivi.

Propositiones de Fluido non Elastico

Axiomata

1. Ex paribus positis paria consectantur.
2. Contingentia corpora se mutuo aequaliter premunt.

---

[1] The remainder of this sentence was added as a marginal note.

Propositio 1

Fluidi non gravitantis eadem intensione quaquaversus compressi, partes omnes se mutuo aequaliter (sive aequali intensione) premunt.

Propositio 2

Et compressio non efficit motum partium inter se.

Utriusque Demonstratio

Ponamus imprimis fluidum a sphaerico limite AB cujus centrum K contineri et uniformiter comprimi [Fig. 42], ejus-

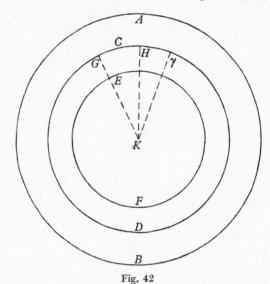

Fig. 42

que portiunculam quamvis $CGEH$ a duabus sphaericis superficiebus CD et EF circa idem centrum K descriptis, una cum conica superficie GKH cujus vertex est ad K terminatam esse. Et manifestum est quod illa $CGEH$ non potest ad centrum K ullatenus accedere, quia propter eandem rationem tota materia inter sphaericas superficies CD et EF undique ad idem centrum accederet,[a] adeoque dimensiones fluidi intra sphaeram EF contenti penetraret.[b] Neque potest ex aliqua parte versus circumferentiam A recederet quia propter eandem rationem tota illa orbita fluida inter superficies CD et EF interjecta pariter recederet,[a] atque adeo dimensiones fluidi

[a] Axiom 1.　　　　　　[b] contra Definitio

inter sphaericas superficies *AB* et *CD* contenti penetraret.[a]
Neque potest ad latera puta versus *H* exprimi, quoniam si
aliam portiunculam *Hγ* ab ijsdem sphaericis superficiebus et
consimili superficie conica quaquaversus terminatam, et huic
GH in *H* contiguam esse subintelligamus; illa *Hγ* propter
eandem rationem ad latera versus *H* exprimetur,[b] adeoque
mutuo contiguarum partium accessu fieret penetratio dimen-
sionum.[c] Constat itaque de qualibet fluidi portiuncula *CGEH*
quod suis limitibus propter pressionem excedere nequit. Et
proinde partes omnes in aequilibrio stabunt. Quod volui
primo ostendere.

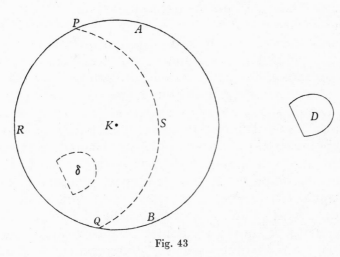

Fig. 43

Dico praeterea quod partes omnes se mutuo aequaliter
premunt, idque eadem pressionis intensione qua superficies
externa premitur. Quod ut pateat concipe *PSQR* esse praefati
fluidi *AB* partem a similibus sphaericarum superficierum seg-
mentis *PRQ* et *PSQ* contentam et compressio ejus juxta in-
ternam superficiem *PSQ* tanta erit ac juxta externam *PRQ*
[Fig. 43]. Hanc enim fluidi partem in aequilibrio stare jam
ostendi, adeoque pares sunt pressionum ejus juxta utramque
superficiem effectus, et inde pares pressiones.[b, d]
Cum itaque sphaericae superficies, qualis est *PSQ* possint
omnifariam in fluido *AB* disponi, et alias quascunque datas
superficies in quibuslibet punctis contingere, sequitur quod

[a] contra Definitio   [b] Axiom   [c] Contra Definitionem.   [d] Definitio.

pressionis partium juxta superficies utcumque positas tanta est intensio quanta fluidum in externa superficie premitur. Quod volui secundo ostendere.

Caeterum cum hujus argumentationis vis in paritate superficierum *PRQ* et *PSQ* fundatur, ne disparitas esse videatur quod altera sit intra fluidum et altera segmentum externae superficiei: juvabit effingere sphaeram integram *AB* esse partem fluidi indefinite majoris, in quo tanquam vase continetur et undique non secus comprimitur quam pars ejus *PRQS* juxta superficiem *PSQ* premitur ab altera parte *PABQS*. Nam nihil interest qua methodo sphaera *AB* comprimitur, dummodo compressio ejus undique statuatur aequabilis.

His de sphaerico fluido ostensis; dico denique quod fluidi *D* quocunque modo terminati et eadem intensione quaquaversum compressi, partes omnes se mutuo aequaliter prement et compressio non efficiet motum partium inter se. Sit enim *AB* fluidum sphaericum indefinite majus et eodem intensionis gradu compressum: sitque δ ejus pars aliqua huic *D* similis et aequalis. Jam e demonstratis patet hanc partem δ aequabili intensione quaquaversum comprimi et pressionis intensionem eandem esse qua sphaera *AB*, hoc est (ex hypothesi) qua fluidum *D* comprimitur. Par itaque est similium et aequalium fluidorum *D* ac δ compressio et proinde pares erunt effectus.[a] At sphaerae *AB*,[b] adeoque fluidi δ in ea contenti partes omnes se mutuo aequaliter prement, pressioque non efficiet motum partium inter se; Quare et idem de fluido *D* verum est.[a] Q.E.D.

Cor. 1. Fluidi partes internae eadem intensione se mutuo premunt qua fluidum premitur in externa superficie.

Cor. 2. Si non eadem sit undique pressionis intensio, fluidum non stabit in aequilibrio. Nam cum stet in aequilibrio propter pressionem undique uniformem, si pressio alicubi augeatur, ibi praepollebit, efficietque ut fluidum ab istis partibus recedat.[c]

Cor. 3. Si motus in fluido pressione non causatur eadem est undique pressionis intensio. Nam si non sit eadem, motus a praepollenti pressione causabitur.[d]

Cor. 4. Quanta intensione fluidum a limitibus ejus pre-

---

[a] Axiom. [b] Secundum jam demonstrata.
[c] Definitio. [d] Corollarium 2.

mitur, tanta limites vicissim premit; et e contra. Quippe cum partes fluidi sint partium contiguarum limites et se mutuo aequali intensione premant; concipe propositum fluidum esse partem majoris fluidi, vel tali parti simile et aequale et similiter compressum, et constabit assertio.[a]

Cor. 5. Fluidum omnes ejus limites, si modo illatam pressionem sustinere valeant, tanta intensione ubique premit, quanta ipsum in quovis loco premitur. Nam alias non ubique premitur eadem intensione.[b] Quo posito cedet intensiori pressioni.[c] Adeoque vel condensabitur,[d] vel contra limites ubi minor est pressio praevalebit.[d]

Schol: Haec omnia de Fluido proposui, non quatenus duro et rigido vase, sed lento et admodum flexibili termino, (puta fluidi exterioris homogenei interna superficie) continetur: eo ut aequilibrium ejus a solo pressionis gradu quaquaversus aequabili causari clarius ostenderem. Sed postquam fluidum in aequilibrio per aequabilem pressionem constituitur, perinde est sive rigido termino sive lento contineri fingas.

TRANSLATION

On the Gravity and Equilibrium of Fluids

It is proper to treat the science of gravity and of the equilibrium of fluid and solid bodies in fluids by two methods. To the extent that it appertains to the mathematical sciences, it is reasonable that I largely abstract it from physical considerations. And for this reason I have undertaken to demonstrate its individual propositions from abstract principles, sufficiently well known to the student, strictly and geometrically. Since this doctrine may be judged to be somewhat akin to natural philosophy, in so far as it may be applied to making clear many of the phenomena of natural philosophy, and in order, moreover, that its usefulness may be particularly apparent and the certainty of its principles perhaps confirmed, I shall not be reluctant to illustrate the propositions abundantly from experiments as well, in such a way, however, that this freer method of discussion, disposed in scholia, may not be confused with the former which is treated in Lemmas, propositions and corollaries.

[a] Axiom.
[b] Corollarium 4.
[c] Corollarium 2.
[d] Contra Hypothesin.

The foundations from which this science may be demonstrated are either definitions of certain words; or axioms and postulates denied by none. And of these I treat directly.

Definitions

The terms *quantity*, *duration* and *space* are too well known to be susceptible of definition by other words.

Def. 1. Place is a part of space which something fills evenly.[a]

Def. 2. Body is that which fills place.[a]

Def. 3. Rest is remaining in the same place.

Def. 4. Motion is change of place.[b]

Note. I said that a body fills place,[a] that is, it so completely fills it that it wholly excludes other things of the same kind or other bodies, as if it were an impenetrable being. Place could be said however to be a part of space in which a thing is evenly distributed; but as only bodies are here considered and not penetrable things, I have preferred to define [place] as the part of space that things fill.

Moreover, since body is here proposed for investigation not in so far as it is a physical substance endowed with sensible qualities but only in so far as it is extended, mobile and impenetrable, I have not defined it in a philosophical manner, but abstracting the sensible qualities (which Philosophers also should abstract, unless I am mistaken, and assign to the mind as various ways of thinking excited by the motions of bodies) I have postulated only the properties required for local motion. So that instead of physical bodies you may understand abstract figures in the same way that they are considered by Geometers when they assign motion to them, as is done in Euclid's *Elements*, Book I, 4 and 8. And in the demonstration of the tenth definition, Book XI, this should be done; since it is mistakenly included among the definitions and ought rather to be demonstrated among the propositions, unless perhaps it should be taken as an axiom.

Moreover, I have defined motion as change of place,[b] because motion, transition, translation, migration and so forth seem to be synonymous words. If you prefer, let motion be transition or translation of a body from place to place.

For the rest, when I suppose in these definitions that space is distinct from body, and when I determine that motion is with respect to the parts of that space, and not with respect to the position of neighbouring bodies, lest this should be taken as being gratuitously contrary to the Cartesians, I shall venture to dispose of his fictions.

I can summarize his doctrine in the following three propositions:

(1) That from the truth of things only one particular motion fits each body (*Principia*, Part II, Art. 28, 31, 32),[1] which is defined as being the translation of one part of matter or of one body from the neighbourhood of those bodies that immediately touch it, and which are regarded as being at rest, to the neighbourhood of others (*Principia*, Part II, Art. 25; Part III, Art. 28).[2]

(2) That by a body transferred in its particular motion according to this definition may be understood not only any particle of matter, or a body composed of parts relatively at rest, but all that is transferred at once, although this may, of course, consist of many parts which have different relative motions. (*Principia*, Part II, Art. 25.)

(3) That besides this motion particular to each body there can arise in it innumerable other motions, through participation (or in so far as it is part of other bodies having other motions). (*Principia*, Part II, Art. 31.) Which however are not motions in the philosophical sense and rationally speaking (Part III, Art. 29),[3] and according to the truth of things (Part II, Art. 25 and Part III, Art. 28), but only improperly and according to common sense (Part II, Art. 24, 25, 28, 31; Part III, Art. 29).[4] That kind of motion he seems to describe (Part II, Art. 24; Part III, Art. 28) as the action by which any body migrates from one place to another.

---

[1] Descartes, *Principia Philosophia*. II, 28: Motion, properly considered, cannot be referred to anything but the bodies contiguous to the moving one. II, 31: How in the same body there can be innumerable diverse motions. II, 32: How also motion properly considered, which is unique for each body, can be regarded as multiple.

[2] II, 25: What motion, properly considered, is. III, 28: That the Earth, properly speaking, does not move, nor any of the Planets, although they are carried by the heaven.

[3] III, 29: That no motion is to be attributed to the Earth, although it is improperly taken as motion according to the vulgar usage, but then it is rightly said that the other planets move.

[4] II, 24: What motion is, according to the vulgar understanding.

And just as he formulates two types of motion, namely particular and derivative, so he assigns two types of place from which these motions proceed, and these are the surfaces of immediately surrounding bodies (Part II, Art. 15),[1] and the position among any other bodies (Part II, Art. 13; Part III, Art. 29).[2]

Indeed, not only do its absurd consequences convince us how confused and incongruous with reason this doctrine is, but Descartes by contradicting himself seems to acknowledge the fact. For he says that speaking properly and according to philosophical sense the Earth and the other Planets do not move, and that he who declares it to be moved because of its translation with respect to the fixed stars speaks without reason and only in the vulgar fashion (Part III, Art. 26, 27, 28, 29).[3] Yet later he attributes to the Earth and Planets a tendency to recede from the Sun as from a centre about which they are revolved, by which they are balanced at their [due] distances from the Sun by a similar tendency of the gyrating vortex (Part III, Art. 140).[4] What then? Is this tendency to be derived from the (according to Descartes) true and philosophical rest of the planets, or rather from [their] common and non-philosophical motion? But Descartes says further that a Comet has a lesser tendency to recede from the Sun when it first enters the vortex, and keeping practically the same position among the fixed stars does not yet obey the impetus of the vortex, but with respect to it is transferred from the neighbourhood of the contiguous aether and so philosophically speaking whirls round the Sun, while afterwards the matter of the vortex carries the comet along with it and so renders it at rest, according to philosophical sense. (Part III, Art. 119, 120.)[5] The philosopher is hardly consistent who uses as the basis of Philosophy the motion of the vulgar which he had rejected a little before, and now rejects that motion as fit for nothing which alone was formerly said to be true and philosophical, according to the nature of things. And

[1] II, 15: How external place is rightly considered as the surface surrounding the body.
[2] II, 13: What external place is.
[3] III, 26: That the Earth is at rest in its heaven, but nevertheless is borne by it. III, 27: That the same is to be understood of all the Planets.
[4] III, 140: On the beginning of the motions of the Planets.
[5] III, 119: How a fixed star is altered into a Comet or Planet. III, 120: How such a star is borne along when first it ceases to be fixed.

since the whirling of the comet around the Sun in his philosophic sense does not cause a tendency to recede from the
centre, which a gyration in the vulgar sense can do, surely
motion in the vulgar sense should be acknowledged, rather
than the philosophical.

Secondly, he seems to contradict himself when he postulates
that to each body corresponds a single motion, according to
the nature of things; and yet he asserts that motion to be a
product of our imagination, defining it as translation from the
neighbourhood of bodies which are not at rest but only seem
to be at rest even though they may instead be moving, as is
more fully explained in Part II, Art. 29, 30.[1] And thus he
thinks to avoid the difficulties concerning the mutual translation of bodies, namely, why one body is said to move rather
than another, and why a boat on a flowing stream is said to be
at rest when it does not change its position with respect to
the banks (Part II, Art. 15). But that the contradiction may
be evident, imagine that someone sees the matter of the
vortex as at rest, and that the Earth, philosophically speaking,
is at rest at the same time; imagine also that someone else
simultaneously sees the same matter of the vortex as moving
in a circle, and that the Earth, philosophically speaking, is
not at rest. In the same way, a ship at sea will at the same
time move and not move, and that without taking *motion* in
the looser vulgar sense, by which there are innumerable
motions for each body, but in his philosophical sense according to which, he says, there is but one in each body, and that
one peculiar to it and corresponding to the nature of things
and not to our imagination.

Thirdly, he seems hardly consistent when he supposes that
a single motion corresponds to each body according to the
truth of things, and yet (Part II, Art. 31) that there really are
innumerable motions in each body. For the motions that
really are in any body, are really natural motions, and thus
motions in the philosophical sense and according to the truth
of things, even though he contends that they are motions in
the vulgar sense only. Add that when a whole thing moves,

---

[1] II, 29: Nor can it [motion] be referred to anything but to those contiguous bodies
which appear to be at rest. II, 30: Why one of two contiguous bodies, which are separated
from each other, is said to move rather than the other.

all the parts which constitute the whole and are translated together are really at rest; unless indeed it is conceded that they move by participating in the motion of the whole, and then indeed they have innumerable motions according to the truth of things.

But besides this, from its consequences we may see how absurd is this doctrine of Descartes. And first, just as he contends with heat that the Earth does not move because it is not translated from the neighbourhood of the contiguous aether, so from the same principles it follows that the internal particles of hard bodies, while they are not translated from the neighbourhood of immediately contiguous particles, do not have motion in the strict sense, but move only by participating in the motion of the external particles: it rather appears that the interior parts of the external particles do not move with their own motion because they are not translated from the neighbourhood of the internal parts: and thus that only the external surface of each body moves with its own motion and that the whole internal substance, that is the whole of the body, moves through participation in the motion of the external surface.[1] The fundamental definition of motion errs, therefore, that attributes to bodies that which only belongs to surfaces, and which denies that there can be any body at all which has a motion peculiar to itself.

Secondly, if we regard only Art. 25 of Part II, each body has not merely a unique proper motion but innumerable ones, provided that those things are said to be moved properly and according to the truth of things of which the whole is properly moved. And that is because he understands by the body whose motion he defines all that which is translated together, and yet this may consist of parts having other motions among themselves: suppose a vortex together with all the Planets, or a ship along with everything within it floating in the sea, or a man walking in a ship together with the things he carries with him, or the wheel of a clock together with its constituent metallic particles. For unless you say that the motion of the whole aggregate cannot be considered as proper motion and as belonging to the parts according to the truth of things, it

[1] Compare with this whole passage the paragraph in the Scholium on space and time (*Principia*, 8), beginning 'Motus proprietas est,...'.

will have to be admitted that all these motions of the wheels of the clock, of the man, of the ship, and of the vortex are truly and philosophically speaking in the particles of the wheels.

From both of these consequences it appears further that no one motion can be said to be true, absolute and proper in preference to others, but that all, whether with respect to contiguous bodies or remote ones, are equally philosophical—than which nothing more absurd can be imagined. For unless it is conceded that there can be a single physical motion of any body, and that the rest of its changes of relation and position with respect to other bodies are so many external designations, it follows that the Earth (for example) endeavours to recede from the centre of the Sun on account of a motion relative to the fixed stars, and endeavours the less to recede on account of a lesser motion relative to Saturn and the aetherial orb in which it is carried, and still less relative to Jupiter and the swirling aether which occasions its orbit, and also less relative to Mars and its aetherial orb, and much less relative to other orbs of aetherial matter which, although not bearing planets, are closer to the annual orbit of the Earth; and indeed relative to its own orb it has no endeavour, because it does not move in it. Since all these endeavours and non-endeavours cannot absolutely agree, it is rather to be said that only the motion which causes the Earth to endeavour to recede from the Sun is to be declared the Earth's natural and absolute motion. Its translations relative to external bodies are but external designations.

Thirdly. It follows from the Cartesian doctrine that motion can be generated where there is no force acting. For example, if God should suddenly cause the spinning of our vortex to stop, without applying any force to the Earth which could stop it at the same time, Descartes would say that the Earth is moving in a philosophical sense (on account of its translation from the neighbourhood of the contiguous fluid), whereas before he said it was resting, in the same philosophical sense.

Fourthly. It also follows from the same doctrine that God himself could not generate motion in some bodies even though he impelled them with the greatest force. For example, if God urged the starry heaven together with all the most remote part of creation with any very great force so as to cause it to revolve about the Earth (suppose with a diurnal motion): yet from this, according to Descartes, the Earth alone and not the sky would be truly said to move (Part III, Art. 38).[1] As if it would be the same whether, with a tremendous force, He should cause the skies to turn from east to west, or with a small force turn the Earth in the opposite direction. But who will imagine that the parts of the Earth endeavour to recede from its centre on account of a force impressed only upon the heavens? Or is it not more agreeable to reason that when a force imparted to the heavens makes them endeavour to recede from the centre of the revolution thus caused, they are for that reason the sole bodies properly and absolutely moved; and that when a force impressed upon the Earth makes its parts endeavour to recede from the centre of revolution thus caused, for that reason it is the sole body properly and absolutely moved, although there is the same relative motion of the bodies in both cases. And thus physical and absolute motion is to be defined from other considerations than translation, such translation being designated as merely external.

Fifthly. It seems repugnant to reason that bodies should change their relative distances and positions without physical motion; but Descartes says that the Earth and the other Planets and the fixed stars are properly speaking at rest, and nevertheless they change their relative positions.

Sixthly. And on the other hand it seems not less repugnant to reason that of several bodies maintaining the same relative positions some one should move physically, while others are at rest. But if God should cause any Planet to stand still and make it continually keep the same position with respect to the fixed stars, would not Descartes say that although the stars are not moving, the planet now moves physically on account of its translation from the matter of the vortex?

Seventhly. I ask for what reason any body is properly said to move when other bodies from whose neighbourhood it is transported are not seen to be at rest, or rather when they

[1] That according to Tycho's hypothesis, the Earth is said to move around its own centre.

cannot be seen to be at rest. For example, in what way can our own vortex be said to move circularly on account of the translation of matter near the circumference, from the neighbourhood of similar matter in other surrounding vortices, since the matter of surrounding vortices cannot be seen to be at rest, and this not only with respect to our vortex, but also in so far as those vortices are not at rest among themselves. For if the Philosopher refers this translation not to the numerical corporeal particles of the vortices, but to the generic space (as he calls it) in which those vortices exist, at last we do agree, for he admits that motion ought to be referred to space in so far as it is distinguished from bodies.

Lastly, that the absurdity of this position may be disclosed in full measure, I say that thence it follows that a moving body has no determinate velocity and no definite line in which it moves. And, what is worse, that the velocity of a body moving without resistance cannot be said to be uniform, nor the line said to be straight in which its motion is accomplished. On the contrary, there cannot be motion since there can be no motion without a certain velocity and determination.

But that this may be clear, it is first of all to be shown that when a certain motion is finished it is impossible, according to Descartes, to assign a place in which the body was at the beginning of the motion; it cannot be said whence the body moved. And the reason is that according to Descartes the place cannot be defined or assigned except by the position of the surrounding bodies, and after the completion of a certain motion the position of the surrounding bodies no longer stays the same as it was before. For example, if the place of the planet Jupiter a year ago be sought, by what reason, I ask, can the Cartesian philosopher define it? Not by the positions of the particles of the fluid matter, for the positions of these particles have greatly changed since a year ago. Nor can he define it by the positions of the Sun and fixed stars. For the unequal influx of subtle matter through the poles of the vortices towards the central stars (Part III, Art. 104), the undulation (Art. 114), inflation (Art. 111) and absorption of the vortices, and other more true causes, such as the rotation of the Sun and stars around their own centres, the generation of spots, and the passage of comets through the heavens, change both the magnitude and positions of the stars so much that perhaps they are only adequate to designate the place sought with an error of several miles; and still less can the place be accurately defined and determined by their help, as a Geometer would require.[1] Truly there are no bodies in the world whose relative positions remain unchanged with the passage of time, and certainly none which do not move in the Cartesian sense: that is, which are neither transported from the vicinity of contiguous bodies nor are parts of other bodies so transferred.[2] And thus there is no basis from which we can at the present pick out a place which was in the past, or say that such a place is any longer discoverable in nature. For since, according to Descartes, place is nothing but the surface of surrounding bodies or position among some other more distant bodies, it is impossible (according to his doctrine) that it should exist in nature any longer than those bodies maintain the same positions from which he takes the individual designation. And so, reasoning as in the question of Jupiter's position a year ago, it is clear that if one follows Cartesian doctrine, not even God himself could define the past position of any moving body accurately and geometrically now that a fresh state of things prevails, since in fact, due to the changed positions of the bodies, the place does not exist in nature any longer.

Now as it is impossible to pick out the place in which a motion began (that is, the beginning of the space passed over), for this place no longer exists after the motion is completed, so the space passed over, having no beginning, can have no length; and hence, since velocity depends upon the distance passed over in a given time, it follows that the moving body can have no velocity, just as I wished to prove at first. Moreover, what was said of the beginning of the space passed over should be applied to all intermediate points too; and thus as the space has no beginning nor intermediate parts it follows that there was no space passed over and thus no determinate

[1] III, 104: Why certain fixed stars disappear, or appear unexpectedly. III, 114: That the same star can alternately appear and disappear. III, 111: A description of the unexpected appearance of a star.

[2] Cf. *Principia*, 7: 'Fieri enim potest ut nullum revera quiescat corpus, ad quod loca motusque referantur.'

motion, which was my second point. It follows indubitably that Cartesian motion is not motion, for it has no velocity, no definition, and there is no space or distance traversed by it. So it is necessary that the definition of places, and hence of local motion, be referred to some motionless thing such as extension alone or space in so far as it is seen to be truly distinct from bodies. And this the Cartesian philosopher may the more willingly allow, if only he notices that Descartes himself had an idea of extension as distinct from bodies, which he wished to distinguish from corporeal extension by calling it generic (*Principia*, Part II, Art. 10, 12, 18).[1] And also that the rotations of the vortices, from which he deduced the force of the aether in receding from their centres and thus the whole of his mechanical philosophy, are tacitly referred to generic extension.

In addition, as Descartes in Part II, Art. 4 and 11 seems to have demonstrated that body does not differ at all from extension, abstracting hardness, colour, weight, cold, heat and the remaining qualities which body can lack, so that at last there remains only its extension in length, width and depth which hence alone appertain to its essence;[2] and as this has been taken as proved by many, and is in my view the only reason for having confidence in this opinion; lest any doubt should remain about the nature of motion, I shall reply to this argument by explaining what extension and body are, and how they differ from each other. For since the distinction of substances into thinking and extended [entities], or rather, into thoughts and extensions, is the principal foundation of Cartesian philosophy, which he contends to be even better known than mathematical demonstrations: I consider it most important to overthrow [that philosophy] as regards extension, in order to lay truer foundations of the mechanical sciences.

Perhaps now it may be expected that I should define extension as substance or accident or else nothing at all. But by

---

[1] II, 10: What space, or internal place is. II, 12: How it differs from [corporeal substance] in the way in which it is conceived. II, 18: How opinion about the vacuum, absolutely considered, is to be emended.

[2] II, 4: That the nature of body does not consist of weight, hardness, colour or the like, but of extension alone. II, 11: How [space] does not differ in itself from corporeal substance.

no means, for it has its own manner of existence which fits neither substances nor accidents. It is not substance; on the one hand, because it is not absolute in itself, but is as it were an emanent effect of God, or a disposition of all being; on the other hand, because it is not among the proper dispositions that denote substance, namely actions, such as thoughts in the mind and motions in body. For although philosophers do not define substance as an entity that can act upon things, yet all tacitly understand this of substances, as follows from the fact that they would readily allow extension to be substance in the manner of body if only it were capable of motion and of sharing in the actions of body. And on the contrary they would hardly allow that body is substance if it could not move nor excite in the mind any sensation or perception whatever. Moreover, since we can clearly conceive extension existing without any subject, as when we may imagine spaces outside the world or places empty of body, and we believe [extension] to exist wherever we imagine there are no bodies, and we cannot believe that it would perish with the body if God should annihilate a body, it follows that [extension] does not exist as an accident inherent in some subject. And hence it is not an accident. And much less may it be said to be nothing, since it is rather something, than an accident, and approaches more nearly to the nature of substance. There is no idea of nothing, nor has nothing any properties, but we have an exceptionally clear idea of extension, abstracting the dispositions and properties of a body so that there remains only the uniform and unlimited stretching out of space in length, breadth and depth. And furthermore, many of its properties are associated with this idea; these I shall now enumerate not only to show that it is something, but what it is.

1. In all directions, space can be distinguished into parts whose common limits we usually call surfaces; and these surfaces can be distinguished in all directions into parts whose common limits we usually call lines; and again these lines can be distinguished in all directions into parts which we call points. And hence surfaces do not have depth, nor lines breadth, nor points dimension, unless you say that coterminous spaces penetrate each other as far as the depth of the surface between them, namely what I have said to be the

boundary of both or the common limit; and the same applies to lines and points. Furthermore spaces are everywhere contiguous to spaces, and extension is everywhere placed next to extension, and so there are everywhere common boundaries to contiguous parts; that is, there are everywhere surfaces acting as a boundary to solids on this side and that; and everywhere lines in which parts of the surfaces touch each other; and everywhere points in which the continuous parts of lines are joined together. And hence there are everywhere all kinds of figures, everywhere spheres, cubes, triangles, straight lines, everywhere circular, elliptical, parabolical and all other kinds of figures, and those of all shapes and sizes, even though they are not disclosed to sight. For the material delineation of any figure is not a new production of that figure with respect to space, but only a corporeal representation of it, so that what was formerly insensible in space now appears to the senses to exist. For thus we believe all those spaces to be spherical through which any sphere ever passes, being progressively moved from moment to moment, even though a sensible trace of the sphere no longer remains there. We firmly believe that the space was spherical before the sphere occupied it, so that it could contain the sphere; and hence as there are everywhere spaces that can adequately contain any material sphere, it is clear that space is everywhere spherical. And so of other figures. In the same way we see no material shapes in clear water, yet there are many in it which merely introducing some colour into its parts will cause to appear in many ways. However, if the colour were introduced, it would not constitute material shapes but only cause them to be visible.[1]

2. Space extends infinitely in all directions. For we cannot imagine any limit anywhere without at the same time imagining that there is space beyond it. And hence all straight lines, paraboloids, hyperboloids, and all cones and cylinders and other figures of the same kind continue to infinity and are bounded nowhere, even though they are crossed here and there by lines and surfaces of all kinds extending transversely,

[1] Presumably Newton is thinking of dropping dye into water, which disturbances in the water cause to spread into swirls and 'shapes'; the boundaries to these disturbances exist before the dye reveals them.

and with them form segments of figures in all directions. You may have in truth an instance of infinity; imagine any triangle whose base and one side are at rest and the other side so turns about the contiguous end of its base in the plane of the triangle that the triangle is by degrees opened at the vertex; and meanwhile take a mental note of the point where the two sides meet, if they are produced that far: it is obvious that all these points are found on the straight line along which the fixed side lies, and that they become perpetually more distant as the moving side turns further until the two sides become parallel and can no longer meet anywhere. Now, I ask, what was the distance of the last point where the sides met? It was certainly greater than any assignable distance, or rather none of the points was the last, and so the straight line in which all those meeting-points lie is in fact greater than finite. Nor can anyone say that this is infinite only in imagination, and not in fact; for if a triangle is actually drawn, its sides are always, in fact, directed towards some common point, where both would meet if produced, and therefore there is always such an actual point where the produced sides would meet, although it may be imagined to fall outside the limits of the physical universe. And so the line traced by all these points will be real, though it extends beyond all distance.

If anyone now objects that we cannot imagine that there is infinite extension, I agree. But at the same time I contend that we can understand it. We can imagine a greater extension, and then a greater one, but we understand that there exists a greater extension than any we can imagine. And here, incidentally, the faculty of understanding is clearly distinguished from imagination.

Should it be further said that we do not understand what an infinite being is, save by negating the limitations of a finite being, and that this is a negative and faulty conception, I deny this. For the limit or boundary is the restriction or negation of greater reality or existence in the limited being, and the less we conceive any being to be constrained by limits, the more we observe something to be attributed to it, that is, the more positively we conceive it. And thus by negating all limits the conception becomes positive in the highest degree. 'End' [*finis*] is a word negative as to sense, and thus 'infinity' [not-end] as it is the negation of a negation (that is, of ends) will be a word positive in the highest degree with respect to our perception and comprehension, though it seems grammatically negative. Add that positive and finite quantities of many surfaces infinite in length are accurately known to Geometers. And so I can positively and accurately determine the solid quantities of many solids infinite in length and breadth and compare them to given finite solids. But this is irrelevant here.

If Descartes now says that extension is not infinite but rather indefinite, he should be corrected by the grammarians. For the word 'indefinite' is never applied to that which actually is, but always relates to a future possibility signifying only something which is not yet determined and definite. Thus before God had decreed anything about the creation of the world (if there was ever a time when he had not), the quantity of matter, the number of the stars and all other things were indefinite; once the world was created they were defined. Thus matter is indefinitely divisible, but is always divided either finitely or infinitely (Part I, Art. 26; Part II, Art. 34).[1] Thus an indefinite line is one whose future length is still undetermined. And so an indefinite space is one whose future magnitude is not yet determined; for indeed that which actually is, is not to be defined, but does either have limits or not and so is either finite or infinite. Nor is it an objection that he takes space to be indefinite in relation to ourselves; that is, we simply do not know its limits and are not absolutely sure that there are none (Part I, Art. 27).[2] This is because although we are ignorant beings God at least understands that there are no limits not merely indefinitely but certainly and positively, and because although we negatively imagine it to transcend all limits, yet we positively and most certainly understand that it does so. But I see what Descartes feared, namely that if he should consider space infinite, it would perhaps become God because of the perfection of infinity. But by no means, for infinity is not perfection except when it is an attribute of

[1] I, 26: Infinity should never be discussed; those things in which we discern no limits should be taken as indefinite only. Such are the extension of the world, the divisibility of the parts of matter, the number of the stars, etc. III, 34: Hence it follows that the division of matter into particles is truly indefinite, although they are imperceptible to us.

[2] What the difference between indefinite and infinite is.

perfect things. Infinity of intellect, power, happiness and so forth is the height of perfection; but infinity of ignorance, impotence, wretchedness and so on is the height of imperfection; and infinity of extension is so far perfect as that which is extended.

3. The parts of space are motionless. If they moved, it would have to be said either that the motion of each part is a translation from the vicinity of other contiguous parts, as Descartes defined the motion of bodies; and that this is absurd has been sufficiently shown; or that it is a translation out of space into space, that is out of itself, unless perhaps it is said that two spaces everywhere coincide, a moving one and a motionless one. Moreover the immobility of space will be best exemplified by duration. For just as the parts of duration derive their individuality from their order, so that (for example) if yesterday could change places with today and become the later of the two, it would lose its individuality and would no longer be yesterday, but today; so the parts of space derive their character from their positions, so that if any two could change their positions, they would change their character at the same time and each would be converted numerically into the other. The parts of duration and space are only understood to be the same as they really are because of their mutual order and position; nor do they have any hint of individuality apart from that order and position which consequently cannot be altered.[1]

4. Space is a disposition of being *qua* being. No being exists or can exist which is not related to space in some way. God is everywhere, created minds are somewhere, and body is in the space that it occupies; and whatever is neither everywhere nor anywhere does not exist. And hence it follows that space is an effect arising from the first existence of being, because when any being is postulated, space is postulated. And the same may be asserted of duration: for certainly both are dispositions of being or attributes according to which we denominate quantitatively the presence and duration of any existing individual thing. So the quantity of the existence of God was eternal, in relation to duration, and infinite in relation to the space in which he is present; and the quantity of the

---

[1] Cf. *Principia*, 7: 'Ut partium Temporis ordo est immutabilis....'

existence of a created thing was as great, in relation to duration, as the duration since the beginning of its existence, and in relation to the size of its presence as great as the space belonging to it.

Moreover, lest anyone should for this reason imagine God to be like a body, extended and made of divisible parts, it should be known that spaces themselves are not actually divisible, and furthermore, that any being has a manner proper to itself of being in spaces. For thus there is a very different relationship between space and body, and space and duration. For we do not ascribe various durations to the different parts of space, but say that all endure together. The moment of duration is the same at Rome and at London, on the Earth and on the stars, and throughout all the heavens. And just as we understand any moment of duration to be diffused throughout all spaces, according to its kind, without any thought of its parts, so it is no more contradictory that Mind also, according to its kind, can be diffused through space without any thought of its parts.

5. The positions, distances and local motions of bodies are to be referred to the parts of space. And this appears from the properties of space enumerated as 1. and 4. above, and will be more manifest if you conceive that there are vacuities scattered between the particles, or if you pay heed to what I have formerly said about motion. To that it may be further added that in space there is no force of any kind which might impede or assist or in any way change the motions of bodies. And hence projectiles describe straight lines with a uniform motion unless they meet with an impediment from some other source. But more of this later.

6. Lastly, space is eternal in duration and immutable in nature, and this because it is the emanent effect of an eternal and immutable being. If ever space had not existed, God at that time would have been nowhere; and hence he either created space later (in which he was not himself), or else, which is not less repugnant to reason, he created his own ubiquity. Next, although we can possibly imagine that there is nothing in space, yet we cannot think that space does not exist, just as we cannot think that there is no duration, even though it would be possible to suppose that nothing whatever

endures. This is manifest from the spaces beyond the world, which we must suppose to exist (since we imagine the world to be finite), although they are neither revealed to us by God, nor known from the senses, nor does their existence depend upon that of the spaces within the world. But it is usually believed that these spaces are nothing; yet indeed they are true spaces. Although space may be empty of body, nevertheless it is not in itself a void; and *something* is there, because spaces are there, although nothing more than that. Yet in truth it must be acknowledged that space is no more space where the world is, than where no world is, unless perchance you say that when God created the world in this space he at the same time created space in itself, or that if God should annihilate the world in this space, he would also annihilate the space in it. Whatever has more reality in one space than in another space must belong to body rather than to space; the same thing will appear more clearly if we lay aside that puerile and jejune prejudice according to which extension is inherent in bodies like an accident in a subject without which it cannot actually exist.

Now that extension has been described, it remains to give an explanation of the nature of body. Of this, however, the explanation must be more uncertain, for it does not exist necessarily but by divine will, because it is hardly given to us to know the limits of the divine power, that is to say whether matter could be created in one way only, or whether there are several ways by which different beings similar to bodies could be produced. And although it scarcely seems credible that God could create beings similar to bodies which display all their actions and exhibit all their phenomena and yet are not in essential and metaphysical constitution bodies; as I have no clear and distinct perception of this matter I should not dare to affirm the contrary, and hence I am reluctant to say positively what the nature of bodies is, but I rather describe a certain kind of being similar in every way to bodies, and whose creation we cannot deny to be within the power of God, so that we can hardly say that it is not body.

Since each man is conscious that he can move his body at will, and believes further that all men enjoy the same power of similarly moving their bodies by thought alone; the free power of moving bodies at will can by no means be denied to God, whose faculty of thought is infinitely greater and more swift. And by like argument it must be agreed that God, by the sole action of thinking and willing, can prevent a body from penetrating any space defined by certain limits.

If he should exercise this power, and cause some space projecting above the Earth, like a mountain or any other body, to be impervious to bodies and thus stop or reflect light and all impinging things, it seems impossible that we should not consider this space to be truly body from the evidence of our senses (which constitute our sole judges in this matter); for it will be tangible on account of its impenetrability, and visible, opaque and coloured on account of the reflection of light, and it will resonate when struck because the adjacent air will be moved by the blow.

Thus we may imagine that there are empty spaces scattered through the world, one of which, defined by certain limits, happens by divine power to be impervious to bodies, and *ex hypothesi* it is manifest that this would resist the motions of bodies and perhaps reflect them, and assume all the properties of a corporeal particle, except that it will be motionless. If we may further imagine that that impenetrability is not always maintained in the same part of space but can be transferred hither and thither according to certain laws, yet so that the amount and shape of that impenetrable space are not changed, there will be no property of body which this does not possess. It would have shape, be tangible and mobile, and be capable of reflecting and being reflected, and no less constitute a part of the structure of things than any other corpuscle, and I do not see that it would not equally operate upon our minds and in turn be operated upon, because it is nothing more than the product of the divine mind realized in a definite quantity of space. For it is certain that God can stimulate our perception by his own will, and thence apply such power to the effects of his will.

In the same way if several spaces of this kind should be impervious to bodies and to each other, they would all sustain the vicissitudes of corpuscles and exhibit the same phenomena. And so if all this world were constituted of this kind of being, it would seem hardly any different. And hence these beings will either be bodies or like bodies. If they are bodies, then we can define bodies as *determined quantities of extension which omnipresent God endows with certain conditions*. These conditions are, (1) that they be mobile; and therefore I did not say that they are numerical parts of space which are absolutely immobile, but only definite quantities which may be transferred from space to space; (2) that two of this kind cannot coincide anywhere; that is, that they may be impenetrable, and hence that when their motions cause them to meet they stop and are reflected in accord with certain laws; (3) that they can excite various perceptions of the senses and the fancy in created minds, and conversely be moved by them, nor is it surprising since the description of the origin [of things?] is founded in this.

Moreover, it will help to note the following points respecting the matters already explained.

1. That for the existence of these beings it is not necessary that we suppose some unintelligible substance to exist in which as subject there may be an inherent substantial form; extension and an act of the divine will are enough. Extension takes the place of the substantial subject in which the form of the body is conserved by the divine will; and that product of the divine will is the form or formal reason of the body denoting every dimension of space in which the body is to be produced.

2. These beings will not be less real than bodies, nor (I say) are they less able to be called substances. For whatever reality we attribute to bodies arises from their phenomena and sensible qualities. And hence we would judge these beings, since they can receive all qualities of this kind and can similarly exhibit all these phenomena, to be no less real, if they should exist. Nor will they be less substance, since they will likewise subsist through God alone, and will acquire accidents.

3. Between extension and its impressed form there is almost the same analogy that the Aristotelians postulate between the *materia prima* and substantial forms, namely when they say that the same matter is capable of assuming all forms, and borrows the denomination of numerical body from its form. For so I suppose that any form may be transferred through any space, and everywhere denote the same body.

4. They differ, however, in that extension (since it is *what*

and *how constituted* and *how much*) has more reality than *materia prima*, and also in that it can be understood, in the same way as the form that I assigned to bodies. For if there is any difficulty in this conception it is not in the form that God imparts to space, but in the manner by which he imparts it. But that is not to be regarded as a difficulty, since the same question arises with regard to the way we move our bodies, and nevertheless we do believe that we can move them. If that were known to us, by like reasoning we should also know how God can move bodies, and expel them from a certain space bounded in a given figure, and prevent the expelled bodies or any others from penetrating into it again, that is, cause that space to be impenetrable and assume the form of body.

5. Thus I have deduced a description of this corporeal nature from our faculty of moving our bodies, so that all the difficulties of the conception may at length be reduced to that; and further, so that God may appear (to our innermost consciousness) to have created the world solely by the act of will, just as we move our bodies by an act of will alone; and, besides, so that I might show that the analogy between the Divine faculties and our own is greater than has formerly been perceived by Philosophers. That we were created in God's image holy writ testifies. And his image would shine more clearly in us if only he simulated in the faculties granted to us the power of creation in the same degree as his other attributes; nor is it an objection that we ourselves are created beings and so a share of this attribute could not have been equally granted to us. For if for this reason the power of creating minds is not delineated in any faculty of created mind, nevertheless created mind (since it is the image of God) is of a far more noble nature than body, so that perhaps it may eminently contain [body] in itself. Moreover, in moving bodies we create nothing, nor can we create anything, but we only simulate the power of creation. For we cannot make any space impervious to bodies, but we only move bodies; and at that not any we choose, but only our own bodies, to which we are united not by our own will but by the divine constitution of things; nor can we move bodies in any way but only in accord with those laws which God has imposed on us. If

anyone, however, prefers this our power to be called the finite and lowest level of the power which makes God the Creator, this no more detracts from the divine power than it detracts from God's intellect that intellect in a finite degree belongs to us also; particularly since we do not move our bodies by our own independent power but through laws imposed on us by God. Rather, if any think it possible that God may produce some intellectual creature so perfect that he could, by divine accord, in turn produce creatures of a lower order, this so far from detracting from the divine power enhances it; for that power which can bring forth creatures not only directly but through the mediation of other creatures is exceedingly, not to say infinitely, greater. And so some may perhaps prefer to suppose that God imposes on the soul of the world, created by him, the task of endowing definite spaces with the properties of bodies, rather than to believe that this function is directly discharged by God. Therefore the world should not be called the creature of that soul but of God alone, who creates it by constituting the soul of such a nature that the world necessarily emanates [from it]. But I do not see why God himself does not directly inform space with bodies; so long as we distinguish between the formal reason of bodies and the act of divine will. For it is contradictory that it [body] should be the act of willing or anything other than the effect which that act produces in space. Which effect does not even differ less from that act than Cartesian space, or the substance of body according to the vulgar idea; if only we suppose that they are created, that is, that they borrow existence from the will, or that they are creatures of the divine reason.

Lastly, the usefulness of the idea of body that I have described is brought out by the fact that it clearly involves the chief truths of metaphysics and thoroughly confirms and explains them. For we cannot postulate bodies of this kind without at the same time supposing that God exists, and has created bodies in empty space out of nothing, and that they are beings distinct from created minds, but able to combine with minds. Say, if you can, which of the views, already well-known, elucidates any one of these truths or rather is not opposed to all of them, and obscures all of them. If we say with Descartes that extension is body, do we not manifestly offer a path to Atheism, both because extension is not created but has existed eternally, and because we have an absolute idea of it without any relationship to God, and so in some circumstances it would be possible for us to conceive of extension while imagining the non-existence of God? Nor is the distinction between mind and body in this philosophy intelligible, unless at the same time we say that mind has no extension at all, and so is not substantially present in any extension, that is, exists nowhere; which seems the same as denying the existence of mind, or at least renders its union with body totally unintelligible, not to say impossible. Moreover, if the distinction of substances between *thinking* and *extended* is legitimate and complete, God does not eminently contain extension within himself and therefore cannot create it; but God and extension will be two substances separately complete, absolute, and having the same significance. But on the contrary if extension is eminently contained in God, or the highest thinking being, certainly the idea of extension will be eminently contained within the idea of thinking, and hence the distinction between these ideas will not be so great but that both may fit the same created substance, that is, but that a body may think, and a thinking being extend. But if we adopt the vulgar notion (or rather lack of it) of body, according to which there resides in bodies a certain unintelligible reality that they call substance, in which all the qualities of the bodies are inherent, this (apart from its unintelligibility) is exposed to the same inconveniences as the Cartesian view. For as it cannot be understood, it is impossible that its distinction from the substance of the mind should be understood. For the distinction drawn from substantial form or the attributes of substances is not enough: if bare substances do not have an essential difference, the same substantial forms or attributes can fit both, and render them by turns, if not at one and the same time, mind and body. And so if we do not understand that difference of substances deprived of attributes, we cannot knowingly assert that mind and body differ substantially. Or if they do differ, we cannot discover any basis for their union. Further, they attribute no less reality in concept (though less in words) to this corporeal substance regarded as being without qualities and forms, than they do to the substance of God, abstracted from his attributes. They conceive of both, when considered simply, in the same way; or rather they do not conceive of them, but confound them in some common idea of an unintelligible reality. And hence it is not surprising that Atheists arise ascribing that to corporeal substances which solely belongs to the divine. Indeed, however we cast about we find almost no other reason for atheism than this notion of bodies having, as it were, a complete, absolute and independent reality in themselves, such as almost all of us, through negligence, are accustomed to have in our minds from childhood (unless I am mistaken), so that it is only verbally that we call bodies created and dependent. And I believe that this preconceived idea explains why the same word, substance, is applied in the schools to God and to his creatures; and also why in forming an idea of body philosophers are brought to a stand and lose their drift, as when they try to form an independent idea of a thing dependent upon God. For certainly whatever cannot exist independently of God cannot be truly understood independently of the Idea of God. God is no less present in his creatures than they are present in the accidents, so that created substance, whether you consider its degree of dependence or its degree of reality, is of an intermediate nature between God and accident. And hence the idea of it no less involves the concept of God than the idea of accident involves the concept of created substance. And so it ought to embrace no other reality in itself than a derivative and incomplete reality. Thus the preconception just mentioned must be laid aside, and substantial reality is rather to be ascribed to these kinds of attributes which are real and intelligible things in themselves and do not need to be inherent in a subject, than to the subject which we cannot conceive as dependent, much less form any idea of it. And this we can manage without difficulty if (besides the idea of body expounded above) we reflect that we can conceive of space existing without any subject when we think of a vacuum. And hence some substantial reality fits this. But if moreover the mobility of the parts (as Descartes imagined) should be involved in the idea of vacuum, everyone would freely concede that it is corporeal substance. In the same way, if we should have an idea of that attribute or power by which God, through the sole action of his will, can create beings, we should readily conceive of that attribute as subsisting by itself without any substantial subject and [thus as] involving the rest of his attributes. But while we cannot form an idea of this attribute nor even of our own power by which we move our bodies, it would be rash to say what may be the substantial basis of mind.

So much for the nature of bodies, in making which plain I judge that I have sufficiently proved that such a creation as I have expounded is most clearly the work of God; and that if this world were not constituted from that creation, at least another very like it could be constituted. And since there is no difference between the materials as regards their properties and nature, but only in the method by which God created one and the other, the distinction between body and extension is certainly brought to light from this. Because extension is eternal, infinite, uncreated, uniform throughout, not in the least mobile, nor capable of inducing change of motion in bodies or change of thought in the mind; whereas body is opposite in every respect, at least if God did not please to create it always and everywhere. For I should not dare to deny God that power. And if anyone thinks otherwise, let him say where he could have created the first matter, and whence the power of creating was granted to God. Or if there was no beginning to that power, but he had the same eternally that he has now, then he could have created from eternity. For it is the same to say that there never was in God an impotence to create, or that he always had the power to create and could have created, and that he could always create matter. In the same way, either a space may be assigned in which matter could not be created from the beginning, or it must be conceded that God could have created it everywhere.

Moreover, so that I may respond more concisely to Descartes' argument: let us abstract from body (as he commands) gravity, hardness and all sensible qualities, so that nothing remains except what pertains to its essence. Will extension alone then remain? By no means. For we may also reject that faculty or power by which they [the qualities] stimulate the perceptions of thinking beings. For since there is so great a distinction between the ideas of thinking and of extension that it is impossible there should be any basis of connection or relation [between them] except that which is caused by divine power, the above faculty of bodies can be rejected without violating extension, but not without violating their corporeal nature. Clearly the changes which can be induced in bodies by natural causes are only accidental and they do not denote a true change of substance. But if any change is induced that transcends natural causes, it is more than accidental and radically affects the substance. And according to the sense of the demonstration, only those things are to be rejected which bodies can be deprived of, and made to lack, by the force of nature. But should anyone object that bodies not united to minds cannot directly arouse perceptions in minds, and that hence since there are bodies not united to minds, it follows that this power is not essential to them: it should be noticed that there is no question here of an actual union, but only of a faculty in bodies by which they are capable of a union through the forces of nature. From the fact that the parts of the brain, especially the more subtle ones to which the mind is united, are in a continual flux, new ones succeeding to those which fly away, it is manifest that that faculty is in all bodies. And, whether you consider divine action or corporeal nature, to remove this is no less than to remove that other faculty by which bodies are enabled to transfer mutual actions from one to another, that is, to reduce body into empty space.

However, as water offers less resistance to the motion of solid bodies through it than quicksilver does, and air much less than water, and aetherial spaces even less than air-filled ones, should we set aside altogether the force of resistance to the passage of bodies, we must also reject the corporeal nature [of the medium] utterly and completely. In the same way, if the subtle matter were deprived of all resistance to the motion of globules, I should no longer believe it to be subtle matter but a scattered vacuum. And so if there were any aerial or aetherial space of such a kind that it yielded without any resistance to the motions of comets or any other projectiles I should believe that it was utterly void. For it is impossible that a corporeal fluid should not impede the motion of bodies passing through it, assuming that (as I supposed before) it is not disposed to move at the same speed as the body (Part II, Epistle 96 to Mersenne).[1]

However it is manifest that all this force can be removed from space only if space and body differ from one another; and thence that they can exist apart is not to be denied before it has been proved that they do not differ, lest a mistake be made by *petitio principii*.

But lest any doubt remain, it should be observed from what was said earlier that there are empty spaces in the natural world. For if the aether were a corporeal fluid entirely without vacuous pores, however subtle its parts are made by division, it would be as dense as any other fluid, and it would yield to the motion of bodies through it with no less sluggishness; indeed with a much greater, if the projectile should be porous, because then the aether would enter its internal pores, and encounter and resist not only the whole of its external surface but also the surfaces of all the internal parts. Since the resistance of the aether is on the contrary so small when compared with the resistance of quicksilver as to be over ten or a hundred thousand times less, there is all the more reason for thinking that by far the largest part of the aetherial space is void, scattered between the aetherial particles. The same may also be conjectured from the various gravities of these fluids, for the descent of heavy bodies and the oscillations of pendulums show that these are in proportion to their densities, or as the quantities of matter contained in equal spaces. But this is not the place to go into this.

Thus you see how fallacious and unsound this Cartesian argument is, for when the accidents of bodies have been rejected, there remains not extension alone, as he imagined, but also the faculties by which they can stimulate perceptions in the mind and move other bodies. If we further reject these faculties and all power of moving so that there only remains a precise conception of uniform space, will Descartes fabricate

[1] 'When I imagine that a body moves in a totally non-resistant medium, what I suppose is that all the particles of the fluid body which surround it have a tendency to move at precisely the same speed as it is doing, and no greater speed, whether yielding to it the place they occupied or going into that which it leaves; and thus there are no fluids which do not resist certain motions. But to suppose a kind of matter which does not at all resist the different motions of some body, one must imagine that God or an angel excites its particles more or less, so that the body which they surround moves more or less quickly.'

any vortices, any world, from this extension? Surely not, unless he first invokes God, who alone can create bodies *de novo* in those spaces (by restoring those faculties, or the corporeal nature, as I explained above). And so in what has gone before I was correct in assigning the corporeal nature to the faculties already enumerated.

And thus at length since spaces are not the very bodies themselves but are only the places in which bodies exist and move, I think that what I laid down about local motion is sufficiently confirmed. Nor do I see what more could be desired in this respect unless perhaps I warn those to whom this is not satisfactory, that by the space whose parts I have defined as places filled by bodies, they should understand the Cartesian generic space in which spaces regarded singularly, or Cartesian bodies, are moved, and so they will find hardly anything to object to in our definitions.

I have already digressed enough; let us return to the main theme.

Definition 5. Force is the causal principle of motion and rest. And it is either an external one that generates or destroys or otherwise changes impressed motion in some body; or it is an internal principle by which existing motion or rest is conserved in a body, and by which any being endeavours to continue in its state and opposes resistance.

Definition 6. Conatus [endeavour] is resisted force, or force in so far as it is resisted.

Definition 7. Impetus is force in so far as it is impressed on a thing.

Definition 8. Inertia is force within a body, lest its state should be easily changed by an external exciting force.

Definition 9. Pressure is the endeavour of contiguous parts to penetrate into each others' dimensions. For if they could penetrate the pressure would cease. And pressure is only between contiguous parts, which in turn press upon others contiguous to them, until the pressure is transmitted to the most remote parts of any body, whether hard, soft or fluid. And upon this action is based the communication of motion by means of a point or surface of contact.

Definition 10. Gravity is a force in a body impelling it to descend. Here, however, by descent is not only meant a motion towards the centre of the Earth but also towards any point or region, or even from any point. In this way if the *conatus* of the aether whirling about the Sun to recede from its centre be taken for gravity, the aether in receding from the Sun could be said to descend. And so by analogy, the plane is called horizontal that is directly opposed to the direction of gravity or *conatus*.

Moreover, the quantity of these powers, namely motion, force, *conatus*, impetus, inertia, pressure and gravity may be reckoned in a double way: that is, according to either intension or extension.

Definition 11. The intension of any of the above-mentioned powers is the degree of its quality.

Definition 12. Its extension is the amount of space or time in which it operates.

Definition 13. Its absolute quantity is the product of its intension and its extension. So, if the quantity of intension is 2, and the quantity of extension 3, multiply the two together and you will have the absolute quantity 6.

Moreover, it will help to illustrate these definitions from individual powers. And thus motion is either more intense or more remiss, as the space traversed in the same time is greater or less, for which reason a body is usually said to move more swiftly or more slowly. Again, motion is more or less in extension as the body moved is greater or less, or as it is acting in a larger or smaller body. And the absolute quantity of motion is composed of both the velocity and the magnitude of the moving body. So force, *conatus*, impetus or inertia are more intense as they are greater in the same or an equivalent body: they have more extension when the body is larger, and their absolute quantity arises from both. So the intension of pressure is proportional to the increase of pressure upon the surface-area; its extension proportional to the surface pressed. And the absolute quantity results from the intension of the pressure and the area of the surface pressed. So, lastly, the intension of gravity is proportional to the specific gravity of the body; its extension is proportional to the size of the heavy body, and absolutely speaking the quantity of gravity is the product of the specific gravity and mass of the gravitating body. And whoever fails to distinguish these clearly, necessarily falls into many errors concerning the mechanical sciences.

In addition the quantity of these powers may sometimes be reckoned from the period of duration; for which reason there will be an absolute quantity which will be the product of intension, extension and duration. In this way if a body [of size] 2 is moved with a velocity 3 for a time 4 the whole motion will be $2 \times 3 \times 4$ or 12 [*sic*!].

Definition 14. Velocity is the intension of motion, slowness is remission.

Definition 15. Bodies are denser when their inertia is more intense, and rarer when it is more remiss.

The rest of the above-mentioned powers have no names.

It is however to be noted that if, with Descartes or Epicurus, we suppose rarefaction and condensation to be accomplished in the manner of relaxed or compressed sponges, that is, by the dilation and contraction of pores which are either filled with some most subtle matter or empty of matter, then we ought to estimate the size of the whole body from the quantity of both its parts and its pores in Definition 15; so that one may consider inertia to be remitted by the increase of the pores and intensified by their diminution, as though the pores, which offer no inertial resistance to change, and whose mixtures with the truly corporeal parts give rise to all the various degrees of inertia, bear some ratio to the parts.

But in order that you may conceive of this composite body as a uniform one, imagine its parts to be infinitely divided and dispersed everywhere throughout the pores, so that in the whole composite body there is not the least particle of extension without an absolutely perfect mixture of parts and pores thus infinitely divided. Certainly it suits mathematicians to contemplate things in the light of such reasoning, or if you prefer in the Peripatetic manner; but in physics things seem otherwise.

Definition 16. An elastic body is one that can be condensed by pressure or compressed within narrower limits; and a non-elastic body is one that cannot be condensed by that force.

Definition 17. A hard body is one whose parts do not yield to pressure.

Definition 18. A fluid body is one whose parts yield to an overwhelming pressure.[1] Moreover, the pressures by which the fluid is driven in any direction whatsoever (whether these are exerted on the external surface alone, or on the internal parts by the action of gravity or any other cause), are said to be balanced when the fluid rests in equilibrium. This is asserted on the assumption that the pressure is exerted in some one direction and not towards all at once.

Definition 19. The limits defining the surface of the body (such as wood or glass) containing the fluid, or defining the surface of the external part of the same fluid containing some internal part, constitute the *vessel of fluid* [*vas fluidi*].

In these definitions, however, I refer only to absolutely hard or fluid bodies, for one cannot ratiocinate mathematically concerning ones partially so, on account of the innumerable circumstances affecting the figures, motions and contexture of the least particles. Thus I imagine that a fluid does not consist of hard particles, but that it is of such a kind that it has no small portion or particle which is not likewise fluid. And moreover, since the physical cause of fluidity is not to be examined here, I define the parts not as being in motion among themselves, but only as capable of motion, that is, as being everywhere so divided one from another that, although they may be supposed to be in contact and at rest with respect to one another, yet they do not cohere as though stuck together, but can be moved separately by any impressed force and can change the state of rest as easily as the state of motion if they move relatively. Indeed, I suppose that the parts of hard bodies do not merely touch each other and remain at relative rest, but that they do besides so strongly and firmly cohere, and are so bound together, as it were by glue, that no one of them can be moved without all the rest being drawn along with it; or rather that a hard body is not made up of conglomerate parts but is a single undivided and uniform body which preserves its shape most resolutely, whereas a fluid body is uniformly divided at all points.

And thus I have accommodated these definitions not to physical things but to mathematical reasoning, after the manner of the Geometers who do not accommodate their definitions

---

[1] Newton originally wrote 'all whose parts are mobile among themselves', but crossed out these words.

of figures to the irregularities of physical bodies. And just as the dimensions of physical bodies are best determined from their geometry (as the measurement of a field from plane geometry, although a field is not a true plane; and the measurement of the Earth from the doctrine of the sphere even though the Earth is not precisely spherical) so the properties of physical fluids and solids are best known from this mathematical doctrine, even though they are not perhaps absolutely nor uniformly fluid or solid as I have defined them here.

Propositions on Non-Elastic Fluids

Axioms

1. From like postulates like consequences ensue.
2. Bodies in contact press each other equally.

Proposition 1. All the parts of a non-gravitating fluid, compressed with the same intension in all directions, press each other equally or with equal intension.

Proposition 2. And compression does not cause a relative motion of the parts.

Demonstration of Both

Let us first suppose that the fluid is contained and uniformly compressed by the spherical boundary *AB* whose centre is *K* [Fig. 42]. Any small portion of it *CGEH* is bounded by the two spherical surfaces *CD* and *EF* described about the same centre *K* and by the conical surface *GKH* whose vertex is at *K*. And it is manifest that *CGEH* cannot in any way approach the centre *K* because all the matter between the spherical surfaces *CD* and *EF* would everywhere approach the same centre for the same reason,[a] and so would penetrate the volume of the fluid contained within the sphere *EF*.[b] Nor can *CGEH* recede in any direction towards the circumference *AB* because all that shell of fluid between *CD* and *EF* would similarly recede for the same reason,[a] and so would penetrate the volume of fluid between the spherical surfaces *AB* and *CD*.[b] Nor can it be squeezed out sideways, say towards *H*,

---

[a] Axiom 1.   [b] Contrary to the definition.

since if we imagine another little section $H\gamma$, terminated in every direction by the same spherical surfaces and a similar conical surface and contiguous to GH at H, this section $H\gamma$ may for the same reason be squeezed out towards H,[a] and so effect a penetration of volume by the mutual approach of contiguous parts.[b] And so it is that no portion of fluid $CGEH$ can exceed its limits because of pressure. And hence all the parts remain in equilibrium. Which is what I wished to demonstrate first.

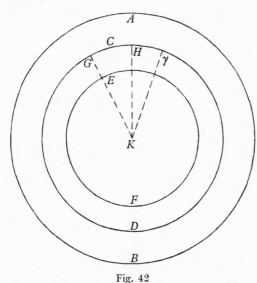

Fig. 42

I say also that all parts press each other equally, and with the same intension of pressure that the external surface is pressed. To show this, imagine that $PSQR$ is a part of the said fluid AB contained by similar spherical segments PRQ and PSQ, and that its compression upon the internal surface PSQ is as great as that upon the external surface PRQ [Fig. 43]. For I have already shown that this part of the fluid remains in equilibrium, and so the effects of the pressures acting on both of its surfaces are equal, and hence the pressures are equal.[c,d]

And thus since spherical surfaces such as PSQ can be described anywhere in the fluid AB, and can touch any other

---

[a] Axiom 1. [b] Contrary to the definition. [c] Axiom. [d] Definition.

given surfaces in any points whatever, it follows that the intension of the pressure of the parts along the surfaces, wherever placed, is as great as the pressure on the external surface of the fluid. Which is the second point I wished to demonstrate.

Moreover, as the force of this argument is based on the equality of the surfaces *PRQ* and *PSQ*, lest it should seem that there is some disparity, in that one is within the fluid and the other is a segment of the external surface, it will help to imagine that the whole sphere *AB* is a part of an indefinitely

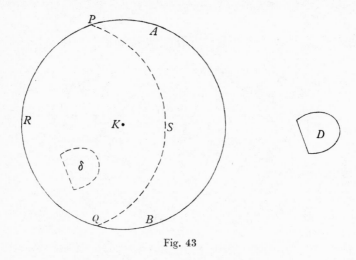

Fig. 43

larger volume of fluid, in which it is contained as within a vessel, and is everywhere compressed just as its part *PRQS* is pressed upon the surface *PSQ* by another part *PABQS*. For the method by which the sphere *AB* is compressed is of no significance, so long as its compression is supposed to be equal everywhere.

Now that these things have been demonstrated for a fluid sphere, I say lastly that all the parts of the fluid *D* (bounded in any manner at all, and compressed with the same intension in all directions) press each other equally and are not made to move relatively by the compression. For let *AB* be an indefinitely greater fluid sphere compressed with the same degree of intension; and let *δ* be some part of it equal and similar to *D*. From what has already been demonstrated it follows that this part *δ* is compressed with an equal intension in all directions and that the intension of the pressure is the same as that of the sphere AB, that is (by hypothesis) as that which compressed the fluid D. Thus the compression of the similar and equal fluids, D and δ, is equal; and hence the effects will be equal.[a] But all the parts of the sphere AB[b] and so of the fluid δ contained in it, press each other equally, and the pressure does not cause a relative motion of the parts. For which reason the same is true of the fluid D.[a] Q.E.D.

Corollary 1. The internal parts of a fluid press each other with the same intension as that by which the fluid is pressed on its external surface.

Corollary 2. If the intension of the pressure is not everywhere the same, the fluid does not remain in equilibrium. For since it stays in equilibrium because the pressure is everywhere uniform, if the pressure is anywhere increased, it will predominate there and cause the fluid to recede from that region.[c]

Corollary 3. If no motion is caused in a fluid by pressure, the intension of the pressure is everywhere the same. For if it is not the same, motion will be caused by the predominant pressure.[d]

Corollary 4. A fluid presses on whatever bounds it with the same intension as the fluid is pressed by whatever bounds it, and vice versa. Since the parts of a fluid are certainly the bounds of contiguous parts and press each other with an equal intension, conceive the aforesaid fluid to be part of a greater fluid, or similar and equal to such a part, and similarly compressed, and the assertion will be evident.[a]

Corollary 5. A fluid everywhere presses all its bounds, if they are capable of withstanding the pressure applied, with that intension with which it is itself pressed in any place. For otherwise it would not be pressed everywhere with the same intension.[e] On which assumption it yields to the more intense pressure.[d] And so it will either be condensed,[f] or it will break through the bounds where the pressure is less.[f]

Scholium. I have proposed all this about fluids, not as

---

[a] Axiom.
[b] According to what has been already demonstrated.
[c] By definition.   [d] Corollary 2.
[e] Corollary 4.
[f] Contrary to the hypothesis.

contained in hard and rigid vessels, but within soft and quite flexible bounds (say within the internal surface of a homogeneous exterior fluid), so that I might more clearly show that their equilibrium is caused only by an equal degree of pressure in all directions. But once a fluid is put into equilibrium by an equal pressure, it is all one whether you imagine it to be contained within rigid or yielding bounds.

2

THE LAWES OF MOTION

MS. Add. 3958, fols. 81–3

This manuscript, written on two sheets folded and stitched to make eight pages, was assigned to Newton's early years in the *Portsmouth Catalogue* (p. 1, no. 5). One can only guess its date, but about 1666 seems plausible. The hand, phraseology and numerous eccentricities of spelling all support this view.

The whole manuscript appears to be holograph, although two styles of writing are mingled. (This may itself be a sign of youth.) The cursive *e* and long *s* (ſ) were not normally used by Newton in later life; here these forms occur together with Newton's usual Greek ε and short *s*. The latter part headed 'Some Observations about Motion' was written a little after the bulk of the paper.

There is a second, nearly identical version of this paper entitled 'On the Laws of Reflection'. It does not include the 'Observations'.

How solitary bodyes are moved.

Sect. 1. There is an uniform extension, space, or expansion continued every way wthout bounds: in wch all bodyes are, each in severall parts of it: wch parts of Space possessed & adequately filled by ym are, their places. And their passing out of one place or part of space into another, through all ye intermediate Space is their motion. Which motion is done wth more or lesse Velocity accordingly as tis done through more or lesse space in equal times or through equall spaces in more or lesse time. But ye motion it selfe & ye force to persevere in yt motion is more or lesse accordingly as ye factus of ye bodys bulk into its velocity is more or lesse. And yt force is equivalent to that motion wch it is able to beget or destroy. *(Of Place motion velocity & force.)*

2. The motion of a body tends one way directly & severall other ways obliqly. As if ye body *A* move directly towards ye point *B* it also moves obliquely towards all ye lines *BC*, *BD*, *BE* &c wch passe through yt point *B*: & shall arrive to ym all at ye same time [Fig. 44a]. Whence its velocity towards ym is in such proportion as its distance from them yt is, as *AB*, *AC*, *AD*, *AE*,&c. *(Wth wt velocity a body moves severall ways at once.)*

3. If a body A move towards B wth the velocity R, & by ye
way hath some new force done to it wch had ye body rested
would have propell'd it towards C wth ye velocity S. Then
making $AB:AC::R:S$, & Completing ye Parallelogram BC ye
body shall move in ye Diagonall AD & arrive at ye point D
wth this compound motion in ye same time it would have
arrived at ye point B wth its single motion [Fig. 44b].

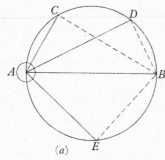 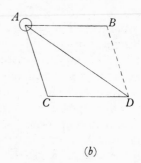

(a)   (b)

Fig. 44

Of centers
& axes of
motion &
ye motion
of those
centers.

4. In every body there is a certaine point, called its center
of motion about wch if ye body bee any way circulated ye
endeavours of its parts every way from ye center are exactly
counterpoised by opposite endeavours. And ye progressive
motion of ye body is ye same wth ye motion of this its center
wch always moves in a streight line & uniformly wn ye body is
free from occursions wth other bodys. And so doth ye common
center of two bodys; wch is found by dividing ye distance twixt
their propper centers in reciprocall proportion to their bulk.
And so ye common center of 3 or more bodys &c. And all ye
lines passing through these centers of motion are axes of motion.

5. The angular quantity of a bodys circular motion &
velocity is more or lesse accordingly as ye body makes one
revolution in more lesse time but ye reall quantity of its circular
motion is more or lesse accordingly as ye body hath more or
lesse power & force to persevere in yt motion; wch motion
divided by ye bodys bulke is the reall quantity of its circular
velocity. Now to know ye reall quantity of a bodys circular
motion & velocity about any given axis EF [Fig. 45]; Suppose
it hung upon ye two end E & F of yt axis as upon two poles:
And yt another globular body of ye same bignesse, whose

center is *A*, is so placed yt ye circulating body shall hit it in ye point *B* & strike it away in ye line *BAG* (wch lyeth in ye same plane wth one of ye circles described about ye axis *EF*) & thereby just loose all its owne motion. Then hath ye Globe gotten ye same quantity of progressive motion and velocity wch ye other had of circular; its velocity being ye same wth yt of ye point *C* wch describes a circle touching ye line *BG*. The Radius *DC* of wch circle I may therefore call ye radius of

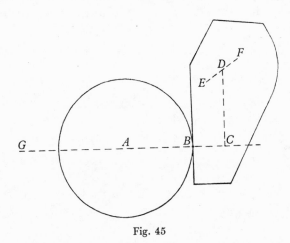

Fig. 45

Circular motion or velocity about yt axis *EF*. And ye circle described wth ye said Radius of Circulation in that plane wch cuts ye axis *EF* perpendicularly in ye center of motion I call ye Equator of circulation about that axis, and those circles wch passe through ye poles, meridians &c.

6. A body circulates about one axis (as *PC*) directly & about severall other axes (as *AC, BC*, &c) obliquely [Fig. 46*a*]. And ye angular quantity of its circulations about those axes (*PC, AC, BC* &c) are as ye sines (*PC, AD, BE*, &c) of ye angles wch those axes make wth ye Equator (*FG*) of ye principall & direct axis (*PC*).

Wth wt velocity a body circulates about severall axes at once.

7. If a body circulates about ye axis *AC* wth ye angular quantity of velocity *R*: & some new force is done to it, wch, if ye body had rested, would have made it circulate about another axis *BC*, wth ye angular quantity of velocity *S* [Fig. 46*b*]. Then in ye plane of ye two axes, & in one of those two opposite angles (made by ye axes) in wch ye two circulations are

How two circular motions are joyned into one.

159

contrary one to another, (as in ye angle ACB), I find such a point P from wch ye perpendiculars (PK, PH) let fall to those axes bee reciprocally proportional to ye angular velocitys about those axes (yt is $PK:PH::R:S$). And drawing ye line PC, it shall bee ye new axis about wch ye compound motion is performed. And ye summe of $\dfrac{CH}{CP} \times R$ & $\dfrac{CK}{CP} \times S$ when ye perpendiculars PH & PK fall on divers sides of ye axis PC, otherwise their difference, is ye angular quantity of circulation

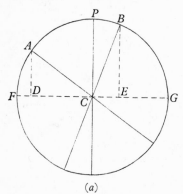 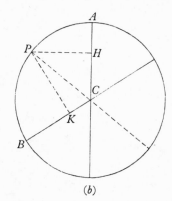

(a)   (b)

Fig. 46

about yt axis: Wch in ye angle $\dfrac{ACP}{BCP}$ tends contrary to ye circulation about ye axis $\dfrac{AC}{BC}$.

In wt cases a circulating body perseveres in ye same state & in wt it doth not.

8. Every body keepes ye same reall quantity of circular motion & velocity so long as tis not opposed by other bodys. And it keeps ye same axis too if ye endeavour from ye axis wch ye two opposite quarters twixt ye Equator & every meridian of motion have, bee exactly counterpoised by the opposite endeavours of ye side quarters; & yn also its axis doth always keepe parallel to it selfe. But if ye said endeavours from ye axis bee not exactly counterpoised by such opposite endeavours: yn for want of such counterpoise ye prevalent parts shall by little & little get further from ye axis & draw nearer & nearer to such a Counterpoise, but shall never bee exactly counterpoised. And as ye axis is continually moved in ye body, so it continually moves in ye space too wth some kind or other of spirall motion; always drawing nearer & nearer to

a center or parallelisme wth it selfe, but never attaining to it. Nay tis so far from ever keeping parallel to it selfe, yt it shall never bee twice in ye same position.

How Bodies are Reflected

9. Suppose ye bodys A & α did move in ye lines DA & $E\alpha$ till they met in the point B [Fig. 47]: yt BC is ye plane wch toucheth them in ye point of contact B: yt ye velocity of ye Body A towards ye said plane of contact is B, & ye motion AB; Some names & letters defined.

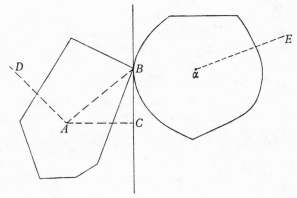

Fig. 47

& yt ye change wch is made by reflection in ye velocity & motion is X & AX. Suppose also yt from ye body A its center of motion two lines are drawne ye one AB to ye point of contact ye other AC to ye plane of Contact: yt ye intercepted line BC is F: that ye axis of motion wch is perpendicular to ye plane ABC & its Equator are called ye axis & Equator of reflected circulation: yt ye radius of yt Equator is G: yt ye reall quantity of velocity about yt axis is D, & ye motion AD: yt ye change wch reflection makes in yt velocity & motion is Y & AY: And yt ye correspondent lines & motions of ye other body α are β, $\alpha\beta$, ξ, $\alpha\xi$, ϕ, γ, δ, $\alpha\delta$, ν & $\alpha\nu$. Lastly for brevity sake suppose yt $\dfrac{1}{A}+\dfrac{1}{\alpha}+\dfrac{F}{AG}+\dfrac{\phi}{\alpha\gamma}=P$. And $2B+2\beta+\dfrac{2DF}{G}+\dfrac{2\delta\phi}{\gamma}=Q$. Observing yt at ye time of reflection if in either body ye center of motion doth move from ye plane of contact, or those parts of it nearest ye point of contact doe circulate from ye plane of contact: yn ye said motion is to bee esteemed negative & ye signe of its velocity B, β, D or δ must bee made negative in ye valor of Q.

The Rule for Reflection.

10. The velocitys B, β, D & δ & they only are directly opposed & changed in Reflection; & yt according to these rules $\dfrac{Q}{AP}=X$. $\dfrac{Q}{\alpha P}=\xi$. $\dfrac{FQ}{AGP}=Y$. & $\dfrac{\phi Q}{\alpha\gamma P}=\nu$. Which mutations $X\,\xi\,Y$ & ν tend all of them from ye plane of Contact. And these four rules I gather thus: The whole velocity of ye two points of contact towards one another perpendicularly to ye plane of contact is $1/2Q$ (arising partly from ye bodys progressive velocity B & β & partly from their circular D & δ): And ye same points are reflected one from another wth ye same quantity of such velocity. So yt ye whole change of all yt their velocity wch is perpendicular to ye plane of contact is Q. Which change must bee distributed amongst ye foure opposed velocitys B, β, D & δ proportionably to ye easinesse (or smallnesse of resistance) wth wch those velocitys are changed, yt is, proportionably to $\dfrac{1}{A}$, $\dfrac{1}{\alpha}$, $\dfrac{F}{AG}$, & $\dfrac{\phi}{\alpha\gamma}$. Soe yt $\dfrac{1}{A}+\dfrac{1}{\alpha}+\dfrac{F}{AG}+\dfrac{\phi}{\alpha\gamma}$:

$$Q::\frac{1}{A}:X::\frac{1}{\alpha}:\xi::\frac{F}{AG}:Y::\frac{\phi}{\alpha\gamma}:\nu.$$ that is $\dfrac{Q}{AP}=X$. $\dfrac{Q}{\alpha P}=\xi$. $\dfrac{FQ}{AGP}=Y$. & $\dfrac{\phi Q}{\alpha\gamma P}$.

The conclusion. In wt method ye precedant rules must be used.

11. Now if any two reflecting bodys A & α, wth ye quantity of their progressive & angular motions; & their position at their meeting & consequently their point & plane of contact &c be given: to know how those bodys shall bee reflected, First find B & β by sec. 2. Then ye lines F & ϕ & ye axis of reflected circulation by Sec 9. & their Radij G & γ by sec 5: Then their angular quantity of velocity about ye axes of reflected circulation by sec 6, & ye reall quantity D & δ by sec 5. Then P & Q by sec 9. Then X, ξ, Y & ν by sec 10. Then ye bodys new progressive determinations & velocitys by sec 3. Then ye angular quantity of yt circulation (Y & ν) wch is generated by reflection by sec 5. And lastly ye new axes & angular quantity of velocity about ym by sec 7.

Some Observations about Motion

Only those bodyes which are absolutely hard are exactly reflected according to these rules. Now the bodyes here amongst us (being an aggregate of smaller other bodyes) have a relenting softnesse & springynesse, wch makes their contact be for some time & in more points then one. And ye touching surfaces during ye time of contact doe slide one upon another more or lesse or not at all according to their roughnesse. And few or none of these bodyes have a springynesse soe strong, as to force them one from another wth ye same vigor that they came together. Besides yt their motions are continually impeded & slackened by ye mediums in wch they move. Now hee yt would prescribe rules for ye reflections of these compound bodies, must consider in how many points ye two bodies touch

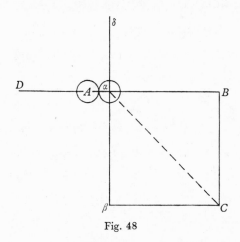

Fig. 48

at their meeting, ye position & pression of every point, wth their planes of contact &c: & how all these are varyed every moment during ye time of contact by ye more or lesse relenting softnesse or springynesse of those bodies & their various slidings. And also what effect ye air or other mediums compressed betwixt ye bodies may have.

2. There are some cases of Reflections of bodies absolutely hard to wch these rules extend not: As when two bodies meet wth their angular point, or in more points then one at once, Or with their superficies. But these cases are rare.

3. In all reflections of any bodies wt ever this rule is true: that ye common center of two or more bodies changeth not its state of motion or rest by ye reflection of those bodies one amongst another.

4. Motion may be lost by reflection. As if two equall Globes A & α wth equall motions from D & δ done in the perpendicular lines DA & $\delta\alpha$, hit one another when the center of ye body α is in ye line DA [Fig. 48]. Then ye body A shall loose all its motion & yet ye motion of α is not doubled. For completing ye square $B\beta$, ye body α shall move in ye Diagonall αC, & arrive at C but at ye same time it would have arrived at β wthout reflection. See ye third section.

5. Motion may be gained by reflection. For if the body α return wth ye same motion back again from C to α. The two bodyes A & α after reflection shall regain ye same equall motions in ye lines AD & $\alpha\delta$ (though backwards) wch they had at first.

3

THE ELEMENTS OF MECHANICKS

MS. Add. 4005, fols. 23–5

This is a holograph manuscript, written on a double folded sheet. Apart from a redrafting of Theorem 9 it is little written over or corrected. It is certainly later in date than the *Principia*.

Theoremes

1. All bodies continue perpetually in their state of resting or moving in right lines unless disturbed by any force & when disturbed the changes of their state are proportional to the force disturbing them: and if this change & ye antecedent motion of the body be represented in situation & quantity by ye sides of a parallelogram, the motion arising from them both shall be represented by ye diagonal of that parallelogram.

2. The actions of two bodies upon one another are always equal & directly contrary, & the resilition of dashing bodies is as their springiness, & the force of a globular body upon a resting plane is as yt part of its velocity wch is perpendicular to the plane.

3. The forces wth wch the parts of one & ye same engine move at one & ye same time are reciprocally proportional to their velocities.

4. In a System of bodies moving variously amongst one another & acting upon one another by any impulses, attractions or other forces & not acted upon by any other bodies, the common center of gravity of them all either rests perpetually or moves uniformly in a right line; & whether it rests or moves the motions of the bodies amongst one another are the same.

5. Falling bodies describe spaces proportional to ye squares of ye times & so do projectiles by their motions of ascent & descent: but their horizontal motions are uniform & with both motions together they describe Parabolas & water spouts up in a Parabolic figure after the manner of Projectiles.

6. Bodies circulating in concentric circles have centrifuge forces proportional to ye radii of ye circles directly & squares

of the times of revolution reciprocally, & are kept in their Orbs by contrary forces of the same quantity.

7. Bodies attracted towards a resting center describe equal areas in equal times & their velocities in approaching that center increase as much as if they made the same approach by falling down directly.

8. Bodies attracted towards a resting center with a force reciprocally proportional to the square of the distance from that center move in Conick Sections wch have a common focus in yt center & their periodical times in Ellipses are in a sesquialterate proportion of their mean distances from that center.

9. The resistance of a globular body in an uniform medium is as the square of the velocity, the surface of ye globe & ye density of the medium.

10. The parts of stagnating water are prest & press all things in proportion to their depth from the surface of the water besides the pressure wch arises from ye weight of ye incumbent Atmosphere. And the parts of all fluids press & are prest in proportion to the incumbent weight. Whence water ascends in Pumps & Mercury in ye Barometer by ye weight of ye Atmosphere & water ascends in Pipes & spouts up to ye height of ye fountain & in a Syphon the water in the deeper legg descends & the gravity of [sic] levity of bodies in water by which they descend or ascend is ye difference between their gravity out of the water & the gravity of so much water as equals them in bulk. And floating bodies dip into the room of so much water as equals them in weight.

11. A leaded string vibrates in the same time whether struck faintly or strongly & so do Pendulums & Springs & all bodies urged by a force proportional to ye space through wch they vibrate. Which is the ground of Pendulum Clocks & Watches & of Sound & Harmony. For a vibrating string shakes the Instrument & the Instrument shakes the air & also other strings, & the air shakes other unison bodies & also ye drum of ye ear for causing sound, & the sound is graver or acuter according to the number of ye vibrations in ye same time. For a leaded string vibrates in time proportional to its length & if it should be leaded by a weight it vibrates in time reciprocally proportional to ye square of the weight & the sweetness of harmony arises from the simplicity of the pro-

portions of these vibrations. If the vibrations be as 1 to 1, 2 to 1, 3 to 1, 4 to 1 or 5 to 1 they create an unison eighth, twelfth, fifteenth or seventeenth; if as 3 to 2, 4 to 3, 5 to 4 or 6 to 5 they create a fift a fourth a third major or a third minor: if as 5 to 3 or 8 to 5 they create a sixt major or sixt minor.

The Mechanical Frame of the World

1. All bodies are impenetrable & have a force of gravity towards them proportional to their matter & this force in receding from ye body decreases in the same proportion that the square of the distance increases & by means of this force the Earth Sun Planets & Comets are round.

2. The Sun is a fixt star & the fixt stars are scattered throughout all the heavens at very great distances from one another & rest in their several regions being great round bodies vehemently hot & lucid & by reason of the great quantity of their matter they are endued with a very strong gravitating power.

3. The Earth is a Planet & the Moon is a Satellite or little secondary Planet & the Planets are round dense opake cold terrestrial bodies illuminated by the Sun & fixt stars & gravitating towards them & towards one another & by means of their gravities revolving in their several Orbs, the Moons about ye Primary Planets & these Planets with their Moons about ye Suns.

4. Comets are a sort of Planets round & opake with very great Atmospheres. By their gravity towards the Sun they revolve about him in very excentric Elliptic Orbs all manner of ways & as often as they descend into the lower parts of their Orbs they become visible to us & then send up tails like a very thin smoke from ye exterior part of their Atmospheres boyed up by ye greater weight of ye Suns Atmosphere into wch they dip.

5. The Comets & Primary Planets revolve about the Sun by ye same laws. They revolve in Ellipses whose lower focus is in the Sun & whose Apsides remain almost immoveable. By radij drawn to ye Sun they describe areas proportional to the times, & the times of their revolutions are in a sesquialterate proportion of their mean distances from ye Sun. And

the same laws are observed by the secondary Planets in their revolutions about the Primary ones. And in general the motions of all the great bodies hitherto observed by Astronomers are exactly such as ought to arise from their mutual gravities in free spaces. For the heavens are empty spaces free to all motions all manner of ways without any sensible resistance.

6. While the Primary Planets revolve in their several Orbs about ye Sun, they revolve also about their several axes, the Earth in 24 hours, Jupiter in 9 hours 56 minutes, Mars in almost 24 hours, the Sun in 27 days. Whence arises day & night in each of the opake Planets & whilst they revolve from west to east the Sun & stars appear to revolve from east to west: These revolutions of the fixt stars are uniform & measure time equally.

7. The axis of the Earth keeps parallel to it self & is inclined 23 1/2 degr to ye Orb in wch the earth revolves about the Sun & thence the Poles of ye earth alternately approach the Sun to cause summer & winter & by the like inclination may summer & winter be caused in any other Planet. Yet the axis of ye Earth keeps not always exactly parallel to it self but changes its position very slowly making an angle with it self of about 20″ yearly & thence arises the Precession of ye Equinox wch amounts to about 50″ yearly.

8. Those parts of ye earth & sea gravitate more towards ye Moon wch are nearest to her & thence ye sea under the Moon rises towards her & opposite to ye Moon it rises from her by the defect of gravity & by rising in both places causes high tyde every 12 hours. The like happens in respect of ye Sun & the solar & lunar Tydes conspiring in the Syzygies become then ye greatest.

Problemes

1. To compound & divide forces & motions.

2. With any given force to move any given weight.

3. To find ye density (or specific gravity) of any body by its weight in Air & Water & by weighing ye same body in several liquors to compare their densities.

4. If a body be thrown with a given force in a given position, to find the line wch the body shall describe & what

shall be ye velocity of ye body in every point of the line & in what time it shall arrive at any point.

5. If any engin emit a body with a given velocity to direct ye engin so as wth ye emitted body to hit a given mark.

6. To draw lines parallel & perpendicular to ye Horizon, find the Meridian & Latitude of your Place, & observe ye altitudes & distances of ye stars & their transits through ye Meridian.

4

GRAVIA IN TROCHOIDE DESCENDENTIA

MS. Add. 3958, fol. 90–1

This holograph manuscript, a double folded sheet, was placed by the compilers of the *Portsmouth Catalogue* (Section I, p. 1, no. 5) among the early papers of Newton. There is no indication, however, that its writing preceded the publication of Christiaan Huygens' *Horologium Oscillatorium sive de motu pendulorum ad horologiam aptato demonstrationes Geometricae*[1] in 1673. This work was necessarily well known to Newton, since he received a copy from the author. As the problems elucidated by Newton are identical with those studied by Huygens (though by different means) it seems more reasonable to suppose that Newton composed this short piece after 1673. He may well have sought demonstrations of these important results that were easier, and to him more natural, than those of Huygens.

We have supplied the three figures, which are not given in the manuscript. Newton was notably careless in his references to the points of these figures.

In Fig: prima [Fig. 49]

Sit DCE Trochoides ad $\frac{1}{2}$ circulum BCY pertinens quae planum horizontale tangat in C insistens ei normaliter. Inque curva DC grave descendat a D ad C dilapsum per puncta δ, P, et π. Et agantur δYS, PVR, $\pi\chi Q$ parallelae ad DE &c.

1. Dico quod gravitatis efficacia sive descendentis acceleratio in singulis descensus locis D, P, O, &c est ut spatium describendum DC, δC, PC, &c. Scilicet obliquitas descensus minuit efficaciam gravitatis ita ut si gravia duo descensura sint ad C alterum B recta per diametrum BC, alterum C [Y] oblique per chordam YC: Minor erit acceleratio gravis Y propter obliquitatem descensus idque in ratione YC ad BC ita ut ambo gravia simul perveniant ad C. Est autem BC parallela curvae in D, ac YC parallela ipsi in δ, ideoque acceleratio gravis in D est eadem cum acceleratione gravis descendentis in

[1] Huygens, *Œuvres Complètes*, **18**, La Haye, 1934.

BC ut et acceleratio gravis in δ eadem cum acceleratione gravis descendentis per YC. Quare descendentis acceleratio in D est ad accelerationem ejus in δ ut BC ad YC, sive ut eorum dupla DC ad δC. Q.E.O.

2. Gravia in Trochoide descendentia, alterum a D, alterum a quolibet alio puncto P, simul pervenient ad C. Nam ut sunt longitudines DC, PC, ita accelerationes sub initio motus in D et P: quare spatia primo descripta puta Dd & Pp erunt in eadem ratione. Unde dividendo est $DC.PC::dC.pC$. Quare accelerationes in d et p permanent in eadem ratione, et etiamnum generabunt velocitates descendentium in eadem ratione, efficientque ut gravia pergant describere spatia $d\delta$ &

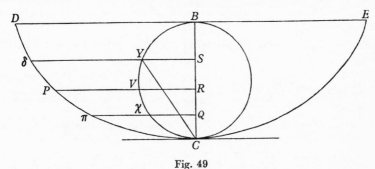

Fig. 49

$p\pi$ in eadem ratione. Adeoque spatia δC et πC erunt in illa ratione idque continuo donec utrumque simul in nihilum evanescat. Quare gravia simul attingent punctum C.

Potuit etiam hoc inde ostendi quod posito $DC.\delta C::PC.\pi C$ sit $D\delta.P\pi::\sqrt{BS}.\sqrt{RQ}::$velocitas post descensum ad profunditatem BS. ad velocitatem post descensum ad profunditatem RQ.

3. Itaque si grave undulet in Trochoide undulationes quaelibet erunt ejusdem temporis.

In Fig: secunda [Fig. 50]

Super Diametrum Trochoidis DE erige perpendiculum $BA = BC$ et a puncto A hinc inde describe duas semi-Trochoides AD, AE tangentes rectam DE in D et E, adeoque ejusdem magnitudinis cum Trochoide DCE. Jam puncto Q in BD ad arbitrium sumpto, fac arcus BT ac DS aequales longitudini DQ et comple parallelogramma $BQPT$ ac $DQRS$. Et constat

1. Quod QP normaliter insistit Trochoidi DC in P, & quod QR tangit Trochoidem DRA in R.

2. Quod QP et QR in directum jacent propter parallelismum rectarum DS, BT. Adeoque omnis recta perpendicularis ad Troch DC tanget Troch AD et contra.

3. Quod est $PQ = TB = DS = \frac{1}{2}$ curvae DR, adeoque $2PQ$ sive $PR = DR$. Et recta $PR +$ cuva $RA =$ toti curvae $DRA =$ Rectae CA.

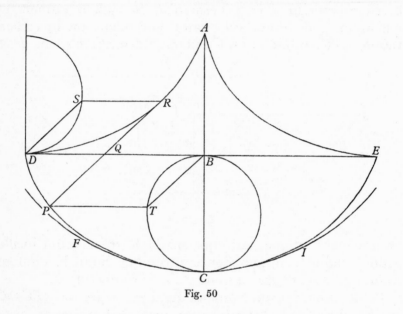

Fig. 50

4. Quare si ARP sit filum datae longitudinis, cui pondus P appenditur ita undulans intra trochoides AD et AE ut filum ab ipsis paululum prohibeatur ne in rectum protendatur, quemadmodum videre est in parte AR, ubi se applicat ad Trochoidem: Tunc pondus P undulabit in Trochoide DCE, adeoque quamlibet utcunque longam vel brevem undulationem in eodem tempore perficiet.

5. Patet etiam quod undulationes in circulo FCI centro A descripto modo sint perbreves (puta 10gr hinc inde vel minus) sunt ejusdem temporis proxime ac in Trochoide DCE. Nam undulatio in utroque casu fit circa centrum A nisi quod filum paululum incurvatur in uno casu; quae curvatura quam parva

sit ex eo percipies quod R tantum supra rectam DE esse debes imaginari quantum P cadit infra.

In Figura tertia [Fig. 51]

1. Stantibus jam ante positis cum gravis a D per P ad C descendentis velocitas in loco quolibet P est ut radix altitudinis BV hoc est ut linea BT: pro designanda illa velocitate exponatur eadem BT.

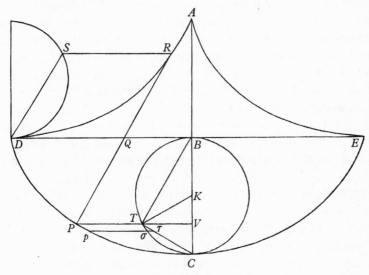

Fig. 51

2. Dein sit Pp particula spatij DC in ejusmodi particulas infinite multas et aequales divisi, et agatur $p\sigma\tau$ parallela ad PTV secans semicirculum in σ et rectam TC in τ. Jam propter parvitatem, curvarum portiunculae Pp ac $T\sigma$ pro rectis haberi possunt, adeoque $P\pi$ [Pp] ac $T\tau$ erunt aequales propter parallelismum et inde $T\tau$ erit datae licet infinite parvae longitudinis; ductaque semidiametero TK, triangula $T\sigma\tau$, TKB erunt similia, siquidem latera unius sunt perpendiculariter posita ad alterius latera correspondentia viz $T\sigma$ ad TK, $T\tau$ ad TB et $\sigma\tau$ ad BK. Quare est $BT.TK::T\tau.T\sigma$. sive $BT \times T\sigma = TK \times T\tau$. Adeoque cum TK ac $T\tau$ pro datis habenda sunt, erunt BT ac $T\sigma$ reciproce proportionalia. Cum itaque tempus et velocitas quibus datum spatium ut Pp describitur sunt reciproce proportionalia, et BT pro velocitate

exponitur exponit potest etiam $T\sigma$ pro tempore. Atque ita si spatij Dp pars quaelibet Pp describitur in parte tempo $T\sigma$, describetur totum spatium Dp in toto tempore $A\sigma\,[B\sigma]$.

3. Hinc posito quod semicircumferentia BTC designat tempus in quo spatium DC percurritur, ut noscas in quo tempore pars DP describetur age $PT\|DB$ et arcus BT designet tempus.

4. Potest etiam tempus per angulum BCT, vel per inclinationem descensus Pp aut per longitudinem DQ designari.

5. Caeterum ut tempora perpendicularis descensus cum temporibus descensus in hac curva conferantur, pone quod grave a B per V ad C descendit, et cum linearum BT quadrata sunt ut lineae BV: exponatur BT pro designando tempore descensus ad V. Adeo ut si grave descendat ad C in tempore BC, descendet ad V in tempore BT.

6. Jam cum descensus aeque a D ac B sub initio sunt ad Horizontem perpendiculares, manifestum est quod utrumque grave D ac B incipit aequaliter descendere, etsi D confestim in obliquum fertur. Adeoque lineae per quas tempora descensuum designantur ita debent inter se constitui ut initialiter exhibeant aequalia tempora descensuum aeque-altorum et postea recte exhibebunt tempora descensuum aeque altorum ut fiunt sensim inaequalia. Et hinc patet arcum et ejus chordam BT (cum sint initialiter aequalia) non modo recte designare haec tempora seorsim, sed et inter se conferre. Ita ut posito quod grave defertur a D ad C in tempore BTC, non modo sequetur quod deferetur ad P in tempore BT, sed etiam quod descendet a B ad V in tempore BT vel a B ad C in tempore BC: et contra.

Tempora etiam descensus ab alijs curvae punctis ut P exhinc noscuntur siquidem partes proportionales in aequalibus temporibus peraguntur.

Sed praecipuum est quod ex dato tempore in quo pendulum datae longitudinis vibrat, datur tempus in quo grave ad datam profunditatem descendit. Nam si grave decidat a B ad C in tempore BC fac $BC^{\mathrm{q}}.BTC^{\mathrm{q}}::BC.A\gamma$, ac decidet ab A ad γ in tempore BTC quod est unius semivibrationis. Posito autem $AC = 1,0000$, calculus dabit $A\gamma$ 1,2337, per cujus quadruplum 4,9348 descendet in tempore unius vibrationis, hoc est per $5AC$ fere, et per 19,7392 sive per $19\tfrac{3}{4}AC$ fere in vibratione replicata.

Nota quod motus gravis a *D* ad *C* descendentis persimilis est motui puncti in rota uniformiter mota quod describit Trochoiden, respectu velocitatis.

2. Quod pendulum ex argento vivo confectum diutius perseverat in motu.

TRANSLATION

The Descent of Heavy Bodies in Cycloids

Figure 1 [Fig. 49]

Let *DCE* be a cycloid derived from the semicircle *BCY*, touching the horizontal plane in *C* and standing upon it normally (figure 49). And in this curve *DC* a heavy body descends from *D* to *C* falling through the points δ, *P*, and π. And δYS, *PVR*, $\pi\chi Q$ are drawn parallel to *DE* etc.

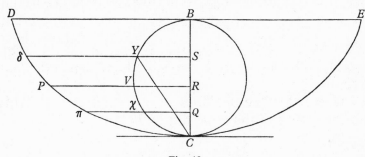

Fig. 49

1. I say that the effectiveness of gravity or the acceleration of the descending body at the various points in the descent *D*, *P*, *O* &c [*D*, δ, *P*,] are as the spaces to be described, *DC*, δC, *PC* &c. Plainly the obliquity of the descent diminishes the effectiveness of gravity so that if two heavy bodies descend to *C*, one (*B*) directly by the diameter *BC*, the other [*Y*] obliquely by the chord *YC*, the acceleration of the heavy body *Y* will be less on account of the obliquity of its descent, and that in the ratio of *YC* to *BC*, so that both heavy bodies may arrive at *C* simultaneously. However, *BC* is parallel to [the tangent to] the curve at *D*, and *YC* is parallel to the same at δ, and so the acceleration of the heavy body at *D* is the same as the acceleration of a heavy body descending *BC* and so also the acceleration

of a heavy body at δ is the same as the acceleration of a heavy body descending along YC. For which reason the acceleration of a descending body at D is to its acceleration at δ as BC is to YC, or as their doubles DC to δC. Q.E.O.

2. When two bodies descend in a cycloid, one from D, the other from any point P, they arrive at C simultaneously. For the accelerations at the beginnings of the motions at D and P are as the lengths DC, PC: for which reason the spaces described in the first instant (say Dd and Pp) will be in the same ratio. Whence, by division, $DC:PC = dC:pC$. Therefore the accelerations in d and p remain in the same ratio, and moreover will generate velocities of descent in the same ratio, and will cause the heavy bodies to describe the spaces $d\delta$ and $p\pi$ in the same ratio. And so the spaces δC and πC will be in that same ratio and so on continually until both vanish to nothing at once. Therefore the heavy bodies reach the point C in the same time.

It would be possible to prove in this way that given

$$DC:\delta C = PC:\pi C,$$

then $D\delta:P\pi = \sqrt{(BS)}:\sqrt{(RQ)}$; that is, they are as the velocity [of a body] after falling from B to S, to the velocity after falling from R to Q.

3. Thus if a body oscillates in a cycloid any oscillations whatever will be of the same duration.

Figure 2 [Fig. 50]

Upon the diameter of the cycloid DE erect the perpendicular $BA = BC$ and from the point A on either side describe two semi-cycloids AD, AE touching the straight line DE in D and E, and thus of the same size as the cycloid DCE (figure 50). Now taking any point Q in BD arbitrarily, make the arcs BT and DS equal to the length DQ and complete the parallelograms $BQPT$ and $DQRS$. And it follows

1. That QP stands normally upon the cycloid DC at P, and that QR touches the cycloid DRA at R.[1]

2. That QP and QR lie in the same straight line because of the parallelism of the straight lines DS, BT. And so all straight

---

[1] Demonstrated by Huygens in *Horologium Oscillatorium*, Pars Tertia, Prop. V; *Œuvres complètes*, **18**, 198–9.

PLATE IV

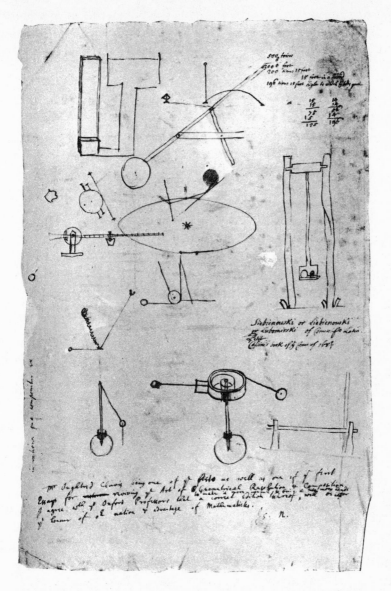

Drawings of apparatus by Newton. Newton was no more than barely competent at making a quick sketch, and a whole sheet of figures is rare in his papers, except for mathematical drawings. These appear to relate to pendulum experiments; a whirling-table for experiments on centrifugal force may be the object sketched at the centre. Perhaps about 1685.

lines perpendicular to the cycloid *DC* touch the cycloid *AD*, and vice versa.

3. Because $PQ = TB = DS =$ one-half the curve *DR*, so $2PQ$ or $PR = DR$.[1] And the straight line $PR+$ the curve $RA =$ the whole curve $DRA =$ the straight line *CA*.

4. Therefore if *ARP* is a thread of a given length, to which the weight *P* is attached, so oscillating between the cycloids *AD* and *AE* that the thread is just prevented from stretching

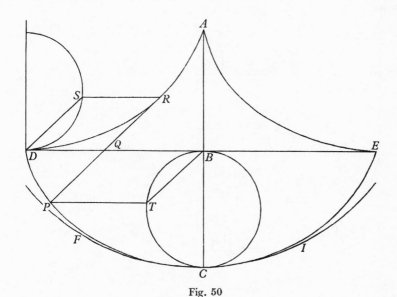

Fig. 50

into a straight line, as may be seen in the portion *AR* where it is applied to the cycloid: then the weight *P* will oscillate in the cycloid *DCE*, and so it will complete any oscillation whether long or short in the same time.[2]

5. It is also clear that oscillations in the circle *FCI* described about the centre *A*, if they are pretty small (say 10° either side or less) are performed in about the same time as those in the cycloid *DCE*. For the oscillation in both cases is made about the centre *A*, except that the thread is a little bent in one case:

[1] Huygens, *ŒC*, **18**, 489, n. 2. This relation between the chord and the cycloidal arc was known to Huygens in 1659, but not published by him (*ŒC*, **14**, 367).

[2] Cf. Huygens, *H. O.* Prop. VI, *ŒC*, **18**, 200–1. Newton sets up the cycloids and proves that the unrolling tangent *PR* always equals *DRA*, i.e. *PRA* can be a 'thread'. Huygens had assumed *PRA* to be a thread and shown that *DCE* was a cycloid.

how slight this bending is you may perceive from the fact that R is to be imagined as much above the line DE as P falls below it.

Figure 3 [Fig. 51]

1. With everything remaining as it was before, the velocity at any point P of a body descending from D through P to C is as the square root of the height BV, that is, as the line BT (figure 51). To designate that velocity let it be expressed by the same BT.

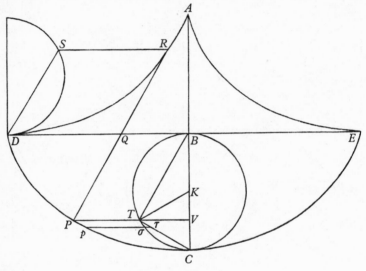

Fig. 51

2. Then let Pp be a small part of the space DC divided into infinitely many equal small parts of the same kind, and $p\sigma\tau$ is drawn parallel to PTV cutting the semicircle in σ and the straight line TC in τ. Now, on account of their smallness, the little curved segments Pp and $T\sigma$ can be taken as straight lines, and so $P\pi$ [Pp] and $T\tau$ will be equal because they are parallel and hence $T\tau$ will be of a given though infinitely small length. When the semidiameter TK is drawn, the triangles $T\sigma\tau$, TKB will be similar, since the sides of one are situated perpendicularly to the corresponding sides of the other, that is, $T\sigma$ to TK, $T\tau$ to TB, and $\sigma\tau$ to BK. Therefore $BT:TK = T\tau:T\sigma$ or $BT \times T\sigma = TK \times T\tau$. And so as TK and $T\tau$ are taken as given, BT and $T\sigma$ will be reciprocally proportional. Since therefore

the time and the velocity with which a given distance such as
Pp is described are reciprocally proportional, and BT can be
used to express the velocity, $T\sigma$ can be taken to express the
time. And so if any part Pp of the space Dp is described in a
fraction of time $T\sigma$, the whole space Dp will be described in the
whole time $A\sigma$ $[B\sigma]$.[1]

3. Hence, having postulated that the semicircumference
BTC designates the time in which the distance DC is passed
over, in order that you may know in what time the portion DP
may be passed over, draw PT parallel to DB, and the arc BT
designates the time.

4. The time may also be designated by the angle BCT, or by
the inclination of descent Pp or by the length DQ.

5. Moreover, in order to compare the times of perpendicular
fall with the times of descent along this curve, suppose that a
heavy body falls from B to C through V, and since the squares
of the various lines BT are as the lines BV, BT may be considered to express the time of descent to V. And so if a heavy
body descends to C in a time BC, it descends to V in a time BT.

6. Now as the descents from D and B are at the beginning
equally perpendicular to the horizon, it is obvious that both
the heavy bodies D and B begin to fall equally, even though D
is immediately borne obliquely. And so the lines by which the
times of descent are designated ought to be so constituted
among themselves that they initially show equal times of
descent of equal distances, and afterwards they will correctly
show times of descent of equal distances gradually becoming
unequal. And hence it appears that the arc and its chord BT,
since they are equal initially, do not only designate these times
separately, but may be compared together. So if it is supposed
that the heavy body is borne from D to C in the time BTC, it not
only follows that it is borne to P in the time BT, but also that it
falls from B to V in the time BT or from B to C in the time BC,
and vice versa.

Also the times of descent from other points of the curve, such
as P, are known from this relation, since proportional parts are
passed over in equal times.

But the outstanding fact is that from the given time in
which a pendulum of given length oscillates, the time in which

---

[1] $B\sigma$, because the arc BT expresses the time to P, etc. (see 3, below).

a heavy body falls to a given depth is given. For if a heavy body falls from B to C in the time BC, $\dfrac{BC^2}{BTC^2} = \dfrac{BC}{A\gamma}$, and it falls from A to γ in the time BTC which is the time of one half oscillation. Suppose however $AC = 1 \cdot 0000$, computation gives $A\gamma = 1 \cdot 2337$, through whose quadruple $4 \cdot 9348$ it descends in the time of one oscillation, that is, through $5\ AC$ nearly, and through $19 \cdot 7392$ or $19\frac{3}{4}\ AC$ nearly in a double oscillation.[1]

Note that as regards velocity the motion of a heavy body falling from D to C is very similar to the motion of the point of the uniformly moved wheel that describes the cycloid.

2. That a pendulum made of quicksilver will continue longer in motion.

---

[1] Since a body falls the distance BC in the time BC, let the distance it falls in a time BTC (the half-period of oscillation of a pendulum of length AC) be $A\gamma$. Then

$$A\gamma = \frac{BTC^2}{BC} = \frac{\pi^2 BC}{4} = \frac{\pi^2 AC}{8}.$$

And the distance fallen during a complete oscillation $= 4A\gamma = \dfrac{\pi^2 AC}{2} = 4 \cdot 9348\ AC$.

Let l be the length of a pendulum beating seconds; the distance fallen in one second $= \frac{1}{2}g = \dfrac{\pi^2 l}{2}$; therefore $g = \pi^2 l$.

Or T (the period of a pendulum of any length L) $= \pi\sqrt{(L/g)}$.

PART III
THEORY OF MATTER

INTRODUCTION

Newton wrote much on the theory of matter, but little of what he wrote was intended for general reading. What has hitherto been known of his ideas on this subject is drawn from the following printed sources, listed below in order of their date of composition:

An Hypothesis explaining the Properties of Light, discoursed of in my several [optical] Papers (before December 1675).[1]

Letter to Henry Oldenburg, 25 January 1675/6.[2]

Letter to Robert Boyle, 28 February 1678/9.[3]

Principia, 1687.[4]

De Natura Acidorum, 1691.[5]

Quaeries in *Opticks*, 1704–17.[6]

Principia, editio secunda, 1713.[7]

These sources by themselves show that Newton's speculations ranged over almost the whole of his active scientific life, from before 1675 until 1713. But the first three—including two documents of the highest significance for the understanding of his thought—were not printed until after his death, and only in one discussion which he himself intended for publication— the *Quaeries*—did Newton allow himself a free hand.

To these published materials we now add the following, listed in the probable order of their composition:

Notes on Hooke's *Micrographia* (Section VI).

De Gravitatione et aequipondio fluidorum (Section II).

De Aere et Aethere (Section III).

Conclusio of *Principia* (Section IV).

Draft of Preface to *Principia* (Section IV).

Drafts of *Scholium Generale* (Section IV).

Only *De Aere et Aethere* is printed in this section; the others we have referred to their more appropriate sections, although

---

[1] Published in Birch, *History*, III, 247–69, 296–305 (reprinted in *Papers & Letters*, 178–235).

[2] Birch, *Boyle*, I, 1744, 74; 1772, cxvii–cxviii (*Papers & Letters*, 254).

[3] *Ibid.* 70–3, cxii–cxvii (*Papers & Letters*, 250–3).

[4] Mathematical theory of the motions and attractive forces of particles, and explanation of phenomena from these; the passage on water (p. 506) is also noteworthy.

[5] John Harris, *Lexicon Technicum*, II, 1710, introduction (*Papers & Letters*, 256–8).

[6] The first sixteen *Quaeries* were given in the first edition of 1704; seven more were added in the first Latin edition of 1706; and the final eight in the second English edition of 1717.

[7] This edition is the first to contain the important *Scholium Generale*.

their relevance to Newton's theory of matter is necessarily discussed in this Introduction.

A clear understanding of Newton's real thoughts about the nature of matter and of the forces associated with material particles has always been (to borrow his own phrase) 'pressed with difficulties'. That a corpuscular or particulate theory was unreservedly adopted by him has long been abundantly evident from many passages in the *Principia*, and from the *Quaeries* in *Opticks*, to mention only discussions fully approved for publication by Newton himself. So far, then, Newton was undoubtedly a 'mechanical philosopher' in the spirit of his age, the spirit otherwise expressed, for example, by Boyle and Locke. But of the exact content and form of his mechanical philosophy it is less easy to be certain. In the *Principia* it is set in a strictly mathematical mould, and Newton for the most part restricts his statements to what is necessary for the development of a mathematical theory of gravitational force: only phenomena of motion are considered. In the *Quaeries*, on the other hand, Newton treats the mechanical philosophy qualitatively, and, having deliberately given to his thoughts a speculative dress in order not to seem utterly committed to them, he roams widely over the phenomena of optics, heat, surface tension, chemistry, and so on, rarely being definite and on occasion being inconsistent. As Newton intended, the *Quaeries* tantalize: we are never sure whether he really means what he says or not.

Interpretations of the *Quaeries*, and of certain passages in the *Principia*, have been further complicated by overemphasis of the positivist element in Newton's scientific method. When the philosopher who declared 'Hypotheses non fingo' is found to be framing hypotheses and formulating conjectures, should these be taken seriously or not? Even if we recognize that Newton did not—for he could not—avoid or mercilessly condemn any entertainment of an hypothesis, it is often tempting to distinguish between Newton the mathematical theorist, and Newton the author of philosophical speculations.

Similar doubts obscure other discussions, especially the *Hypothesis of Light* and the *Letter to Boyle*: again, when touching on fundamental explanations, Newton seems to don a cloak of elusiveness. Of the former he wrote almost in terms of impatience. Considering, he said, that an hypothesis would illustrate his optical papers, or at least that they were felt to need such an explanation by some great virtuosos in the Royal Society whose heads ran much upon hypotheses: 'I have not scrupled to describe one, as I could on a sudden recollect my thoughts about it; not concerning myself whether it shall be thought probable or improbable, so it do but render the papers I send you, and others I sent formerly, more intelligible.' And he made it clear that he was not to be supposed to accept this hypothesis himself, nor to be seeking adherents for it. This is as much as to say: if the virtuosi must amuse themselves with hypotheses, I will supply one, as good as any other.[1] As for the *Letter to Boyle*, it begins with an apology for reproducing notions so ill digested 'that I am not well satisfied my self in them; and what I am not satisfied in, I can scarce esteem fit to be communicated to others; especially in natural philosophy, where there is no end of fancying'. And the letter ends yet more discouragingly: 'For my own part, I have so little fancy to things of this nature, that, had not your encouragement moved me to it, I should never, I think, have thus far set pen to paper about them.'[2]

These pieces, together with *De natura acidorum*, are so brief and so *ad hoc* in their composition, and so little related to the main stream of Newton's mathematical and experimental enquiry, that they have defied accurate assessment. Historians have been inclined to take Newton at his word and dismiss them as occasional pieces not truly representative of his permanent convictions. Although Newton was thirty-three in 1675 and had by this time performed original work of outstanding quality—indeed, no truly creative thought entered his mind after this date—it has been tempting to suppose that these papers were juvenilia, products of a state of mind which the author of the *Principia* had outgrown. That Newton only renounced almost total public silence in 1717 with the final group of *Quaeries* seemed to indicate that his thoughts were uncertain until that time. Even after going into print (with all the safeguards of the *Quaery* form) Newton was still undecided, apt to plan further discussion and then draw back. For he informed Roger Cotes, editor of the second edition of

[1] Birch, *History*, III, 248–9.   [2] Birch, *Boyle*, 1744, I, 70, 73.

the *Principia*, that 'I intended to have said much more about the attraction of the small particles of bodies, but upon second thoughts I have chose rather to add but one short Paragraph about that part of Philosophy'.[1] This was with reference to the final passage of the General Scholium, which Newton ultimately rewrote so mysteriously that its meaning has escaped commentators ever since.[2] Once more Newton could not speak his mind openly and plainly; once more his compulsion to utter only unchallengeable truths or perplexing generalities prevailed over his sense of the importance of explaining his theory of matter.

Yet, as the documents here published for the first time prove, it was not lack of continuity in his ideas nor want of an impulse to publish them that caused Newton to hover between silence and oracular utterance. Newton planned and drafted more versions of his theory of matter, both before writing the *Principia* and in conjunction with it, than have been known hitherto. The various versions with their similar phrases and identical examples form a continuous chain over the years 1675–1713 to show that the basic fabric of Newton's theory of matter remained always the same. Moreover, despite all his overt professions, Newton was genuinely anxious to discuss his theory of matter in detail and publicly. Three times he tried to find a place for it in the *Principia*: in the *Conclusio*, in the *Preface* and in the *General Scholium*; and three times he rejected it. Somewhere, he seems to have felt, some notice should be given of the microscopic architecture of nature side by side with the majestic system of celestial motions unfolded by mathematical analysis. But all that at length emerged after painful reflection was a cautious hint in the printed version of the Preface to the first edition, to be followed years later by the oracular but confusing conclusion to the General Scholium in the second edition. No one can be sure of knowing the reasons for the ultimate suppression of his cherished ideas. The most plausible is Newton's persistent fear of committing himself to some position which might be made to seem foolish. It may also have struck him that the

---

[1] 2 March 1712/13. Letter LXXV in Joseph Edleston, 'Correspondence of Sir Isaac Newton and Professor Cotes', London, 1850, 147.

[2] See below, p. 207, and the drafts of the General Scholium, Section IV, no. 8.

juxtaposition of the mathematical positivism of the *Principia* (where he was so careful to deny that when he spoke of gravitational *attraction* he meant attraction at all) and speculations about half-a-dozen other attractive forces (even less well known than gravity) would be particularly strange. Especially since the *Principia* was the work which Newton was the most anxious to make immune from attack. His second thoughts were no doubt tactically wise, and helped avoid bitter controversy; yet, as we shall try to show, the annexing of an essay in the mechanical philosophy to the *Principia mathematica* would have been by no means as paradoxical as it might seem.

Newton's adherence to the mechanical philosophy was a very early development. In his student days he had read with care the *Principia Philosophiae* of Descartes, the *Origin of Forms and Qualities* among many works of Boyle, and the writings of Henry More and Robert Hooke, with others of the generation somewhat senior to his own.[1] His annotations portray him as thoroughly imbued with the mechanical philosophy; and among his earliest original notes there occur, for example, speculations 'Of Attomes'. A very little later—perhaps about 1665—there is a sketch of an attempt to trace colour to the reflection and absorption of light in the pores of solid bodies.[2] Probably to this same period—it can hardly be very much later—belongs the paper printed in Section II (no. 1) which also indicates the mechanical outlook, and which contains some traces of the later Newtonian theory of matter. The first surviving document fully devoted to this theory we believe to be *De Aere et Aethere*, in which it emerges in a tolerably complete form. The existence of this document, in fact, belies Newton's own statement to Boyle that his ideas would never have been committed to paper but for the latter's persuasion, or the earlier claim to Oldenburg, that it was only to gratify the virtuosi that he had written an hypothesis of light. The well-organized, though summary, treatment in two chapters suggests that it was intended as a synopsis of a more ambitious work. From internal evidence it seems to have been written between 1673, when Boyle's calcination experiments which it mentions were described in his *New Experiments to Make Fire*

---

[1] A. R. Hall, 'Sir Isaac Newton's Note-Book, 1661–1665', *Cambridge Historical Journal*, IX, 1948, 239–50.   [2] *Ibid.* 243, 248.

and Flame Stable and Ponderable,[1] and 1675 when Newton wrote the *Hypothesis on Light*. Comparison of *De Aere et Aethere* with parallel passages in the *Hypothesis* and in the *Letter to Boyle* indicates that while the same observations and examples were used in all three, as they were to be used many times more until Newton's ideas crystallized in print in the *Quaeries*, the explanations offered of the observed phenomena represent a less mature stage of development than those of the *Hypothesis*.

The feature of *De Aere et Aethere* that immediately distinguishes it from Newton's other discussions of phenomena of attraction and repulsion (the examples here are capillary attraction, the lack of cohesion in a dry powder, the difficulty of pressing two surfaces together, the walking of flies on water—all to be used repeatedly again) is that he here finds the cause of the effect in the repulsive force of air particles. As he says, 'from the way these effects occur those who philosophize rightly know that air seeks to avoid the parts of these bodies; and so, since the air is more rare in these [pores] than in wider spaces, the water can penetrate into them'. Similarly, the refraction of light in passing through or near bodies is attributed to the rarefaction of the air in or near them. In later hypotheses Newton attributes such properties and effects to the particles of aether, not to particles of air. In the *Hypothesis* air is still said to be repelled from bodies, being more dense in large spaces than in small ones, and capillary attraction is regarded as caused by this repulsion of air; but the aether is now made the cause of cohesion, of the abnormal standing of mercury in the Torricellian experiment (a phenomenon not mentioned in *De Aere et Aethere*),[2] of the refraction of light and so on.[3] Later again, in the *Letter to Boyle*, Newton makes no mention of air at all in this connexion; the aether is made responsible for all the phenomena of attraction and repulsion, even capillary attraction.[4]

If we suppose that Newton's ideas changed consistently, the *Hypothesis* of 1675 must mark a half-way stage in his thinking,

[1] Birch, *Boyle*, 1772, III, 706 ff.

[2] This anomalous effect was first observed by Huygens in December 1661, when the experiment was made with water (*Œuvres Complètes*, vol. XVII, 262, 320). The same effect with mercury, standing 50 or more inches in a tube, was first demonstrated by Hooke in October 1663 (*ibid.* vol. IV, 438–9).

[3] Birch, *History*, III, 252, 256.　　　　[4] Birch, *Boyle*, 1744, I, 70.

and thus be later than *De Aere et Aethere*; in modifying his hypothesis Newton has transferred the repulsive force from the particles of air to the particles of aether in certain cases, as he was later to do in all cases. At any rate, the explanations of *De Aere et Aethere* are almost identical with those that Newton offered later, save that the word 'air' here appears instead of the later 'aether'. Perhaps he broke off the paper when he began to consider the aether more attentively in his second chapter. Indeed, certain expressions in the first chapter, *De Aere*, already suggest that repulsive forces between particles of air and of matter may result from some properties of an intervening medium. (But there is also the equally fascinating suggestion that particles consist of a solid nucleus surrounded by a tenuous sphere that admits other bodies with difficulty). So the way was open for a shift to aetherial explanations.

In *De Aere et Aethere* the particles of air are, so to speak, active agents in phenomena in virtue of their intrinsic repulsive force. In the *Letter to Boyle*, this is no longer the case: particles of air or of other matter, even light-rays, are passively subjected to the force exerted by aether particles, to which the intrinsic repulsive force has now been transferred. Some economy of explanation is gained by the change. For instead of supposing that material particles are endowed with a variety of forces, gravitational, chemical, electrical and so forth, it may be possible to reduce all these to one force in the aether—but unfortunately for economy, Newton did not succeed in this. On the other hand, there has necessarily been a multiplication of entities in a fashion to invite the slash of Ockham's razor. One does not make much progress by supposing that repulsion between material particles is caused by the repulsion between aetherial particles, for this leaves the latter just as unexplained as the former was previously. Following the general line of argument familiar to the seventeenth century mechanical philosophy, one may reasonably enough maintain that all material bodies, including airs derived from them, are composed of particles; and argue further with Newton, in his first great extension of the mechanical philosophy, that since composite bodies exert forces on each other, such as that of gravity, their component particles must also exert the same forces on each other, so that the force between two bodies is the sum of enormously numerous corpuscular forces.[1] These concepts are, loosely, inferred from the phenomena or at least are directly related to the phenomena. This is not so with the aether, if it is introduced to account for the forces with which the particles appear to be endowed. For the aether cannot be inferred from the phenomena: it can only be imagined. Suppose, for instance, that we consider an iron bar to be made of iron particles: then it seems to follow that when the bar as a whole is magnetized, each particle must be magnetic as well. But there is a much wider gap of imagination to be crossed before one can agree that the magnetism of each particle is caused by an aether.

This was a perpetual problem in Newton's philosophy of matter. He could adopt either of two kinds of language. He could speak of forces between the material particles (of whose existence he was confident) as being the causes of phenomena; or he could speak of the forces between the aetherial particles, which in turn acted on the material particles, as being the true causes of phenomena. Sometimes, but not always, he could translate from one language to the other. Thus he would sometimes speak of gravity as though it were a force produced by the repulsive effect of material particles upon aether particles. Or again, compare the language of *Quaery* 1 in *Opticks*:

Do not bodies act upon light at a distance, and by their action bend its rays...?

with that of *Quaery* 19:

Doth not the refraction of light proceed from the different density of [the] aetherial medium in different places, the light receding always from the denser parts of the medium?

and with *Quaery* 31:

Have not the small particles of bodies certain powers, virtues, or forces by which they act at a distance, not only on the rays of light...but also upon one another for producing a great part of the phenomena of nature?

---

[1] Cf. General Scholium (Section IV, no. 8, MS. A): 'For the force of the Sun is composed of the forces of all its particles and the forces of all the particles are propagated... through all the orbs of the Planets....' Cf. also the Third Rule of Reasoning in Philosophy, *Principia*, Book III, second edition.

In the 1st and 31st quaeries the material particles exert a force directly upon light; in the 19th they bring about a density-gradient in the aether by repelling its particles (such a one as Newton earlier discussed in the *Letter to Boyle*), which is the true cause of refraction. (In the first case, refraction without the presence of a material body is impossible; in the second case it occurs whenever there is a density-gradient in the aether, from any cause.) In instances such as these the two kinds of language are nearly interchangeable, and it might even be supposed that the second provides a wider explanation than the first. It would be tempting, perhaps, to consider that when Newton attributed forces to material particles, he always meant that these forces were produced by an aether. But this, as we shall show, was not the case; and in any event the aether-version of the particulate theory of matter as Newton developed it cannot be considered as a profounder theory underlying the force-version, for it takes a totally different view of the ultimate properties of material particles.

Using this criterion, Newton's writings on the theory of matter fall into two groups. He used aether-language in the *Hypothesis* of 1675, the *Letter to Boyle*, and certain of the *Quaeries*. He attributed forces directly to material particles, without the interposition of an aether, in *De Aere et Aethere*, the whole text and printed preface of the *Principia*, the suppressed *Conclusio* and draft preface, and certain other *Quaeries*. It is significant that when dealing with some of the most obscure—because not directly observable—phenomena of attraction and repulsion, in chemical reaction, Newton never introduced the aether at all, but always spoke of forces between the reactive particles.[1] On the other hand, Newton was particularly careful not to commit himself to the aetherial hypotheses of 1675 and 1678; the aether-language is always qualified by a cautionary note, whereas when he spoke of forces exerted by material particles he felt no such disclaimer to be necessary.

One should hesitate, therefore, before thinking that New-

---

[1] There would appear to be an exception in the discussion of solution in the *Letter to Boyle*; but in the hypothetical explanation offered there, any reference to the role of the aether is quite superfluous.

ton's theory of matter consisted of nothing but a series of hypotheses about the aether and its importance in phenomena. Rather, the solid part of this theory consisted of the view that phenomena result from the motions of material particles, and that these motions are the result of the interplay of forces between the particles. This he suggests again and again, both in print and in previously unpublished drafts. Nothing could be more emphatic than the statement of *Quaery* 31, free from conjectural disguises and disclaimers about hypotheses:

it seems probable to me that God in the beginning formed matter in solid, massy, hard, impenetrable, moveable particles...it seems to me farther that these particles have not only a *vis inertiae*, accompanied with such passive laws of motion as naturally result from that force, but also that they are moved by certain active principles, such as is that of gravity, and that which causes fermentation, and the cohesion of bodies.

Nothing that Newton wrote furnishes authority for going beyond this, nor for fathering upon him the opinion that aether was the ultimate cause of everything, as was mistakenly and ridiculously done by Bryan Robinson in the eighteenth century.[1] It is true that Newton, going beyond the bounds of his theory, would speculate on the hypothesis that all the forces of Nature might originate in the properties of an aether, but this speculation is no more essential to the theory proper (as described in so many passages where there is no mention of an aether) than Darwin's hypothesis of pangenesis is to the theory of evolution, or (for that matter) Maxwell's aether is to Maxwell's equations.

Without seeking to cramp Newton's thought into a Procrustean bed of positivism, to which indeed his theory of matter is ill adapted, it seems unnecessary to run to the opposite extreme and (with Lord Keynes) make Newton a *magus* whose scientific thinking was at the mercy of inexplicable whims and medieval fancies. Nor can we agree with I. B. Cohen, who has suggested that, although 'Newton presented his thoughts on the aether with some degree of tentativeness' (an understatement indeed!) 'he did so over so long

[1] *A Dissertation on the Æther of Sir Isaac Newton*, 1743; *Sir Isaac Newton's Account of the Æther, with some additions by way of an appendix*, 1745. The latter was occasioned by the first appearance in print of the *Letter to Boyle*.

a period of time that the conclusion is inescapable that a belief in an aetherial medium, penetrating all bodies and filling empty space, was a central pillar of his system of nature'.[1] For, since Newton always presented his aetherial hypotheses tentatively, even at the height of his scientific prestige, there is no reason to suppose that he regarded them as other than tentative: he was certainly not equally coy about his theory of particulate motions and forces, which in a far more real and effective sense he looked upon as offering a key to the understanding of the inner mysteries of nature. It may be that in a sense Newton pointed the way up a blind alley. The very fact that he speculated at all on the aether as a mechanism to account for the forces attributed to material particles gratified the prejudice of an age that, lacking any concept of field-theory, loathed the notion of action at a distance and saw in the push-and-pull mechanism of an aether the only escape from it. Faced with a choice between a universe of Cartesian, billiard-ball mechanism rewritten in Newtonian terms and a universe requiring the inconceivable concept of action at a distance, the seventeenth, eighteenth and nineteenth centuries preferred the former. But, because this was so, and because Newton himself shared the general contempt for the notion of action at a distance, we should not suppose that Newton was unaware of the distinction between an hypothesis and a theory; nor should we conclude that his speculations on the aether were the foundation of his theory of matter, when in fact they were at most no more than hypothetical ancillaries to it.

From the phenomena of physics, chemistry and even physiology, Newton concluded that matter can act upon matter in ways other than the direct mechanical impact of Cartesian physics. Whether an aether be imagined or not this must be so, for if a non-mechanical type of action be denied to the particles of ordinary matter, then it must be allowed to the particles of aether (which, for Newton, are widely dispersed in a vacuum). It is this type of action that imposes a belief in forces, associated with matter though certainly not inherent in it; and these forces must be accepted as ultimate facts of nature, of the way the world was created. But what, more

[1] *Papers & Letters*, General Introduction, 7.

specifically, does the word 'force', or as he otherwise termed it, power, virtue, or active principle mean for Newton? Unfortunately, he gives no specific answer. The forces between particles are certainly of the same nature as those between macroscopic bodies;[1] they can be qualified as gravitational, magnetic, electric, 'and there may be more attractive powers than these'; they are identifiable from experimental phenomena. But precisely what their nature is Newton never did declare, because he could not discover it.

Confronting this difficulty, Alexandre Koyré takes the view that, as Newton knew that forces could not be explained in terms of aetherial mechanisms, he held 'them to be non-mechanical, immaterial and even "spiritual" energy extraneous to matter'.[2] As Newton wrote to Bentley, 'It is inconceivable, that inanimate brute Matter should, without the Mediation of something else, *which is not material*, operate upon, and affect other Matter, without mutual contact. . . .'.[3] Forces would thus require, for Newton, 'in the last analysis, the constant action in the world of the Omnipresent and All-powerful God'.[4] M. Koyré's position appears to be this: Newton was much too intelligent not to perceive that the mechanical hypotheses of forces lead to infinite regress; therefore he must on the contrary have believed that forces are non-mechanical, quasi-spiritual.

Undoubtedly Newton was a teleologist for whom the celestial system was the product of divine design, who believed that God had created particles with such properties, and moved by such forces, as were necessary to create the phenomena intended in the divine plan. Just as Newton rejected Descartes' contention that God could not create extension without matter, so he would have denied that God could not create matter without forces. Therefore forces were certainly not innate in matter. If asked why the planetary orbits have certain parameters and not others, or why particles have certain forces and not others, Newton would reply: Because God made them so. God could have made our world

---

[1] *Quaery* 31, 376.

[2] *From the Closed World to the Infinite Universe*, Baltimore, 1957, 209.

[3] *Four Letters from Sir Isaac Newton to Doctor Bentley*, London, 1756, 25; *Papers & Letters*, 302. Italics added.

[4] Koyré, *op. cit.* 217.

differently if he wished.[1] So, to discourse of God 'from the appearances of things, does certainly belong to Natural Philosophy' for natural philosophy teaches 'what is the first Cause, what power he has over us, and what benefits we receive from him'.[2] Moreover, Newton seemed to require God's activity not in the first creation alone, but continually. Why should not the matter in the universe congregate together, unless a divine power prevented it?[3] How could the action of comets and planets upon each other avoid a disturbance of the celestial harmony, unless the same power preserved it?[4] And how could the quantity of motion in the universe be hindered from decreasing?[5]

For attributing such a ceaseless activity to God Newton was criticized by Leibniz and in turn defended by Samuel Clarke who, nevertheless, did not challenge the accuracy of Leibniz's understanding of Newton's views.[6] M. Koyré believes that the metaphysical opinions of Newton's champion literally represent those of Newton himself, as is indeed probable.[7] Clarke maintains that there is no true distinction between 'natural' and 'miraculous' things, both being the work of God; the former are regular and common, the latter irregular and rare.[8] Nothing, even a miracle (allowed by both Clarke and Leibniz), is more the result of divine intervention than anything else, since everything that is depends always on God's actual government of the world.[9]

The notion of the world's being a great machine, going on without the interposition of God, as a clock continues to go without the assistance of a clockmaker; is the notion of materialism and fate; and tends, (under pretence of making God a *supra-mundane intelligence*), to

---

[1] *Quaery* 31, 403–4.

[2] General Scholium, Cajori, 546; *Quaery* 31, 405.

[3] *Quaery* 28, 369. *Four Letters*, 29; *Papers & Letters*, 306. However, in a draft of the *Scholium Generale* (Section IV, no. 8, MS. C) Newton supposes that the stars are too remote from each other to experience a centripetal tendency.

[4] *Quaery* 31, 402.

[5] *Quaery* 31, 397–9. These three points imply design, of course. They do not exclude the possibility that the remedies for these three perils to the harmony of the universe were incorporated in it from the beginning by divine wisdom.

[6] H. G. Alexander, *The Leibniz–Clarke Correspondence*, Manchester, 1956.

[7] Koyré, *op. cit.* 301.

[8] Alexander, *op. cit.* 23–4, 35: 'There is nothing more extraordinary in the alterations [God] is pleased to make in the frame of things, than in his continuation of it.'

[9] *Ibid.* 117 and *passim*.

exclude providence and God's government in reality out of the world.[1]

Clarke argues that the Newtonian conception of forces involves nothing extraordinary when God's relation to the world is properly understood; forces are divinely produced because the whole of nature is divinely maintained; they are not mechanically produced because the universe is not a machine:

> That one body should attract another without any intermediate means, is indeed not a miracle, but a contradiction: for 'tis supposing something to act where it is not. But the means by which two bodies attract each other, may be invisible and intangible, and of a different nature from mechanism; and yet, acting regularly and constantly, may well be called natural; being much less wonderful than animal-motion, which yet is never called a miracle.
>
> If the word *natural forces*, means here mechanical; then all animals, and even men, are as mere machines as a clock. But if the word does not mean, mechanical forces; then gravitation may be effected by regular and natural powers, though they be not mechanical.[2]

This allows no more spiritual quality to forces than to the very existence of matter. For a long time Newton had believed that matter itself depends immediately for its being on God. In the document printed as no. 1 of Section II he argues against Cartesian mechanism precisely on the ground that the Cartesian identification of matter with extension implies that God did not create matter out of nothing, since extension is certainly eternal. For God, who is eternal, is in space, which is extension: 'If we say with Descartes that extension is body, do we not offer a path to Atheism, both because extension is not created but has existed eternally, and because we have an absolute idea of it without any relationship to God?' Extension, in fact, only became matter when it was endowed by God with the properties of impenetrability, motion, inertia and so forth by which we identify substance; but extension itself, Newton insists, 'is as it were an emanent effect of God'. Nor is this creation of matter a mere historical act, one that required God's will for its accomplishment but not for its continuance. That there was a moment of creation as a consequence of divine will (since the universe did not always

---

[1] Alexander, *op. cit.* 14.   [2] Alexander, *op. cit.* 53.

exist) does not entail the subsequent independent reality of matter. On the contrary, Newton writes, the notion of substance being 'real' is unintelligible and atheistic; for independent reality belongs only to God:

And hence it is not surprising that atheists arise ascribing that to corporeal substances which solely belongs to the divine. Indeed, however we cast about we find no other reason for atheism than this notion of bodies having as it were a complete, absolute and independent reality in themselves, such as almost all of us, through negligence, are accustomed to have in our minds from childhood (unless I am mistaken), so that it is only verbally that we call bodies created and dependent.

The significance of Newton's argument is clear, and fits with his insistence on the omnipresence of God (in *space*, not in a spiritual or moral sense merely): matter exists by a continued act of divine will. For the universe to cease to exist, God would not have to will its annihilation; he would only have to cease willing its continued existence. And if matter in its merely passive properties of impenetrability and inertia exists only by the continued exertion of divine will, so too matter in its active properties—that is, particles acting on one another through the mediation of forces whose laws God defines—must exist equally through the continued exertion of divine will. Thus the gravitational pull of the Sun on the Earth is no more a spiritual property of the matter in them, than is the fact that two bodies ('determinate quantities of extension') cannot be in the same place at the same time. Equally, of course, it is no less a spiritual property. No phenomenon of the physical universe can be more (or less) immediately dependent on God than another (if that is what the word 'spiritual' signifies) because all phenomena depend continuously on God. All are 'eminently contained in him'. Natural forces are certainly immaterial, in that they are not the result of mechanical impacts (like Descartes' gravity and magnetism), but they are nevertheless physical, in that they are subject to law and open to experimental investigation in a way that miracles and spiritual powers are not.

This is a difficult, perhaps an impossible, metaphysic. Clarke, who had some insight into it, was nevertheless forced by Leibniz into seeking for a middle position where none appeared to exist.

Attempting to avoid equally the soulless mechanism of which he strove to convict Leibniz, and the unscientific spiritualism of which Leibniz accused him, he seeks to define a world in which all is under divine government and yet subject to law and regularity. Neither he nor his mentor Newton (who said nothing about the question openly) were able to give a clear idea of forces that, although within the natural order of things, were at the same time neither material and mechanical, nor miraculous and spiritual. Leibniz presented the nature of forces as a metaphysical problem, in which Clarke (and Newton) were invited to say that they were *either* spiritual *or* mechanical, one or the other. Clarke wanted a *tertium quid*, which he could not define, because Newton could not solve the force in physical terms. Leibniz was right: Newton's conception of force could not be justified by Clarke's metaphysics; it could only be justified by its empirical usefulness in physics.

The content of this theory, discernible in published work such as the *Quaeries*, is made still clearer in the suppressed *Conclusio* to the *Principia*. Here the motions of the particles in hot, fermenting and growing bodies (such, in other words, as exhibit the principal phenomena of change) are said to be strictly analogous to 'the greater motions that can easily be detected', and these motions offer the chief clue to 'the whole nature of bodies as far as the mechanical causes of things are concerned'. The same reasoning applies to the lesser motions as to the greater one, and just as the latter 'depend upon the greater attractive forces of larger bodies', so do the former upon 'the lesser forces, as yet unobserved, of insensible particles'. In chemical phenomena the rapid motion of particles is made especially evident, but Newton suggests that the attractive force actively involved in chemical reaction is the same as that responsible for the static cohesion of particles; and the interplay of cohesive force with particulate shape yields the varying characteristics of fluidity, hardness, elasticity, and so on that different bodies exhibit. Particles both repel and attract: the repulsive force acts more strongly at greater distances while the attractive force, diminishing more rapidly with distance, preponderates when particles are in close proximity. For this reason contiguous particles in a solid cohere; but as soon as the particles are separated, whatever the means, they fly further apart, forming an air or vapour whose particles are mutually repellent. If the particles were sufficiently heavy and dense, as Newton had explained in *De Aere et Aethere*, they would constitute permanent air, and in fact he expressly stated that metals were the most apt to do this. (Presumably this idea derived at least in part from an erroneous interpretation of Boyle's production of a permanent air—hydrogen—by the action of strong mineral acids on steel filings. If the permanently elastic part of air was metallic, this would explain why it would not support respiration or combustion, and why some air was always left in a closed vessel after the respirable or combustible part had been used up). Many other phenomena are discussed, and then Newton pauses a moment to consider his purpose in entering into so much detail:

I have briefly set these matters out, not in order to make a rash assertion that there are attractive and repulsive forces in bodies, but so that I can give an opportunity to imagine further experiments by which it can be ascertained more certainly whether they exist or not. For if it shall be settled that they are true [forces] it will remain for us to investigate their causes and properties diligently, as being the true principles from which, according to geometrical reasoning, all the more secret motions of the least particles are no less brought into being than are the motions of greater bodies which as we saw in the foregoing [books] derived from the laws of gravity.

After this he turns to a problem that often excites his interest, that of order and pattern in the arrangement of particles. They are not, he says, thrown together like a heap of stones, for 'they coalesce into the form of highly regular structures almost like those made by art, as happens in the formation of snow and salts'. This occurs, he suggests, because the individual particles join up into long elastic rods, the rods in turn forming retiform corpuscles, and so on until visible bodies take shape. Bodies assembled in this regular way will transmit light and the vibrations of heat, and permit variation of density through chemical change. 'Thus', writes Newton, 'almost all the phenomena of nature will depend upon the forces of particles if only it be possible to prove that forces of this kind do exist'.

Special mention should be made of the application of the particulate theory of matter to chemical phenomena to be found in the *Conclusio*, for some ideas expressed there are not found elsewhere. As Newton continually emphasized, he believed that 'the matter of all things is one and the same'. The differences in substances could be explained by variations in the arrangement and the proximity of their particles; these were both subject to alteration by chemical and physical change, especially by the processes which Newton called fermentation. Thus, water 'can be transformed by continued fermentation into the more dense substances of animals, vegetables, stones', earths, and even ultimately metals, because the fermentation causes the particles to be brought more closely together. Conversely, the particles of solid substances if separated by some means may revert to a less dense aggregation under the action of a repulsive force, the opposite of the cohesive force acting in the solid state; they may thus constitute either vapours (which, as Newton explained in *De Aere et Aethere*, are transient forms) or true airs (permanent gases) such as those yielded from the corrosive action of acids on metals. When these particles are of the smallest size they constitute the matter of light, so that the body emitting them is said to be luminous. (Luminosity may also result from a cold fermentation, as well as from the action of heat—both agitate the particles, though in different ways.)

In his remarks on combustion Newton echoes the ideas of earlier seventeenth-century mechanical philosophers—Boyle, Hooke, Lower, Mayow, even Beccher—and in turn foreshadows the phlogistic theory. Burning, he says, requires a sulphureous or fatty matter, which is converted into vapour in the flame. This is apparently because sulphur abounds in a corrosive, acid spirit capable of agitating the parts of other bodies enough to heat them to a point where they are ready to burn. Newton is rather confused on this point, especially since he claims that spirit of nitre and spirit of vitriol differ only by the phlegm in the spirit; this is perhaps why his 'sulphureous spirit' so closely resembles the 'nitro-aerial' particles and spirit of Hooke and Mayow. The sulphureous spirit abounds in air, Newton says, and hence air is necessary for the maintenance of fire; perhaps also the Sun's heat derives from its sulphureous atmosphere. Again, the sulphureous spirit, perhaps from the air, joins with the fixed parts of vegetable salts (thereby driving off their own spirits) to form Sal Alkali, which because of its consequent sulphureous content then runs *per deliquium*, like oil of vitriol which is indubitably sulphureous. Yet again, sulphur, presumably from the air, joins with nitre to form Sal Alkali, with spirit of nitre driven off; and when nitre and sulphur are ignited together in gunpowder, the expulsion of the spirit of nitre occurs explosively. (Newton says that gunpowder is made as unstable as aurum fulminans if Sal Alkali is added to it, because the sulphureous spirit combines even more eagerly with it than with nitre—though he has already explained that this Alkali is formed from the sulphureous spirit.) Such bodies as metals, however, neither abound in volatile parts nor have pores wide enough to admit the sulphureous spirit, and hence they neither fume nor burn. In all the above Newton seems, in his own way and not without much muddled thinking, to be trying to continue the idea, so much discussed in the previous generation, of air as a powerful menstruum, dissolving bodies through the mechanical action of particles contained in it.

The theory of matter in the *Conclusio* could, indeed, almost be reconstructed from scattered passages in the published writings. It is presented in compressed form in the well-known sentence of the printed Preface: 'I wish we could derive the rest of the phenomena of Nature by the same kind of reasoning from mechanical principles, for I am induced by many reasons to suspect that they may all depend upon certain forces by which the particles of bodies, by some causes hitherto unknown, are either mutually impelled towards one another, and cohere in regular figures, or are repelled and recede from one another', which is a perfect summary of Newton's thinking on the subject. And a fuller account of this theory as a part of the *Principia* itself would hardly have been inappropriate. The *Principia* is, for the most part, a treatise on the one force of nature, the gravitating force which ultimately resides in the particles of matter, whose laws

Newton was able to determine. Having determined these laws, and believing that real theoretical physics was mathematical physics and not a parlour-game of hypotheses, Newton went on to work out mathematically the consequences of the laws of gravity and to show that these corresponded with observable phenomena.[1] This was Newton's second great advance in the mechanical philosophy, arising from his conception of particulate forces: that which rendered it mathematical. None of his predecessors had succeeded in this, or even attempted it, though many had seen its desirability. But the 'mechanical principles' of the *Principia* require that the forces acting on particles and the motions produced by them be exactly calculated. In the *Principia* dynamics and the mechanical philosophy are united—but only with respect to the force of gravity. This, as Newton conceived it, was to be the general pattern of theoretical physics; if only it were possible to extend this union of dynamics and the mechanical philosophy until it embraced the operation of other forces, such as those of electricity or chemical reaction, and thereby effected a precise correspondence between theory and phenomena such as he had achieved for the force of gravity, then indeed these phenomena of nature would be rationally understood.[2] This extension of a philosophy that was at once mathematical and mechanical Newton did not expect to accomplish himself, and the suggestions he made in *Opticks* and elsewhere were intended for the guidance of others. Yet perhaps the best guide he could have provided, in some such explanation of his widest conception of matter and its forces as he drafted in the *Conclusio*, he denied them. This remains inexplicable. The sentence in the Preface already quoted survived as an indication of his hopes; was he afraid of influencing posterity too much by offering more explicit directions?

Before the final version of the Preface went to the printer, Newton again contemplated a more open statement. In this draft, printed in Section IV (no. 3), he proposes

the inquiry whether or not there be many forces of this kind never yet perceived, by which the particles of bodies agitate one another

---

[1] This is, of course, a summary of the intellectual architecture of the book; how Newton formed his ideas and developed his theory of gravitation would be a very different story.

[2] We have discussed this view of the *Principia* more amply in our article on 'Newton's Mechanical Principles', *Journal of the History of Ideas*, **20**, 1959, 167–78.

and coalesce into various structure. For if Nature be simple and pretty conformable to herself, causes will operate in the same kind of way in all phenomena, so that the motions of smaller bodies depend upon certain smaller forces just as the motions of larger bodies are ruled by the greater force of gravity.

As before, particles are said to exhibit both repulsive and attractive forces, the former having a longer range, and in general the discussion is much like that of the *Conclusio*, with the same examples, though of course on a smaller scale. Once more, however, Newton changed his mind and omitted all but the simple statement of his hopes already quoted.

There the matter rested until Newton composed the *Quaeries* in *Opticks*, nearly twenty years later, demonstrating both the continuity of his thought by the essential similarity between the *Quaeries* and the earlier drafts, and the liveliness of his interest in these oft-considered problems by the vigour of this fresh attempt to express his mind. The *Quaeries* reflect Newton's firmly-held theory of matter as well as those speculative thoughts about the aether that he had previously expounded in the *Hypothesis* of 1675 and the *Letter to Boyle*. Thus, setting aside those *Quaeries* devoted to problems of physical and physiological optics, numbers 8–12 and 29–31 correspond fairly closely in their content to the *Conclusio* and the draft Preface, while numbers 17–22 deal with the aetherial hypothesis.[1] *Quaeries* 23 and 24 have a special relevance to the draft of the *Scholium Generale* that will be mentioned later.

The *Quaeries* show, however, that Newton's theory of matter had made no progress since 1687, or even earlier, for its roots are visible in the chapter *De Aere* written before 1675. His conception of the production of natural phenomena from the motion of material corpuscles caused by the action of a variety of forces, illustrated by many observations and experiments, had not been given greater definition. He was not even sure whether there was one basic force, or a pair of repulsive and attractive forces, or as many forces as there were classes of phenomena—gravitational, magnetic, electrical, optical, chemical and physiological—involving such forces. For the

---

[1] The *Quaeries* on aether were first printed in the second English edition of the *Opticks* in 1717, which extends the total to 31. The additional *Quaeries* of the Latin edition of 1706 did not treat of the aether.

further development of this theory three requirements had to be met. It was necessary to be able to analyse the motions of particles mathematically: some of the methods of doing this had been established in the *Principia*. Secondly, it was necessary to know more, from experiments, of the nature of the force or forces: in comparison the discovery of the laws of gravitation and of the way in which these explained the phenomena of astronomy and tidal motion had been a relatively straightforward task. Newton recognized that the experimental insight available to him was far too shallow to carry a theoretical superstructure. 'I have least of all undertaken the improvement of this part of philosophy', he wrote in the *Conclusio*. Too much, indeed, was founded on his reiterated assumption (found in the Third Rule of Reasoning in the *Principia*) that 'Nature is always simple and conformable to herself'. That this principle could be relied on in the study of particulate forces had never been experimentally (or mathematically) proved, though it seemed to be confirmed in the case of gravity. And thirdly, it was necessary to resolve the philosophic doubt concerning action at a distance, always lurking in the attribution of forces to material particles. In the *Principia*, when speaking of gravity as an attractive force, Newton several times asserted that he used such terms as attraction only in a loose or popular sense, not considering how the motion so described was produced. 'Attraction' at this level was a description of an observable effect, seen when an apple falls or iron is drawn to a magnet, without causative implications. By extension Newton applied the same word 'attraction' to, for example, chemical phenomena where the motions of invisible particles could not, like those of the iron or the apple, be actually seen but were inferred. It now described an unobservable effect and in such cases again when revealing his theory of matter, Newton was careful to assert that his language was not to have implications fastened upon it; he was not considering the cause of the particle's motion, though he could call this cause a force because forces are the causes of motions.[1]

[1] *Conclusio*: 'The force of whatever kind by which distant particles rush towards one another is usually, in popular speech, called an attraction. For with common folk I call every force by which distant particles are impelled mutually towards one another, or come together by any means and cohere, an attraction.'

This last presented him with a supreme conceptual difficulty. In the *Principia* Newton had penetrated as deeply into the nature of the gravitational force as science seemed to permit; he had a far more complete understanding of this force and its effects than he could hope to attain of any other. Yet he had failed to discover the cause of gravity or any approach to such a discovery: he could say only that it existed, in proportion to the quantity of matter. Far less, therefore, could he hope to elucidate the cause or origins of the forces that seemed to operate in optics and chemistry, whose laws and phenomena were quite unknown. When pressed—or soliciting himself—to declare the cause of gravity and the manner of its action between material bodies (to avoid the charge of countenancing action at a distance) Newton could only fall back on speculative hypotheses, as Descartes and Huygens had done before him. Yet such hypotheses are far from central to Newton's theory of gravitation; indeed he always maintained that his theory had no need of them.

The hypothesis explaining gravity by variation of density in the aether, outlined in the *Letter to Boyle*, is familiar enough. The fragment printed as no. 5 in Section IV shows Newton, rather surprisingly, preferring an alternative one. In this draft, probably written about 1690, he argues in favour of the vacuum, and of his own view that the volume of such empty space even in solid bodies is far greater than the volume of the solid particles in them.[1] He rejects 'the vulgar sophism [of Descartes] generally opposed to the concept of the vacuum, by which bodies are defined by extension', and in passing refers approvingly to Nicholas Fatio de Duillier's explanation of gravity by the motion of aetherial particles through the vacuum. He calls this 'the unique hypothesis by which gravity can be explained'.

Fatio (1664–1753), a Swiss by birth, possessed mathematical abilities that were considerable though not of the highest order. He became one of the most strongly opinionated of Newton's early supporters and launched the latter's dispute with Leibniz. Not long after his arrival in England Fatio reported to Huygens on the then unpublished *Principia*; he was soon elected to the Royal Society and, being already

[1] Cf. *On the Gravity and Equilibrium of Fluids*, Section II, no. 1, p. 89.

205

familiar with Huygens' ideas on the subject, took an early opportunity to address the Society on the explanation of gravity by aetherial action. In these discourses (27 June, 4 July 1688) Fatio expounded an hypothesis that he attributed to Huygens in which gravity was caused by the rotation of an 'Aether, or such like subtle matter,...so as to pass every-way, in great-Circles, about the earth at the rate to surround the Globe in 1 h. 25''.[1] Three years later Fatio described his own aetherial hypothesis:

That in the whole Universe there is dispersed a very-subtle Matter, infinitely little and exceedingly agitated in right-lines; the Reflections of each particle whereof against gross Bodies and against any great mass of matter floating in them, lessening somewhat the swiftness of the particles. This he conceived would drive down all manner of Bodies towards one another or towards any greater Mass;

in accordance with Newton's inverse-square law.[2]

The development of this hypothesis in detail, which is not without absurdity, is beside the point here. Its author proudly noted that

Sir Isaac Newton's Testimony is of the greatest weight of any. It is contained in some Additions written by himself at the End of his own printed Copy of the first Edition of his Principles, while he was preparing it for a second Edition, And he gave me leave to transcribe that Testimony. There he did not scruple to say *That there is but one possible Mechanical cause of Gravity, to wit that which I had found out:* Tho he would often seem to incline to think that Gravity had its Foundation only in the arbitrary Will of God.[3]

So far as can be ascertained, the fragment just mentioned contains the only evidence supporting Fatio's story. There is, of course, no reference to him or his hypothesis in the second edition of the *Principia*, and none can be found in the interleaved copy of the first edition in the Cambridge University Library in which Newton wrote many corrections adopted in the later editions. Yet Rouse Ball was certainly mistaken in

---

[1] Bernard Gagnebin, *De la Cause de la Pesanteur. Mémoire de Nicholas Fatio de Duillier présenté à la Royal Society le 26 Février 1690. Notes and Records of the Royal Society,* vol. vi, 1949, 106–60.

[2] Gagnebin, *loc. cit.* 115.

[3] *Ibid.* 117. This is interesting as showing that Newton's friends were aware of his hesitation between a mechanical, that is aetherial, explanation of gravity, and the view that gravity is a direct product of the divine will.

supposing that Newton never approved of Fatio's hypothesis, though he must have turned against it at some time after writing the relevant fragment.[1] It is much easier to understand this, than it is to understand why Newton should ever have favoured Fatio's hypothesis at all; for action at a distance is no more incomprehensible than Fatio's aether. In fact one might have thought that this hypothesis would be just the facile piece of imagination that Newton would deplore.

The remainder of this draft is mainly of interest for the light it sheds on Newton's ideas about the rarity of material bodies, and the importance he attached to the regularity of the pattern of particle arrangement within them. It emphasizes once again that any hypothetical aether that Newton was prepared to entertain in order to facilitate the explanation of natural forces was itself a mechanical fluid. Its effects were wrought by impact and it was totally unlike the Cartesian aether in that it did not fill space, but on the contrary left nearly all space vacuous. As Newton had written in an earlier paper (Section II, no. 1):

if the aether were a corporeal fluid entirely without vacuous pores, whatever the division of the more subtle parts, it would be as dense as any other fluid, and it would yield to the motion of bodies through it with no less sluggishness; rather indeed with a greater, if the projectile should be porous, because then the aether would enter its internal pores, and encounter and resist not only the whole of its external surface but also the surfaces of all the internal parts. Since the resistance of the aether is on the contrary so small when compared with the resistance of quicksilver as to be over ten or a hundred thousand times less, there is all the more reason for thinking that part of the aetherial space is void, scattered between the aetherial particles.[2]

Without doubt the most puzzling of Newton's declarations of his aetherial hypothesis is that concluding the *Scholium Generale* of the second edition of the *Principia*, so often quoted in the form due to the translator, Andrew Motte, which reads as follows:

And now we might add something concerning a certain most subtle spirit which pervades and lies hid in all gross bodies; by the force and

---

[1] Rouse Ball, 125–6.   [2] Compare *Opticks*, *Quaery* 22.

action of which spirit the particles of bodies attract one another at near distances, and cohere, if contiguous; and electric bodies operate to greater distances, as well repelling as attracting the neighbouring corpuscles; and light is emitted, reflected, refracted, inflected, and heats bodies; and all sensation is excited, and the members of animal bodies move at the command of the will, namely, by the vibrations of this spirit, mutually propagated along the solid filaments of the nerves, from the outward organs of sense to the brain, and from the brain into the muscles. But these are things that cannot be explained in a few words, nor are we furnished with that sufficiency of experiments which is required to an accurate determination and demonstration of the laws by which this electric and elastic spirit operates.

A partial gloss on this sentence is provided by *Quaeries* 23 and 24 (published later), which ask,

Is not vision perform'd chiefly by the vibrations of this medium, excited in the bottom of the eye by the rays of light, and propagated through the solid, pellucid and uniform capillamenta of the optick nerves into the place of sensations?...Is not Animal Motion perform'd by the vibrations of this medium, excited in the brain by the power of the will, and propagated from thence through the solid, pellucid and uniform capillamenta of the nerves into the muscles, for contracting and dilating them?

These *Quaeries* at least suggest that the 'spirit' of the *Scholium Generale* is none other than the aetherial medium which was so often the subject of Newton's conjectures. It has not previously been noted, however, that the qualification of this 'spirit' as 'electric and elastic' does not appear in any Latin edition of the *Principia*; these words appear to have been interpolated by Motte. His authority for the epithets is unknown, but the words agree very well with a set of twelve propositions written out by Newton in one draft of the *Scholium Generale* (Section IV, no. 8, MS. C). This draft is in a very rough state: Newton set down some ideas in a group of paragraphs without ever welding them together into a coherent exposition, as he often did in first tackling a difficult topic. The propositions are interspersed among the paragraphs, to which they are in no way stylistically connected, yet it is obvious enough that they correspond almost exactly to what Newton wrote in the passage just quoted; perhaps he had some thought of writing a complete explanation of his ideas (as he suggested in his letter to Cotes about the General Scholium) based upon these propositions, to which parallels can easily be found in the *Conclusio* and the *Quaeries*. The difference lies in Prop. 2, 'That attraction is of the electric kind', a notion that Newton suppressed in the *Scholium Generale*, though Motte nearly revived it.

The propositions are Newton's last private thoughts upon the origin of the natural forces by which particles are moved and visible phenomena occasioned, and they are very hard to interpret. The electric spirit is the cause of cohesion, for it causes a strong attraction between contiguous particles; at the same time it causes repulsion at greater distances. It permeates all bodies, emits and bends light, and when vibrating rapidly causes the sensation of heat. Physiologically it is the vehicle by which sensation is transmitted to the brain, and by which that organ commands the muscles; it also effects nutrition. This is virtually to allege that the electric spirit effects all the phenomena of nature; yet Newton's propositions do not say what it really is, nor how it operates. The statement that the electric spirit emits light reminds one of the experiment with the frictional machine described in *Quaery* 8, where the glass globe becomes luminous when rubbed, and it suggests that Newton understands 'electric' in the normal sense. Seemingly, the electric spirit was the 'fluid' (as the eighteenth century would have said) collected by friction on glass and sulphureous materials. Newton appears to suggest that whereas in experiments in electricity large forces give rise to conspicuous effects of attraction, repulsion and luminous discharge, in the minute world of material particles electric forces might exist normally without excitation, though such 'attraction without friction extends only to small distances'.[1] This undetectable electric force might cause the invisible motions of the particles that are sensed as heat, light or chemical change. But it is hard to understand how Newton could imagine that the electric force between particles which is a force of attraction at microscopic distances could become a force of repulsion between bodies at macroscopic distances, if that is what he means. And although it is easy to see how

[1] Cf. *Quaery* 31: '...and perhaps electrical attraction may reach to such small distances, even without being excited by friction'.

electrification could be associated with attraction and repulsion, and light and hence heat, its connexion with animal physiology seems obscure. Nor does this hypothesis really solve his difficulties in giving a true explanation of natural forces; for to convert the aether of earlier writings into an electric spirit may avoid the implications of the word 'aether', so inevitably associated with Cartesian and plenist speculations, and it may offer an analogy. But it leaves the cause of these forces no less mysterious than before, and no less removed from any experimental verification.

Newton's own experiments, described in the *Hypothesis of Light*, had shown him that the electric spirit is both attractive and repulsive, and this is undoubtedly why he conceived it as playing such a major role in nature. He was, of course, ignorant of any concept of electric charge or of the distinction between positive and negative electrification; all he knew was that his tiny scraps of paper would leap up to a rubbed piece of glass, and then be repelled from it.[1] His general theory of matter postulated the existence of both repulsive and attractive forces in all phenomena except gravitation, which was unique; as he put it vividly in *Quaery* 31, 'as in algebra, where affirmative quantities vanish and cease, there negative ones begin; so in mechanicks where attraction ceases, there a repulsive virtue ought to succeed'. Electrical phenomena offered a particularly significant combination of the two kinds of force. (It was a more significant example than the case of magnetism, which was specialized and limited to one kind of matter). This duality of force remained a perpetual problem. In his mathematical treatment of the force of gravity in the *Principia* it did not concern him, except in Book II, Proposition XXII. Here, in deriving Boyle's Law from the supposition that the particles of an elastic fluid repel each other with a force inversely proportional to the distance between them, Newton simply ignored the gravitational attraction of the particles and explained that he was not asserting the physical truth of the supposition that elastic fluids consist of particles repelling each other in this way. (In fact, he always explained the elasticity of airs in this fashion, but he did not believe that all airs or vapours were permanently elastic.) In other cases,

[1] Birch, *History*, III, 250–1.

in the qualitative discussions of cohesion, capillary attraction, solution and so forth, Newton could do no more than merely assert the simultaneous existence of dual forces without further reconciling them; it was without doubt one of the merits of the aether hypothesis that it seemed to offer some possibility of doing this. It was perhaps his London acquaintance with Francis Hawksbee (who made the electrical machine mentioned in *Quaery* 8), that influenced Newton to consider the universal importance of the electrical duality which had not struck him so forcibly in earlier years. But, like so much else in his thinking on the theory of matter and natural forces, the idea was not fully worked out, and all that remained was a hint to influence later Newtonians.[1]

However hard he struggled, Newton could not devise a theory which would overcome the supreme deficiency which he always recognized: lack of experimental information. His aetherial hypotheses, early or late, could do nothing to remedy this. As a mechanical philosopher Newton knew that no theory of matter could be firmly established until the forces effecting phenomena were thoroughly understood from experimental investigation of the phenomena themselves, while as a mathematical physicist he required such a theory to have mathematical rigour. When Newton wrote adversely of hypotheses he was (though condemning himself) methodologically correct in the sense he meant: for the theory of matter could not be advanced by framing hypotheses that were neither verifiable nor falsifiable, especially at a time when even an elementary theory of the transmission of light, the strength of bodies, and the formation of chemical compounds was still totally lacking. He was aware that his own theory of interparticulate forces was defective and incomplete, except perhaps in the case of gravity. It could not even explain adequately how water was variously a solid, a liquid and a vapour; yet this was apparently a relatively easy problem for the mechanical philosophy to solve. As for fundamental mechanical causes—the true cause of the forces of gravity, cohesion, optical refraction, chemical attachment and so on—

---

[1] As Bryan Robinson developed an extended aetherial hypothesis on the basis of Newton's hints, Roger Boscovich developed an extended theory of point particles endowed with inertia and surrounded by alternating zones of attraction and repulsion in his *Theoria philosophiae naturalis*, 1758, which was also Newtonian in inspiration.

insight into these was almost impossibly remote. Hence Newton's remark:

> To tell us that every species of things is endow'd with an occult specifick quality by which it acts and produces manifest effects, is to tell us nothing; But to derive two or three general principles of motion from phaenomena, and afterwards to tell us how the properties and actions of all corporeal things follow from those manifest principles, would be a very great step in philosophy, though the causes of those principles were not yet discover'd: And therefore I scruple not to propose the principles of motion above mention'd, they being of very great Extent, and leave their causes to be found out.[1]

This is primarily Newton's justification for using the words 'attraction' and 'repulsion': they were not occult qualities to him. One may translate his statement: if a few more phenomena could be verifiably accounted for in terms of attractive and repulsive forces, as I have accounted for gravity, that would be a great deal; let the elucidation of the causes of those forces come afterwards, when it may. Hypotheses about the causes—which could only be finally discovered *after* the first objective was attained—were of merit in the meantime only in so far as they suggested new experimental enquiries; otherwise they were as useless as Hooke's hypothesis of colour or Descartes' of gravity, which were formulated before the basic theories of colour and gravity were known.

And yet—after all this, and after allowing Clarke to underline his metaphysical opinions, Newton still continued to face both ways. The *Quaeries* of 1717 confuse his role as the mechanical philosopher who wrote the *Principia* with his role as a maker of mechanical hypotheses. (This distinction remains clear in the suppressed drafts printed in this volume.) One is left with an enigma. Newton appears to have been in some part of his mind a Cartesian *malgré lui*: conscious of the folly of aetherial speculations (which, for him, had neither physical basis nor metaphysical justification) he could not wholly resist playing that beguiling game, even if it meant inventing an aether 49×10^{10} times more elastic than air in proportion to its density.[2] Leibniz seemed to show that the heads of the

---

[1] *Quaery* 31, 401–2.   [2] *Quaeries* 21 and 22.

virtuosi still ran upon mechanical explanations, as they had forty years before. In response Newton could devise plausible mechanical explanations, though he knew that they could not contain the ultimate cause of natural forces. One remembers his *cri de cœur* to Bentley: 'You sometimes speak of Gravity as essential and inherent to Matter. Pray do not ascribe that Notion to me. . . .' From this peril the aetherial hypothesis, the concept of force as mass multiplied by acceleration, a shock-wave in a line of billiard-balls, offered an escape. . . or at least a reprieve.

'I have not yet disclosed the cause of gravity, nor have I undertaken to explain it, since I could not understand it from the phenomena', as he said on one occasion.[1] In one obvious sense this is true, and that sense knocks the bottom out of aetherial hypotheses. In another sense it is false: Newton knew that God was the cause of gravity, as he was the cause of all natural forces, of everything that exists and happens. That his statement could be both true and false was Newton's dilemma: in spite of his confident expectations, physics and metaphysics (or rather theology) did not smoothly combine. In the end, mechanism and Newton's conception of God could not be reconciled. The *tertium quid* demanded by Clarke's arguments was not really available. Newton's mind must make the enormous leap from particles and forces (the proximate causes of phenomena) to the First Cause—as though leaping a chasm were a proof of its non-existence. For aetherial hypotheses offered no solid bridge. Forced to choose, Newton preferred God to Leibniz.

---

[1] *Scholium Generale*, Section IV, no. 8, MS. A.

DE AERE ET AETHERE

MS. Add. 3970, fols. 652–3

This is described in the *Portsmouth Catalogue* (Section I, xiii, no. 7) as 'Fragments on Light and Heat'. The papers in this box are mainly concerned with optics, but folios 652–3 are an exception. They consist of a double foolscap sheet, closely written on all four sides. The hand is Newton's and there is much crossing out and interlineation in his usual style. The top of the first sheet has been water-soaked and the ink has consequently faded so much as to make the beginning quite illegible. Newton ceased writing shortly before the end of the fourth page; as there was room to write about another line and a half, it appears that, as on other occasions, Newton simply abandoned the draft. It was, however, originally intended as a formal piece of work, and is carefully illustrated with sketches, which we here reproduce, though in one case (p. 217) there is a sketch which Newton never discussed. As indicated in the Introduction to this section, we believe that this draft was written between 1673 and 1675.

Cap. 1. De Aere

De rerum natura scriptores a... caled... ...libus aere scilicet & naturis aereis ut sensu doce[or] progrediar. Inter aeris proprietates insignis est ejus ingens rarefactio et condensatio & haec tribus potissimum fit causis, relaxatione, compressione, calore & vicinia corporum. Prior caeteras totamque naturam aeream pandet, eaque pluribus modis deprehenditur. Primo quod in fistula angustissima cujus inferior extremitas in aquam stagnantem immergitur, aqua interior altius quam exterior ascendit, eoque altius quo fistula sit angustior, ita ut in summe angusta ad plures digitos ascenderit. Id quod non fit ubi aqua et fistula in Vitro aeris vacuo collocantur. Deinde quod Ampulla vitrea cineribus compressis plena multum aquae stagnantis, in quam os immergitur ebibet nullo aere egrediente. Tum quod in Charta vel panno ex quo fit Philtrum aqua sponte ascendit. Quinetiam restem etsi po[n]dere ingenti distensum aqua tanta vi ingreditur ut intumescentem faciat attollere pondus. Et inde haec omnia evenire novunt recti philosophantes quod aer poros seu intervalla partium istorum corporum exilia refugit et in iis rarior existens quam in spatiis amplioribus aquam sinit ingredi; minus premens superficiem ingredientis quam Aer externus premit superficiem stagnantis. & sic non sustinens pressionem externi. Hinc esse etiam quod aqua stagnans ascendit paululum juxta latera vasis, aere latera refugiente; quodque aqua superficiebus omnium rerum adhaerere solet, sive ut loqui solemus, humectare omnia. Denique haec aeris in vicinia corporum rarefactio per refractionem se prodit. Nam jubare per foramen *A* in cubiculum tenebrosissimum immisso [Fig. 52], & imagine solis ad magnam distantiam e vitro convexo *B* projecta, omne jubar ad usque latitudinem vigesimae trigesimae vel quadragesimae circiter partis digiti, aciebus parallelis duorum pluriumve cuneorum

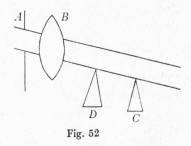

Fig. 52

C, D, deinceps dispositorum intercipias, videbis imaginem ab *E* versus *F* nonnihil recedere ob refractionem scilicet quam lux transiens per Medium juxta acies cuneorum rarefactum patitur.

Quinetiam non aer tantum corpora, sed corpora se mutuo *Corpora etiam mutuum contactum fugere* refugiunt. Nam si super vitrum nonnihil convexum & politissimum, puta ultimum vitrum Telescopii longissimi, aliud minus vitrum planum & aeque politum collocetur, senties vim aliquam requiri qua ad contactum adducantur, et cessante ista vi ab invicem sponte recedere atque eo magis quo vitrum superius sit minus grave: videbis enim urgente vi plures colorum circulos concentricos emergere et ultimo maculam nigram in centro eorum quae indicio est contactum plenarium eo loci, tunc coepisse, et vi cessante macula ista et colores deinceps evanescent ordine contrario quo apparuere. Quod si vitra plana sint, etsi non ultra semissens digiti lata, possis quidem eo usque comprimere, ut colores appareant sed ad contactum plenarium adducere, ita ut macula ista nigra appareat, vix aut ne vix quidem sufficit integra vis unius hominis. [Et tamen aer intermedius propter angustiam spatii ita rarefit ut ubi colores apparuerint vitra sane absque vi haud exigua divelli nequeunt, aere externo magis comprimente, quam internus vi elastica per rarefactionem dibilitata valet dispellere] Eodem modo partes vitri, Chalybis aut cujusvis corporis diffracti non possunt etsi vi maxima urgente ad priorem contactum reduci ita ut cohaereant sicut prius, nam contactus plane restitutus, procul dubio restitueret cohaesionem. Plumbum etiam vel stannum liquefactum & in ferreium vas infusum non ita assequetur contactum ut figefactum adhaereat ferro. Pulveres etiam liquoribus innatantes contactum liquorum fugiunt & aegre submerguntur etsi satis graves, uti sunt ramenta metallorum. Et similiter muscae et alia exigua animalia super aquam pedibus non madefactis cedentem ambulare solent. Denique particulae pulveris alicujus etsi in se invicem incumbere videntur tamen si se plene contingerent arctius cohaererent, uti solent ubi ad contactum per effusam ac deinceps evaporatam aquam adducuntur.

Suspiciones de causa hujus fugae [De causa hujus fugae plures possunt esse opiniones vel quod Medium interfluum aegre recedit aut patitur se nimis coarctari, vel quod Deus naturam quandam incorpoream creavit quae conatur corpora respuere & reddere minus constipata, vel quod de natura corporea sit non tantum habere nucleum durum et impenetrabilem sed et sphaeram quandam fluidissimam ac tenuissimam circumfluam quae aegre admittit alia corpora [De his jam nihil disputo Sed cum aer aeque fugiat corpora ac corpora se mutuo, hinc recte videor colligere quod aer ex particulis corporum a contactu divulsis & ab invicem vi quadam haud exigua renitentibus constat] Vel denique quod ex his duae vel plures aliquando concurrunt. De his jam nihil][1]

Atque hoc fundamento facile intelliguntur omnes aeris proprietates

Quemadmodum quod tam mire rarefit & condensatur per gradus pressionis. Norunt Philosophi per Experimentum Torricellianum quodnam sit pondus totius incumbentis Atmosphaerae quo aer hic juxta terram comprimitur, atque experimento probavit Hookius quod duplum vel triplum pondus comprimit aerem in dimidium vel trientium sui spatii & contra quod sub dimidia vel triente vel etiam centesima milesimave parte ejus ponderis expanditur ad duplum triplum centuplum vel mille-

Quomodo condensatur per pressionem, & rarefit per diminutionem pressionis

---

[1] This whole paragraph was ultimately crossed out; the final interpolation was crossed out first.

cuplum ejus spatii, quod sane si partes aeris se mutuo contingerent vix fieri posse videatur, sed si per principium aliquod
in distantes agens se mutuo fugiunt, ratio suadet ut in duplum
centrorum distantia sit dimidia vis fugae in tripla triens & sic
in aliis & inde facili computo colligitur quod expansio aeris
reciproce erit ut vis compressionis.

Praeterea quod aer calore expanditur non mirabimur si
consideremus quod partes ejus calore agitati debent contre-
Quomodo rarefit
per calorem mescere et contremescentes abigere hinc inde
vicinas partes. Sint *A, B, C* [Fig. 53], tres
particulae in statu quietis et si *B* a causa calorifica movea-

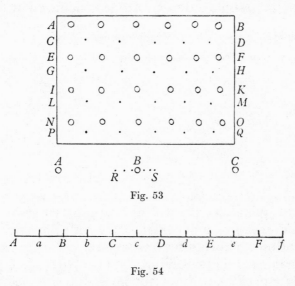

Fig. 53

Fig. 54

tur versus *A*, ad usque *R*, abiget *A* ad majorem distantiam et
eadem actione vicissim resiliens versus *S* abiget etiam *C* & sic
alternis vicibus motu tremulo repercutiens *A* et *C*, atque *A* et
C similiter repercutiens proximos, omnes per spatium majus
pro quantitate motus diffundentur. Non tamen necesse est
tremor iste (in quo calorem consistere suppono) semper sit in
linea recta sed potest particula in linea curva gyrare. Neque
tantum rarefactio per calorem ex hoc motu tremulo pro-
ficiscatur sed et motus

Eodem principio facile explicatur motus iste undulatorius
quo soni propagantur. Sit enim *A* centrum undulationum
[Fig. 54], et particulae *A, B, C, D, E, F*, tremant in lineis *Aa*,

Bb, Cc, Dd, &c ita ut ubi *A* sit in *A,* sit *B* in medio *Bb, C* in *c,* *D* in medio *Dd* & *E* in *E,* et maximae aeris expansiones erunt in mediis *Bb* & *Ff,* maximae contractiones in mediis sed[1]

Praeterea quod aer dumadmodum fluidus et subtilis videatur difficilius quam aqua vel oleum per poros corporum serpit ita ut in vesica contineatur quam liquor facile pervadit exinde est quod particulae ejus distantiam suam ab omnibus servantes nolunt accedere latera pororum   Exiguae quidam sunt, sed quoad facultatem permeandi angustias subtilitas ex tota sphaera quam quaeque sibi vendicat aestimari debet.   Atque hinc fit etiam quod aer

Quare Medium segne et minime subtile.   Quodque contactum corporum impedit et contingentia facit cohaerere.

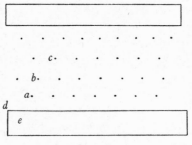

Fig. 55

ad motum quemlibet praeter praefatum tremorem continuandum minime sit idonaeus quodque difficillissime ex intervallo duorum politorum marmorum vel vitrorum exprimitur aut expressus denuo ingreditur.   Nam si particula *a* [Fig. 55] ad cavitatem aliquam insensibilem *e* sita non potest egredi nisi proprius accedendo ad latera cavitatis et *b* non potest egredi nisi proprius accedendo ad *a,* neque *c* nisi proprius accedendo ad [*b*] quam sinit earum vis fugae ab invicem; necesse est ut haereant et caeteras omnes interiores detineant inclusas, nisi motu contritiones marmore superiore super inferiorem labente expellantur.   Unde fit quod corpora solida vix nisi per fusionem & particulae pulverum vix nisi per aquam interfusam & evaporatam ad contactum plenum adducantur, & ad contactum tandem adductae arctissimae cohaereant, aere circumfluo comprimente ut nihil jam de simili vi aetheris loquar.

[1] This paragraph also was crossed out.

Ex iisdem denique principiis generatio aeris facile innotescit. Etenim ad hanc nihil aliud requiritur quam actio
Quomodo generatur vel motus aliquis qui divellat exiguas partes corporum, siquidem divulsae fugient se mutuo ad instar aliarum particularum aeris. Et hinc est quod omnis agitatio vehemens (ut corrasio, fermentatio, ignitio, & calor ingens) substantiam aeream generet quae se in liquoribus per ebullitionem prodit, quodque vehementiori actione substantia illa in majori copia generetur. Sic limatura plumbi aesis aut ferri in *AF* dissoluta ingentem ebullitionem efficit, quae tamen in aceto vel in eadem *AF* per misturam aquae satis debilita sine ebullitione dissolvitur. Sic et Nitrum fusum et injecto carbone accensum, multum quidem aereae substantiae spirat remanente etiam multo sale fixo in fundo; sed si actio ignis acceleretur per debitam misturam sulphuris & carbonum ut fit in faciendo pulvere bombardico, substantia pene omnis mistorum per vehementiam agitationis in aeream formam convertitur, cujus subitanea expansione pro natura aeris oritur vis ingens istius pulveris. Caeterum substantiae aereae
Triplex genus aeris ex triplici genere substantiarum ortum. Aer permanens & ponderosus ex metallis, exhalatio ex terrestri praesertim parte substantiarum vegetabilium ac vapor ex liquoribus. pro natura corporum ex quibus generantur valde diversae sunt. Metalla per corrosionem dant verum aerem permanentem, substantiae vegetabiles & animales per corrosionem fermentationem aut ignitionem dant aerem non diu durantem qualis est exhalatio & substantiae volatiles per calorem rarefactae dant aerem minime omnium durantem quam vaporem dicimus. In his etiam magna est differentia vapor aquae citius decidit quam vapor spiritus vini & vapor quorundam spirituum salinosorum et citius quam vapor aquae ut in vapore maris ubi spiritus salis non nisi ad altitudinem paucorum pedum ascendens decidit redditque aerem ibi crassiorem & minus transparentem & vaporem aquae dulcem linquit. Est et diversitas in pondere aeris nam aer ex substantiis vegetabilibus & animalibus vi ignis generatus tanto levior est aere reliquo circumfluo ut non tantum ipse [non] descendat sed et implicatos fumos crassiores secum abtollat. Sic et vapores omnis generis supra reliquum aerem ascendere conantur. Et non dubium est quin vapores & exhalationes habeant varios gradus gravitatis; nam tot esse aeris diversitatis videntur, quot sunt substantiarum in terra ex

quibus ortum ducit. Atmosphera itaque ex multiplici aere componitur sed qui tamen in tria summa genera distingui potest: vapores qui liquoribus orti minime permanentes & levissimi esse videntur; exhalationes quae ex substantiis crassioribus & fixioribus in regno praesertim vegetabili ortum ducentes sunt mediae naturae; & aer proprie dictus cujus permanentia & gravitas indicio est ipsum nihil aliud esse quam congeriem particularum metallicarum quas corrosiones subterraneae ab invicem quotidie discutiunt; Id quod ex eo etiam confirmatur quod aer hisce (sicut exigit fere indestructibilis natura metallica) neque igni conservando neque usui animalium in respiratione inservit sicut inserviunt exhalationes quaedam ex tenerioribus rerum vegetabilium vel salium substantiis ortae.

Cap. 2. De Aethere

Et quemadmodum Corpora hujus terrae dissiliendo in exiguas particulas convertuntur in aere, sic et particulae illae in minores actione aliqua violenta dissilire possunt & converti in aerem subtiliorem quam ante quem si sit adeo subtilis ut poros vitri Crystalli aliorumque corporum terrestrium penetrat spiritum aeris vel aethereum nominare possumus. Tales vero spiritus dari constat experimentis hisce Boylianis quod metalla in vitro Hermetice clauso aliquamdiu fusa ut pars in scoriam vertatur graviora evadunt. aucta scilicet subtillissimo spiritu salinoso qui per poros vitri ingrediens calcinat metallum & in scoriam vertit. Quodque in vitro aeris vacuo pondus pendulum non multo diutius motum oscillatorium conservat quam in aperto aere, cum tamen motus iste non deberet cessare si exhausto aere nihil subtilius in vitro restaret quod suffocaret motum ponderis. Hujus etiam generis quisquam credo agnoscet effluvia magnetica qui viderit limaturam ferri his effluviis a polo lapidis ad polum circulantibus in lineas curvas quasi Meridianas dispositam. Sic et attractio vitri, succini, gagatis, caerae resinae & similium, per ejusmodi materiam tenuissimam fieri videtur

Dari varia genera substantiarum longe subtiliorum aere

TRANSLATION

Chapter 1. On Air

...Among the properties of air its great rarefaction and condensation are remarkable. Of these there are three chief causes: expansion, compression, heat and the proximity of bodies. The former accounts for the rest, and for the whole nature of air, and it may be demonstrated in several ways. As first that water ascends within a very narrow pipe whose lower end is immersed in stagnating water higher than the external level, and ascends the higher in proportion to the narrowness of the pipe, so that it will rise several inches in the narrowest pipes. This is a thing that does not happen when the water and the pipe are placed in an exhausted glass vessel. And next, that when a glass jar filled with compressed ashes stands with its mouth immersed in stagnating water, it imbibes much of the water although no air escapes. And again, that water rises spontaneously in that paper or sheet of which filters are made. A rope, even when stretched by a heavy weight, is so swollen by the force of the absorption of water that the weight is lifted up. And those who philosophize rightly know that all of these effects occur because the air seeks to avoid the pores or intervals between the parts of these bodies; and so, since in these [pores] the air is more rare than in wider spaces, the water can penetrate into them, [the air in the pores] pressing the surface of the incoming [water] less than the external atmosphere presses the surface of the stagnant water, and thus not sustaining the pressure of the external air. For this reason standing water creeps little by little up the sides of vessels, the air withdrawing from their sides; and water commonly clings to the surface of all substances, or as we ordinarily say, wets everything. And lastly this rarefaction of air in the neighbourhood of bodies reveals itself by the refraction [of light]. For light being admitted into a very dark room through the hole A [Fig. 52], and the image of the sun thrown to a great distance from the convex lens B, if you intercept all the light where [the beam] is the twentieth or thirtieth or fortieth part of an inch wide on the parallel sharp edges of two or more wedges C, D, placed there, you will see the image somewhat deflected from E towards F on account

221

of the refraction which the light undergoes in passing through the rarefied medium close to the edges of the wedges.

Bodies also seek to avoid mutual contact.
Moreover air does not only seek to avoid bodies, but bodies also tend to fly from each other. For if you place upon a somewhat convex lens, very highly polished, such as the objective of a very long telescope, a second lens which is smaller, plane, and equally well polished, you will find that some effort is required to bring them into contact, and that when the pressure is removed they will spring apart spontaneously, the more so as the upper lens is less heavy; for you will see several coloured concentric circles appear as the pressure is applied, and finally a black spot at their centre which is the indication of complete contact at that spot, then begun; when the

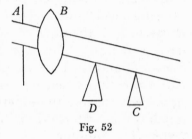

Fig. 52

pressure is removed that spot and the colours then vanish in the reverse order from that of their appearance. When plane glasses are used, even though they are not more than half an inch wide, you can so compress them that the colours appear, but the whole strength of one man is scarcely or not quite sufficient to bring them into complete contact, so that the black spot appears. [However, the intervening air because of the minuteness of the space becomes so rare that where the colours have appeared the glasses cannot be separated without the use of considerable force. The external air exerts a greater pressure than the internal air, with its elastic force weakened by rarefaction, is able to overcome.] In the same way the parts of glass, steel or any broken body cannot even when pressed with the greatest force be reduced to their former contact so that they cohere as before, for were complete contact restored without doubt their cohesion would be restored [also]. Further, lead and tin when melted and poured into an iron vessel do not attain contact in such a way that the casting adheres to the iron. Also powders floating on liquids avoid contact with the liquid and are submerged with difficulty even when they are fairly heavy, like filings of metal. Similarly, flies and other small creatures are wont to walk on the yielding [surface of] water without wetting their feet. Lastly, if the particles of any powder even though they seem to lie one upon another did fully touch they would cohere strongly, as they do when they are brought into contact by moistening with water and then drying.

[Many opinions may be offered concerning the cause of this repulsion. The intervening medium may give way with difficulty or not suffer itself to be much compressed. Or God may have created a certain incorporeal nature which seeks to repel bodies and make them less packed together. Or it may be in the nature of bodies not only to have a hard and impenetrable nucleus but also [to have] a certain surrounding sphere of most fluid and tenuous matter which admits other bodies into it with difficulty. [About these matters I do not dispute at all. But as it is equally true that air avoids bodies, and bodies repel each other mutually, I seem to gather rightly from this that air is composed of the particles of bodies torn away from contact, and repelling each other with a certain large force.] Or lastly it may be that two or more of these causes sometimes operate together. About these matters [I dispute] not at all. Upon this foundation all the properties of air are easily understood.

Suspicions about the cause of this repulsion.

In just the same remarkable manner [air] rarefies and is condensed according to the degree of pressure. The whole weight of the incumbent atmosphere by which the air here close to the Earth is compressed is known to philosophers from the Torricellian experiment, and Hooke proved by experiment[1] that the double or treble weight compresses air into the half or third of its space, and conversely that under a half or a third or even a hundredth or a thousandth part of that [normal] weight [the air] is expanded to double or treble or even a hundred or a thousand times its normal space, which would hardly seem to be possible if the particles of air were in mutual contact; but if by some principle acting at a distance [the particles] tend to recede mutually from each other, reason persuades us that when the distance between their centres is doubled the force of recession will be halved, when trebled the force is reduced to a third

How it is condensed by pressure and rarefies by diminution of the pressure.

---

[1] Cf. Newton's notes on *Micrographia*, Section VI, p. 400.

and so on, and thus by an easy computation it is discovered that the expansion of the air is reciprocal to the compressive force.[1]

How it rarefies by heat.   Moreover we need not wonder that air is expanded by heat if we consider that its parts when agitated by heat must vibrate, and, by vibrating, propel hither and thither the neighbouring parts. Suppose A, B, C, three particles in a state of rest [Fig. 53], and if B is set in motion by heat towards A, as far as the point R, it drives A away to a greater distance and, by the same action in reverse

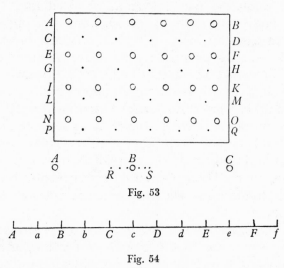

Fig. 53

Fig. 54

springing back towards S, drives away C and so on, B alternately repelling A and C with a vibrating motion and A and C similarly repelling their neighbouring [particles so that] all are scattered through a wider space proportionate to the quantity of motion. It is not necessary, however, that this vibration (of which I suppose heat to consist) should always be in a straight line, for the particles may revolve in curves. Nor need the rarefaction through heat arise only from this vibratory motion, for the motion...

On the same principle the undulatory motion by which sounds are propagated is easily explained. For suppose A the centre of the undulation [Fig. 54], and that the particles A, B,

[1] Cf. *Principia*, Book II, Proposition XXII, for the proof that if the force is inversely as the distance, the density is as the compression.

C, D, E, F, vibrate along the lines Aa, Bb, Cc, Dd, etc. so that when A is at A, B is in the middle of Bb, C at c, D in the middle of Dd and E at E, and the greatest expansions of the air will be at the midpoints of Bb and Ff, and the greatest contractions at the midpoints but. . . .

Furthermore, that air although fluid and subtle seems to creep through the pores of bodies with greater difficulty than water or oil, so that it may be contained in a bladder which a liquid easily pervades, results from the fact that its particles, keeping their distance from all [others], are reluctant to approach the sides of the pores. They are indeed very small; but as to the

Why the medium is sluggish and anything but subtle. And that it impedes the contact of bodies, and makes bodies in contact cohere.

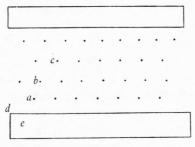

Fig. 55

faculty of permeating narrow places, [their] subtlety ought to be estimated from the whole sphere which each claims for itself. And thus it is that air is not at all inclined to any motion except the before-mentioned continuous vibration and that it is very difficult to force it out from the space between two smooth marble or glass [surfaces], or when forced out to admit it again. For if a particle a [Fig. 55] situated in any imperceptible cavity e cannot move outwards unless by approaching nearer the sides of the cavity, and b cannot go out except by approaching near a, nor c unless by approaching nearer c [b] than their mutual force of repulsion permits; it is necessary that they adhere and that they keep all the remaining ones within enclosed, unless they are expelled as grindings by the sliding motion of the upper marble on the lower one. Whence it is that solid bodies are scarcely brought to full contact unless by fusion, and particles of powder scarcely except by infusion in water and evaporation; yet when at last brought into contact they cohere most strongly, compressed by the surrounding air, to say nothing now of the similar force of the aether.

How it is generated. Lastly, from these principles the generation of air is easily learned. For this nothing else is required save a certain action or motion which tears apart the small parts of bodies; since when separated they mutually flee from one another, like the other particles of air. And thus it is that every vehement agitation (like friction, fermentation, ignition and great heat) generates the aerial substance which reveals itself in liquids by ebullition; and the more vehement the action the more copiously that substance is generated. So filings of lead, brass, or iron dissolved in Aqua Fortis produce a great ebullition; these, however, dissolve without ebullition in vinegar or in the same Aqua Fortis sufficiently weakened by mixing with water. So nitre, melted and ignited by charcoal thrown upon it, emits much of the aerial substance, with much fixed salt remaining in the bottom; but if the action of the fire is hastened by the due admixture of sulphur and charcoal as is used in making gunpowder, almost all the substance of the mixt is changed by vehement agitation into an aerial form, the huge force of this powder arising from its sudden expansion, as is the nature of air. Moreover, aerial substances

The threefold kind of air arises from the threefold kind of substances. Permanent and heavy air from metals, exhalation especially from the earthy part of vegetable substances and vapour from liquids. are very different according to the nature of the bodies from which they are generated. Metals by corrosion give true permanent air; vegetable and animal substances by corrosion, fermentation or burning give an air of short duration like an exhalation; and volatile substances rarefied by heat give an air least lasting of all, which we call a vapour. Among these there are great differences, the vapour of water condenses more quickly than the vapour of spirit of wine and the vapour of certain saline spirits, and more quickly than the vapour of water,[1] as in the vapour of the sea where the spirit of salt condenses only when rising to the height of a few feet, and makes the air there thicker and less transparent and leaves [behind] the sweet vapour of water. There is a diversity in the weight of air, for

[1] ? [salt] water....

226

air generated by the force of fire from vegetable and animal substances is so much lighter than the rest of the surrounding air [i.e. atmosphere] that not only does it not descend itself but it also carries away the associated thicker fumes with it. So vapours of every kind seek to rise above the rest of the air. And there is no doubt but that vapours and exhalations have various degrees of gravity; for there appear to be as many kinds of air as there are of substances on the Earth from which its origin comes. So the atmosphere is composed of many kinds of air, which nevertheless can be divided into three chief kinds: vapours, which arising from liquids seem to be the least permanent and the lightest; exhalations, which arise from thicker and more fixed substances, especially in the vegetable kingdom, are of a middle nature; and air properly so called whose permanence and gravity are indications that it is nothing else than a collection of metallic particles which subterranean corrosions daily disperse from each other. This is confirmed by the fact that this latter air serves (as the almost indestructible nature of metals demands) neither for the preservation of fire nor for the use of animals in breathing, as do serve some of the exhalations arising from the softer substances of vegetable matter or salts.

Chapter 2. On the Aether

And just as bodies of this Earth by breaking into small particles are converted into air, so these particles can be
Various kinds of substances are shown to be much subtler than air. broken into lesser ones by some violent action and converted into yet more subtle air which, if it is subtle enough to penetrate the pores of glass, crystal and other terrestrial bodies, we may call the spirit of air, or the aether. That such spirits exist is shown by the experiments of Boyle in which metals, fused in a hermetically sealed glass for such a time that part is converted into calx, become heavier. It is clear that the increase is from a most subtle saline spirit which, coming through the pores of the glass, calcines the metal and turns it into calx. And that in a glass empty of air a pendulum preserves its oscillatory motion not much longer than in the open air, although that motion ought not to cease unless, when the

air is exhausted, there remains in the glass something much more subtle which damps the motion of the bob. I believe everyone who sees iron filings arranged into curved lines like meridians by effluvia circulating from pole to pole of the [load-]stone will acknowledge that these magnetic effluvia are of this kind. So also the attraction of glass, amber, jet, wax and resin and similar substances seems to be caused in the same way by a most tenuous matter of this kind. . . .

PART IV

MANUSCRIPTS RELATED TO THE *PRINCIPIA*

INTRODUCTION

The way in which the *Principia* came to be written and published has been described by a number of writers, the most complete accounts being those of Rigaud, Edleston, and Rouse Ball. These accounts were largely based upon Newton's correspondence, which will at last be rendered accessible to all historians in the edition prepared under the direction of the Royal Society. The manuscripts in the Portsmouth Collection relating to the preparation of the first and subsequent editions of the *Principia* have, by contrast, been neglected, although they throw interesting light on Newton's procedure and the working of his mind.[1]

Most of these manuscripts—scattered throughout the Portsmouth Collection—consist of drafts, at all stages from the rough to the final, some of them dealing with material ultimately incorporated into the *Principia*, some covering topics rejected from the final version. In addition, there are in the Cambridge University Library manuscript copies of two series of lectures by Newton, one corresponding very closely to the earlier part of Book I of the *Principia*, the other to *De Systemate Mundi*, first printed in 1728.

Many of these draft passages, although of interest to an editor of the *Principia*, are of insufficient general interest to be considered here. They elucidate the evolution of the finished form of the book without adding to our understanding of Newton's thought, or of his purposes in writing. Others indicate, however, that the *Principia* might have emerged as a work different from—and perhaps even more astonishing than—that which ultimately appeared.

It has long been recognized that the *Principia* was an expansion of a relatively short paper, entitled by Rigaud *Propositiones de Motu*, which we suggest was written in the late summer of 1684, and which was subsequently submitted to the Royal Society. In this paper are to be found versions of propositions included in Books I and II of the *Principia*, and of various topics more fully developed in Book III (cf. below,

---

[1] W. W. Rouse Ball was the first to make use of the Newton papers after their deposition in the Cambridge University Library, and subsequent scholars owe him a considerable debt.

no. 1). From it Newton worked up a more elaborate course of lectures which he delivered in the autumn of 1684. And by the winter of 1685, perhaps as a result of pressure from Edmond Halley, he was committed to writing a treatise on a larger scale than he had originally contemplated; as his work proceeded it grew still further.

Newton's own statement that 'the book of the Principles was written in seventeen or eighteen months' is somewhat disingenuous, for in fact he was working on the propositions on elliptical motion in August 1684, and the third book was not complete until March 1687: that is, two years and seven months later. He proceeded at amazing speed, however, during the first twelve or eighteen months of his work: Book I was taken at least as far as Prop. XXXIX by December 1684 (though all the geometry was not necessarily completely developed); propositions LXX–LXXXIV, on the attractive forces of spherical bodies, were probably discovered in the spring of 1685;[1] and the whole book was nearly complete by the summer of 1685 at the latest, since Newton speaks of the second book as having been finished that summer.[2]

At this point Halley and others in the Royal Society presumably knew of Book I only, for the second and third books were not mentioned by Newton before the early summer of 1686. It was therefore of Book I alone that Halley spoke to the Royal Society on 21 April 1686, and it was Book I alone that was presented to the Society a week later by Dr Vincent. Even more strange, perhaps, is that Samuel Pepys can have seen no more than Book I when he gave his imprimatur as President of the Royal Society on 5 July 1686.

Moreover, the second and third books, on which Newton was, in fact, working through the summer and autumn of 1685, and on which he continued to work into the spring of 1687, were originally very different from those ultimately published. From Newton's remarks in his letter to Halley of 20 June 1686 and from his apology in the Preface to the *Principia* for a certain incoherence in the book, it seems that having completed Book I he turned next, in the summer of

---

[1] Rouse Ball, 61; Halley to Newton 20 June 1686 says simply 'last year' (*ibid.* 157).

[2] To Halley, 20 June 1686 (*ibid.* 158). However, he returned to Book I in the winter of 1685–6, enlarging 'it with divers propositions, some relating to comets, others to other things' (*ibid.* 158).

1685, to considering the application of its dynamical theory to various phenomena of astronomy, including the motions of the moon and the orbits of comets. And at about the same time he found himself increasingly occupied with problems of fluid mechanics, on which he had composed some propositions a year earlier.

Perhaps, at this stage, Newton intended Book I to be succeeded directly by another in which the truth of his discoveries should be made evident from their correspondence with the celestial phenomena. This would have been a logical plan— more logical than that of the *Principia* as printed, where Book II serves for the most part only to disrupt a coherent exposition: few of its propositions are relevant to what has gone before in Book I, or what follows in Book III. The demonstration of the frame of the world upon mathematical principles was originally (as Newton explained at the opening of the ultimate Book III) to have been 'composed in a popular method'. It is plausible to suppose that he wrote for this purpose—beginning in the summer of 1685—the tract later published (in 1728) as *De Systemate Mundi* (or the *System of the World*). In the Michaelmas Term 1687 Newton in fact delivered five lectures, the surviving manuscript of which agrees in all but occasional verbal detail with the first 27 sections of *De Systemate Mundi*.[1] The manuscript is holograph, with the date of the first lecture given as 29 September. It is headed *De Motu Corporum Liber Secundus*, but the word *Secundus* has been crossed out and a stop added after *Liber*. Now *De Motu Corporum* is the subtitle of Books I and II of the *Principia*, and was at one time considered by Newton as a possible title for the whole work: he subsequently decided that *Philosophiae Naturalis Principia Mathematica* was a title more likely to attract buyers.[2] Why should one not suppose that Newton proceeded with this book as economically as he had with Book I? That is

---

[1] MS. Dd-IV-18. The five lectures cover sections 1–6, 7–12, 13–17, 18–22, 23–7, respectively. In numbering the sections Newton omitted 17, hence sections 17–27 are numbered in the MS. 18–28. Cf. Rouse Ball, 64–5, who also thought that *De Systemate Mundi* might be the original form of Book III. The year 1687 is found on Cotes' copy of these lectures, not on the original MS. It is possible that Cotes (who was five years old in 1687) made a mistake, and that these lectures were given in 1686, for which year no lectures of Newton have been found. This would have been the natural time for them.

[2] Newton to Halley, 20 June 1686; Rouse Ball, 159. The proceeds, of course, were to go to Halley.

to say, that he wrote it up first for delivery as a course of lectures, and gave it a title intended to indicate that it followed immediately upon the already completed Book I, from which he had continued to draw his lectures in the autumn of 1685.

In the midst of this activity he reconsidered his design, realizing that the headlong pace of past months could not be maintained. He told Halley that in the 'autumn last [1685] I spent two months in calculations [on the theory of comets] to no purpose for want of a good method', and the theory of lunar motions was even more intransigent.[1] He decided (as recorded in the Preface to the *Principia*) to defer 'publication till I had made a search into those matters [the problems treated in Books II and III] and could put forward the whole together'. These words in their full context imply a decision to treat fluid mechanics at length, in addition to the astronomical problems, in the proposed sequel to Book I. But we do not know what precise progress was made between the autumn of 1685 and June 1686.

On the 20th of that month, Newton wrote to Halley announcing the preparation of three books, apparently thus giving the first definite news of the two later ones.[2] Presumably by then the ultimate Book II was well in hand, but it had not yet taken its final shape, and could still be described as 'short'.[3] In the same letter Newton announced and killed the third book:

The third I now design to suppress. Philosophy is such a litigious lady, that a man had as good be engaged in lawsuits, as have to do with her. I found it so formerly, and now I am no sooner come near her again, but she gives me warning.

The reasons for this design have never been quite clear.

---

[1] Rouse Ball, 158.

[2] Cf. Halley to Newton, 7 June 1686, 'I hope you will please to bestow the second part, or what remains of this [Book I], upon us as soon as you shall have finished it, for the application of this Mathematical part to the system of the world, is what will render it acceptable to all Naturalists, as well as Mathematicians; and much advance the sale of ye book.' (Rouse Ball, 156.) Halley obviously did not know whether the application was to be developed in a separate Book or not, and expected it to be popular, so as to 'render it acceptable to those, that will call themselves Philosophers without Mathematicks which are by much the greater number' (29 June 1686, *ibid.* 164).

[3] 'Some new propositions' had been thought on since the preceding year, which Newton said he could as well let alone—one supposes that they were in fact included. Perhaps he meant the propositions on the speed of sound, and so forth.

Newton's only explanation is offered at the beginning of the ultimate Book III:

I had, indeed, composed the third Book in a popular method, that it might be read by many; but afterwards considering that such as had not sufficiently entered into the principles could not easily discern the strength of the consequences, nor lay aside the prejudices to which had been for many years accustomed, therefore, to prevent the disputes which might be raised upon such accounts, I chose to reduce the substance of this Book into the form of Propositions (in the mathematical way), which should be read by those only who had first made themselves masters of the principles established in the preceding Books....[1]

The decision was, in fact, made when Newton was irritated by Hooke's claim to priority in the discovery of the law of gravitation, which occasioned the outburst against 'lawsuits'. But the suppression of Book III could hardly have quietened Hooke, or any scientific critic of Newton's theories, though Newton may have believed that it would do so. In any case, for the moment an astronomical book, finished or unfinished, was laid aside. *De Systemate Mundi*, or the early portion of it, remained available for a course of lectures (presumably undergraduates were not fearsomely litigious) and was otherwise suppressed till after Newton's death, though manuscript copies of it circulated. It seems plausible to suppose that Newton now pressed on with Book II (in the latter half of 1686), for according to his letter to Halley of 18 February 1687 it was complete by the preceding November or December. As he had received no word from Halley through the autumn and winter, however, he had kept the manuscript in his own hands. In his letter to Halley Newton refers to Wallis's paper on motion in a resisting medium (read 26 January 1687) and to the Royal Society's resolution to inquire whether Newton intended to print his own propositions on such problems: they were, he wrote, inserted 'at the beginning of the second book with divers others of that kind', but it is unlikely that he meant that he had added them during the past two weeks.[2] Finally, the manuscript of Book II was sent up to London about 1 March.

The ultimate Book III followed shortly afterwards, reaching

---

[1] Cajori, 397.   [2] Newton to Halley, 18 February 1687 (Rouse Ball, 169–70).

London on 28 March 1687. Calmed by Halley's letter of 29 June 1686, and seeking 'how best to compose the present dispute' with Hooke, Newton may have started pretty rapidly thereafter on the recomposition of the astronomical book 'in the mathematical way' and accomplished this, as well as the completion of Book II, in the autumn and winter following. It seems more natural to believe that matters happened thus, than that Newton first wrote *De Systemate Mundi*, then abandoned its more popular approach and rewrote it, then decided to suppress Book III altogether, and finally changed his mind once more to include it. Indeed, his expressed intention of preventing those incapable of mastering Book I from following Book III may well have been specially aimed at Hooke, of whose mathematical attainments Newton had a low opinion.

Drafts in the Portsmouth Collection contain preliminary studies of many problems discussed in each of the three books of the *Principia*, as well as of problems that were not included. They confirm the usual assumption that Newton used the method of fluxions in handling some problems, though he seems generally to have been happier with synthetic geometry. (We have discussed this at greater length in the first section, on mathematics.) They show how, repeatedly and critically, he revised his writing of controversial passages—passages where he touched on such diverse topics as matters of priority, God as the Creator and First Cause, the theory of matter and the origin of physical forces. The last topic particularly exercised his mind: he was not only impelled to make clear his belief that attractive force was not an esoteric conception, nor a consequence of inexplicable action at a distance, but was under a kind of compulsion to make plain and to illustrate his conception of the nature of physical forces, including gravitation. He tried to express himself in a *Conclusion* (No. 7) and renounced his draft; he tried again in a draft of the *Preface* (No. 3) and struck out the revealing passages; and he tried yet again in drafts of the *Scholium Generale* of the second edition, only to abandon the attempt for the third time. Newton never succeeded in giving more than hints of his profoundest thoughts on physical structure within the framework of the *Principia Mathematica*; what he would like to have said may be seen from the drafts printed here.

1

EARLY DRAFTS OF PROPOSITIONS IN MECHANICS

Several draft statements of Definitions, Axioms and Laws appropriate to the science of mechanics, and of propositions relating to the motions of the bodies *in vacuo* and in fluids, exist in the archives of the Royal Society and the Portsmouth Collection (MS. Add. 3965). All these correspond to passages in the *Principia*, as indicated in the following pages, and all of them antedate the first draft of Book I of the *Principia* itself, that is, the series of lectures given by Newton in the Michaelmas Term of 1684.[1]

The exact date at which these drafts were written is uncertain. There is nothing to indicate that any of them was written before Halley's famous visit to Newton at Cambridge in August 1684 when, according to Conduitt's story, he asked Newton 'what would be the curve described by the planets on the supposition that gravity diminished as the square of the distance. Newton immediately answered, an Ellipse. Struck with joy and amazement, Halley asked him how he knew it? Why, replied he, I have calculated it; and being asked for the calculation, he could not find it, but promised to send it to him'.[2] The demonstration was sent to Halley in November 1684, but this paper and any accompanying letter do not appear to have survived, although Halley seems to speak of it as having been registered by the Royal Society.[3]

Newton's letters and the records of the Royal Society refer to other drafts besides this, but they cannot be identified with certainty now. There was what Halley referred to at a meeting of the Royal Society as 'a curious treatise *De Motu*', seen by him on a second visit to Cambridge in November 1684; and there is a letter from Newton to the Secretary of the Society, of 23 February 1685, in which he thanks the latter for registering 'my notions about motion' which he proposed to complete in a few weeks after a visit to Lincolnshire. These various allusions have been the subject of conjecture by Rigaud, Edleston, Brewster and Rouse Ball, but it is now impossible to decide which (if any) of the surviving drafts was intended and how much was communicated by Newton to his friends in London at any given date between August 1684 and March 1685.

---

[1] These lectures, as written out for deposition in the University Library, are still preserved there in more than one version (MS. Dd-9-46). Cf. Rouse Ball, 28–9, 59–60.
[2] Rouse Ball, 26.   [3] *Ibid.* 163. There is no record of a registration at this time.

In any case these matters are trivial. For comparison of the various pre-*Principia* drafts makes it perfectly clear from internal evidence that *all* of them precede the Michaelmas Term lectures, which themselves are headed in Newton's own hand 'Octob. 1684'. The drafts are preliminary and fragmentary studies, whereas the lectures correspond very closely in language and organization to what was later printed in the *Principia*.

One may suppose that events moved somewhat as follows. Immediately after Halley's visit in August 1684, Newton must have returned to his study of motion in an ellipse which he proceeded to work out, in his customary fashion, through four or more different drafts. By October 1684 this part of his work—later incorporated into the *Principia*—had reached an almost definitive form. At the same time he was drafting the Definitions, Axioms and Laws which had to precede the mathematical propositions, and these too had reached a nearly definitive form by October, when he began to lecture and chose (naturally) the material on which he had been working throughout the latter part of the summer. Halley in November 1684 may have seen either one of the earlier drafts, or the manuscript lectures; it matters little which, since their content is the same. In sending evidence of his discoveries to the Royal Society, however, Newton very rationally retained his own lecture notes and dispatched instead an amanuensis' copy of one of the drafts written during the preceding summer. This copy, still at the Royal Society and printed by Rigaud under the title *Propositiones de Motu*,[1] was received in London at an unknown date in the winter of 1684/5.

There are four drafts (A, B, C, D) in the Portsmouth Collection, MS. Add. 3965, which we attribute to the summer of 1684:

A. fols. 21–4   Latin, holograph. Title, 'De Motu Corporum'.
B. fols. 55–62 bis   Latin, holograph. Title, 'De motu corporum in gyrum'.
C. fols. 63–70   Latin, copy. No title.
D. fols. 40–54   Latin, copy. Title, 'De Motu sphaericorum Corporum in fluidis'.

A is a draft, with many alterations and interlineations, of the *Definitiones* (as printed in the *Principia*, pp. 1–5), breaking off a little before the end of the final paragraph.

B, C, and D are successive drafts, similar to the so-called *Propositiones de Motu* of the Royal Society. B seems to be the earliest version, corrected by Newton at some later stage. C is a rather rough copy

---

[1] Appendix, 1–19. Mr J. W. Herivel argues that this was made from draft C, later returned to Newton.

of this, incorporating some fresh thoughts. D is a neatly written revision in the hand of an amanuensis, perhaps Humphrey Newton, and gives the longest and most polished version. In the text that follows we have used D, as the best authority, comparing it with the Royal Society copy as printed by Rouse Ball, and especially with the holograph B. The result gives the most complete version of Newton's achievement in the summer of 1684, before he wrote the lectures that (in their corresponding passages) are in turn directly antecedent to the text of the *Principia* itself.

Note added in Proof

Relying on Conduitt's reference to *May* as the month of Halley's visit to Newton in 1684, although Halley himself two years later gave it as *August*, and on a document (now lost) printed by Brewster (I, 471) in which Newton stated, much later, that some propositions of the *Principia* were discovered in June and July 1684, Mr J. W. Herivel suggests that Halley's visit to Newton occurred in May. This conjecture has no other effect than to allow more time for Newton's composition of the *De Motu* drafts before the Michaelmas lectures began. However, it by no means appears improbable that Newton could have proceeded as far as he did in the two months of August and September alone.

MS. A. De Motu Corporum

Definitiones

1. *Quantitas materiae* est quae oritur ex ipsius densitate et magnitudine conjunctim. Corpus duplo densius in duplo spatio quadruplum est. Hanc quantitatem per nomen corporis vel massae designo.

2. *Quantitas motus* est quae oritur ex velocitate et quantitate materiae conjunctim. Motus totius est summa motuum in partibus singulis, adeoque in corpore duplo majore aequali cum velocitate duplus est et dupla cum velocitate quadruplus.

3. *Materiae vis insita* est potentia resistendi qua corpus unumquodque quantum in se est perseverat in statu suo vel quiescendi vel movendi uniformiter in directum. Estque corpori suo proportionalis, neque differt quicquam ab *inertia* massae nisi in modo conceptus nostri. Exercet vero corpus hanc vim solummodo in mutatione status sui facta per vim aliam in se impressam estque Exercitium ejus *Resistentia* et

Impetus respectu solo ab invicem distincti: Resistentia quatenus corpus reluctatur vi impressae, *Impetus* quatenus corpus difficulter cedendo conatur mutare statum corporis alterius. Vulgus insuper resistentiam quiescentibus & impetum moventibus tribuit: sed motus et quies ut vulgo concipiuntur respectu solo distinguuntur ab invicem: neque vere quiescunt quae vulgo tanquam quiescentia spectantur.[1]

4. *Vis impressa* est actio in corpus exercita ad mutandum statum ejus vel quiescendi vel movendi. Consistit haec vis in actione sola neque post actionem permanet in corpore. Est autem diversarum originum, ut ex impetu, ex pressione, ex vi centripeta.

5. *Vis Centripeta* est vel actio vel potentia quaelibet qua corpus versus punctum aliquod tanquam ad centrum trahitur, impellitur, vel utcunque tendit. Hujus generis est gravitas, qua corpus tendit ad centrum terrae: vis magnetica, qua ferrum petit centrum magnetis, et vis illa, quaecunque sit, qua Planetae retinentur in orbibus suis et perpetuo cohibentur ne abeant in eorum tangentibus. Est autem vis centripetae quantitas triplex; *absoluta, acceleratrix,* et *motrix. Quantitas absoluta* (quae et *vis absoluta* dici potest) major est ad unum centrum, minor ad aliud, nullo habito respectu ad distantias et magnitudines attractorum corporum; uti virtus magnetica major in uno magnete, minor in alio. *Quantitas* seu *vis acceleratrix* est velocitati proportionalis quam dato tempore generat; uti virtus magnetis ejusdem major in minori distantia, minor in majori: vel vis gravitans major prope terram minor in regionibus superioribus. *Quantitas* seu *vis motrix* est motui proportionalis quem dato tempore producit; uti pondus majus in majori corpore, minus in minore. Ita se habet igitur *vis motrix* ad *vim accelerationem* ut *motus* ad *celeritatem.* Namque oritur *quantitas motus* ex *celeritate* ducta in corpus mobile et *quantitas vis motricis* ex *vi acceleratrice* ducta in idem corpus. Unde juxta superficiem terrae, ubi gravitas acceleratrix in corporibus universis eadem est, gravitas motrix seu pondus est ut corpus: at si longius recedatur a terra inque regiones ascendatur ubi gravitas acceleratrix fit minor, pondus pariter minuetur, eritque semper ut corpus in gravitatem acceleratricem ductum. Porro *attractiones* & impulsus eodem

---

[1] This paragraph is much corrected and interlineated.

sensu *acceleratrices* & *motrices* nomino. Voces autem attractionis, impulsus vel *propensionis* cujuscunque in centrum indifferenter & pro se mutuo usurpo, has vires non *physice* sed
mathematice tantum considerando. Unde caveat Lector ne per
hujusmodi voces cogitet me speciem vel modum actionis
causamve aut rationem physicam alicubi definire   [end]

TRANSLATION

On the Motion of Bodies

Definitions

1. The *quantity of matter* is that which arises from its density
and bulk conjointly. A body of a double density in a double
space is quadruple. This quantity I designate under the name
of body or mass.

2. The *quantity of motion* is that which arises from the velocity
and quantity of matter conjointly. The motion of the whole
is the sum of the motions in the individual parts, and therefore in a body twice as great with equal velocity it is double,
and with twice the velocity, quadruple.

3. The *innate force of matter* is the power of resisting by which
a body continues, as much as in it lies, in its normal state,
either of resting or of moving uniformly in a straight line. And
it is proportional to the body [or, by def. 1, mass] to which it
belongs, nor does it differ at all from the *inertia* of the mass,
except in our manner of conception. Indeed, a body only
exerts this force in the change of its state made by means of
another force impressed upon it, and the Exercise of it is
Resistance and *Impetus*, which are mutually distinct only in this
respect, that it is Resistance so far as the body struggles
against the impressed force, *Impetus* in so far as the body, not
easily yielding, endeavours to change the state of another
body. Moreover common folk ascribe resistance to bodies at
rest, and impetus to those in motion: but motion and rest as
commonly conceived are only distinguished with respect to
each other: nor do those bodies truly rest, which are commonly considered as resting.

4. The *impressed force* is the action exerted on a body to
change its state of rest or motion. This force consists in the

action only and does not remain in the body after the action [is completed]. It is moreover of diverse origins, as from impetus, from pressure, from centripetal force.

5. The *Centripetal Force* is any action or potential by which a body is drawn, impelled or in any way tends towards any point as to a centre. Of this kind is gravity, by which a body tends towards the centre of the earth: the magnetic force, by which iron goes towards the centre of the magnet, and that force, whatever it may be, by which the Planets are preserved in their orbits and perpetually confined there, lest they go astray along the tangents to these [orbits]. And the quantity of centripetal force is threefold: *absolute, accelerative* and *motive*. The *absolute quantity* (which may be called the *absolute force*) is greater towards one centre, less towards another, without respect to the distances and bulks of the attracted bodies; thus the magnetic power is greater in one magnet, less in another. The *accelerative quantity* or *force* is proportional to the velocity which it generates in a given time; so the power of the same magnet is greater at a less distance, less at a greater; or the force of gravity is greater near the earth, less in higher regions. The *motive quantity* or *force* is proportional to the motion which it produces in a given time; so weight is greater in a greater body, less in a lesser one. So the *motive force* is to the *accelerative force* as *motion* to *celerity*. For the *quantity of motion* arises from the *celerity* multiplied by the moving body [i.e. its mass] and the *quantity of motive force* from the *accelerative force* multiplied by the same body. Whence near the surface of the earth, where the accelerative gravity is the same in all bodies, the motive gravity or weight is as the body [or mass]: but if one should go further away from the earth and ascend into regions where the accelerative gravity is less, the weight will diminish as well, and will always be as the body multiplied by the accelerative gravity. I further call *attractions* and impulses in the same sense *accelerative* and *motive*. Indeed I employ the words attraction, impulse or *propensity* of any kind to a centre indifferently and one for another, considering these forces not *physically* but only *mathematically*. Whence the Reader is to take care not to think that I anywhere define the kind or manner of an action, its cause or physical reason.

MSS. B, C, D. De Motu sphaericorum Corporum in fluidis[1]

Definitiones

Def. 1. Vim centripetam appello qua corpus attrahitur vel impellitur versus punctum aliquod quod ut centrum spectatur.

Def. 2. Et vim corporis seu corpori insitam qua id conatur perseverare in motu suo secundum lineam rectam.

Def. 3. Et resistentiam quae est medii regulariter impedientis.

Def. 4. Exponentes quantitatum sunt aliae quaevis quantitates proportionales expositis.[2]

[Hypotheses][3]

Lex 1. Sola vi insita corpus uniformiter in linea recta semper pergere si nil impediat.

Lex 2. Mutationem status movendi vel quiescendi proportionalem esse vi impressae et fieri secundum lineam rectam qua vis illa imprimitur.

Lex 3. Corporum dato spatio inclusorum eosdem esse motus inter se sive spatium illud quiescat sive moveat id perpetuo et uniformiter in directum absque motu circulari.

Lex 4. Mutuis corporum actionibus commune centrum gravitatis non mutare statum suum motus vel quietis. Constat ex Lege 3.

Lex 5. Resistentiam medii esse ut medii illius densitas et corporis moti sphaerica superficies et velocitas conjunctim.

[Lemmata][4]

Lemma 1. Corpus viribus conjunctis diagonalem parallelogrammi eodem tempore describere quo latera separatis.

---

[1] B: title, 'De Motu corporum in gyrum.'   [2] Not in B or C.

[3] B and C: 'Hypotheses'; in D this word is altered to 'Lex'. B reads:

1. Resistentiam in proximis novem propositionibus nullam esse, in sequentibus esse ut corporis celeritas et medii densitas conjunctim.

2. Corpus omne sola vi insita uniformiter secundum rectam lineam in infinitum progredi, nisi aliquid extrinsecus impediat.

3. Corpus in dato tempore viribus conjunctis eo ferri, quo viribus divisis in temporibus aequalibus successive.

4. Spatium quod corpus urgente quacunque vi centripeta ipso motus initio describit, esse in duplicata ratione temporis.

C follows B with the following exceptions: 1. ...ut medii densitas et celeritas conjunctim. 2. ...lineam rectam....

[4] B has no lemmas; C has lemmas 3 and 4 only, and they are so numbered.

Si corpus dato tempore vi sola *m* ferretur ab *A* ad *B* [Fig. 56] et vi sola *n* ab *A* ad *C*, compleatur parallelogrammum *ABCD*, et vi utraque ferretur id eodem tempore ab *A* ad *D*. Nam quoniam vis *m* agit secundum lineam *AC* ipsi *BD* parallelam, haec vis per Legem 2, nihil mutabit celeritatem accedendi ad lineam illam *BD* vi altera impressam. Accedet igitur corpus eodem tempore ad lineam illam *BD* sive vis *AC* imprimatur

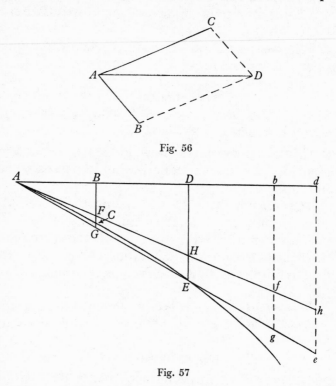

Fig. 56

Fig. 57

sive non, atque adeo in fine illius temporis reperietur alicubi in linea illa *BD*. Eódem argumento in fine temporis ejusdem reperietur alicubi in linea *CD*, et proinde in utriusque lineae concursu *D* reperiri necesse est.[1]

Lemma 2. Spatium quod corpus urgente quacunque vi centripeta ipso motus initio describit, esse in duplicata ratione temporis.

Exponantur tempora per lineas *AB*, *AD* [Fig. 57]. Datis *Ab*, *Ad* proportionales, et urgente vi centripeta aequabili

[1] Cf. *Principia*, Lex III, Corol. I, 13.

exponentur spatia descripta per areas rectilineas *ABF*, *ADH*, perpendiculis *BF*, *DH*, et recta quavis *AFH* terminatas ut exposuit Galileus. Urgente autem vi centripeta inaequabili exponantur spatia descripta per areas *ABC*, *ADE* curva quavis *ACE* quam recta *AFH* tangit in *A*, comprehensas. Age rectam *AE*, parallelis *BF*, *bf*, *dh* occurrentem in *G*, *g*, *e* et ipsis *bf*, *dh* occurrat *AFH* producta in *f* et *h*. Quoniam area *ABC* major est area *ABF* minor area *ABG*, et area curvilinea *ADEC* major area *ADH* minor area *ADEG*, erit area *ABC*, ad aream *ADEG* major quam area *ABF* ad aream *ADEG*, minor quam area *ABG* ad aream *ADH*, hoc est, major quam area *Abf* ad aream *Ade*, minor quam area *Abg* ad aream *Adh*. Diminuantur jam lineae *AB*, *AD* in ratione sua data usque dum puncta *A*, *B*, *D* coeunt et linea *Ae* conveniet cum tangente *Ah*, adeoque ultimae rationes *Abf* ad *Ade* et *Abg* ad *Adh* evadent eaedem cum ratione *Abf* ad *Adh*. Sed haec ratio est dupla rationis *Ab* ad *Ad* seu *AB* ad *AD* ergo ratio *ABC* ad *ADEC* ultimis illis intermedia jam fit dupla rationis *AB* ad *AD* id est ratio ultima evanescentium spatiorum seu prima nascentium dupla est rationis temporum.[1]

Lemma 3. Quantitates differentiis suis proportionales sunt continue proportionales.

Ponatur *A* ad *A−B*, ut *B* ad *B−C* & *C* ad *C−D* &c et dividendo fiet *A* ad *B* ut *B* ad *C*, et *C* ad *D* &c[2]

Lemma 4. Parallelogramma omnia circa datam ellipsin descripta, esse inter se aequalia.

Constat ex Conicis.[3]

Theoremata[4]

1. Gyrantia omnia radiis ad centrum ductis, areas temporibus proportionales describere.

2. Corporibus in Circumferentiis Circulorum uniformiter gyrantibus vires centripetas esse ut arcuum simul descriptorum quadrata, applicata ad radios circulorum.

Coroll. 1. Vires Centripetae sunt ut celeritatum quadrata Applic. ad Radios.

---

[1] Cf. *Principia*, Book I, Section I, Lemma X, 32.
[2] Cf. *ibid*. Book II, Section I, Lemma I, 237.
[3] Cf. *ibid*. Book I, Lemma XII, 47.
[4] This list of titles occurs only in C, where it precedes lemmas numbered 3 and 4.

2. Et reciproce ut quadrata Temporum periodicorum applicata ad Radios.

3. Unde si quadrata Temporum Periodicorum sunt ut Rad. Circulorum, vires centripetae sunt aequales et vice versa.

4. Et si Quadrata Temporum Periodicorum sunt ut Quadrata Radiorum Vires centripetae sunt reciproce ut radii.

5. Si quadrata Temporum Periodicorum sunt ut Cubi Radiorum Vires centripetae sunt reciproce ut quadrata Radiorum.

3. Corporis circa punctum gyrando Vim centripetam esse reciproce ut solidum *a quadrato distantiae, in quadratum perpendiculi ad Distantiam a puncto peripheriae*, applicato ad parallelam ab eodem puncto in peripheria ad Tangentem Ductam.[1]

Coroll. Hinc si detur Figura quaevis, et in ea punctum, ad quod vis centripeta dirigitur, inveniri potest lex vis centripetae, quae corpus in figurae illius perimetro gyrare faciet, nimirum computando solidum illud huic vi reciproce proportionale. Exempla dantur in problematis sequentibus.

Problemata

1. Gyrat corpus in Circumferentia Circuli requiritur lex Vis centripetae tendentis ad punctum aliquod in circumferentia.

2. Corpus gyrat in Ellipsi veterum requiritur Lex Vis Centripetae tendentis ad Centrum Ellipseos.

3. Gyrat corpus in Ellipsi requiritur lex vis centripetae tendentis ad Umbilicum Ellipseos.

Theor. 4. Posito quod Vis Centripeta sit reciproce proportionalis quadrato distantiae a centro, Quadrata Temporum periodicorum in Ellipsibus sunt ut Cubi Transversorum Axium.

Prob. 4. Posito quod Vis Centripeta sit reciproce proportionalis quadrato Distantiae et cognita vis illius quantitate requiritur Ellipsis quam Corpus describit de loco dato, cum data celeritate, secundum datam rectam emissum.

Prob. 5. Posito quod vis Centripeta sit reciproce propor-

---

[1] This form of words is modified in the Theorem as given below.

tionalis quadrato distantiae a centro, spatia definire, quae corpus recta cadendo datis temporibus describit.

6. Corporis sola vi insita per medium similare resistens delati motum definire.

7. Posita uniformi vi Centripeta motum corporis in medio similari recta Ascendentis ac Descendentis definire.

Beneficio duorum novissimorum problematum innotescunt motus projectilium in aere nostro ex hypothesi quod aer iste similaris sit quodque gravitas uniformiter et secundum lineas parallelas agat.

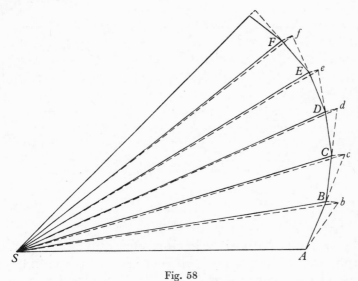

Fig. 58

De motu corporum in mediis non resistentibus

Theorema 1. Gyrantia omnia radiis ad centrum ductis areas temporibus proportionales describere.

Dividatur tempus in partes aequales, et prima temporis parte describat corpus vi insita rectam *AB* [Fig. 58]. Idem secunda temporis parte, si nihil impediret, recta pergeret ad *c*, describens rectam *Bc* aequalem ipsi *AB*, adeo ut radiis *AS*, *BS*, *cS* ad centrum actis, confectae forent areae aequales *ASB*, *BSc*. Verum ubi corpus venit ad *B*, agat vis centripeta impulsu unico at magno, faciatque corpus a recta *Bc* deflectere et pergere in recta *BC*. Ipsi *BS* parallela agatur *cC* occurrens *BC* in *C*, et, completa secunda temporis parte, corpus reperietur in *C*. Junge *SC*, et triangulum *SBC* ob parallelas *SB*, *Cc*, aequale erit triangulo *SBc*, atque adeo etiam triangulo *SAB*. Simili argumento, si vis centripeta successive agat in *C*, *D*, *E*, &c. faciens corpus singulis temporis momentis singulas describere rectas *CD*, *DE*, *EF*, &c, triangulum *SDC*, triangulo *SBC*, et *SDE* ipsi *SCD*, et *SEF* ipsi *SDE* aequale erit. Aequalibus igitur temporibus aequales areae describuntur. Sunto jam haec triangula numero infinita, et infinite parva, sic ut singulis temporis momentis singula respondeant triangula, cogente vi centripeta sine remissione, et constabit propositio.[1]

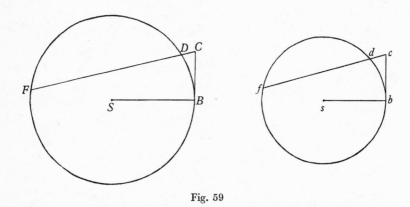

Fig. 59

Theorema 2. Corporibus in circumferentiis circulorum uniformiter gyrantibus, vires centripetas esse ut arcuum simul descriptorum quadrata applicata ad radios circulorum.

Corpora *B*, *b* in circumferentiis circulorum *BD*, *bd* gyrantia, simul describant arcus *BD*, *bd* [Fig. 59]. Sola vi insita describerent tangentes *BC*, *bc* his arcubus aequales; vires centripetae sunt quae perpetuo retrahunt corpora de tangentibus ad circumferentias, atque adeo hae sunt ad invicem ut spatia ipsis superata *CD*, *cd*; id est productis *CD*, *cd* ad *F* et *f* ut $\dfrac{BC^{\text{quad}}}{CF}$ ad $\dfrac{bc^{\text{quad}}}{cf}$, sive ut $\dfrac{BD^{\text{quad}}}{\frac{1}{2}CF}$ ad $\dfrac{bd^{\text{quad}}}{\frac{1}{2}cf}$. Loquor de spatiis *BD*, *bd* minutissimis, inque infinitum diminuendis, sic ut pro $\frac{1}{2}CF$, $\frac{1}{2}cf$ scribere liceat circulorum radios *SB*, *sb* quo facto constabit propositio.

---

[1] B reads: describens lineam *Bc*—unico sed magno—triangula, agente vi centripeta sine intermissione & constabit propositio (cf. *Principia*, Book I, Prop. I, 37).

Cor. 1. Vires centripetae sunt ut celeritatum quadrata applicata ad radios.[1]

Cor. 2. Et reciproce ut quadrata temporum periodicorum applicata ad radios.

Cor. 3. Unde si quadrata temporum periodicorum sunt ut radii circulorum, vires centripetae sunt aequales; et vice versa.

Cor. 4. Si quadrata temporum periodicorum sunt ut quadrata radiorum vires centripetae sunt reciproce ut radii.

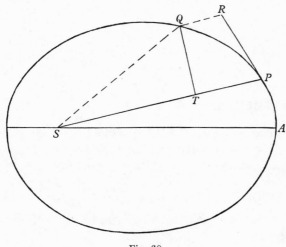

Fig. 60

Cor. 5. Si quadrata temporum periodicorum sunt ut cubi radiorum vires centripetae sunt reciproce ut quadrata radiorum.[2]

Schol. Casus Corollarii quinti obtinet in corporibus coelestibus. Quadrata temporum periodicorum sunt ut cubi distantiarum a communi centro circum quod volvuntur. Id obtinere in Planetis majoribus circa Solem gyrantibus inque minoribus circa Jovem jam statuunt Astronomi.[3]

Theorema 3. Si corpus P circa centrum S gyrando describat lineam quamvis curvam APQ [Fig. 60]; et si tangat recta PR curvam illam in puncto quovis P, et ad tangentem ab alio

---

[1] B reads: Hinc vires centripetae—ad radios circulorum.
[2] B reads: quadrata radiorum. Et vice versa. (Cf. *Principia*, Book I, Prop. IV and corols. 41–2.)
[3] B reads: circa Jovem et Saturnum.... The Scholium occurs in D and B only.

249

quovis puncto Q agatur QR distantiae SP parallela ac demittatur QT perpendicularis ad distantiam SP; dico quod vis centripeta sit reciproce ut solidum $\dfrac{SP^{\text{quad}} \times QT^{\text{quad}}}{QR}$, si modo solidi illius ea semper sumatur quantitas quae ultimo fit ubi coeunt punta P et Q.

Namque in figura indefinite parva $QRPT$, lineola QR dato tempore est ut vis centripeta, et data vi ut quadratum temporis, atque adeo neutro dato ut vis centripeta et quadratum temporis conjunctim, id est, ut vis centripeta semel, et area SPQ tempori proportionalis seu duplum ejus $SP \times QT$ bis.[1] Applicetur hujus proportionalitatis pars utraque ad lineolam QR, et fiet unitas ut vis centripeta et $\dfrac{SP^{\mathrm{q}} \times QT^{\mathrm{q}}}{QR}$ conjunctim, hoc est vis centripeta reciproce ut $\dfrac{SP^{\mathrm{q}} \times QT^{\mathrm{q}}}{QR}$. Q.E.D.

Corol. Hinc si detur figura quaevis, et in ea punctum ad quod vis centripeta dirigitur, inveniri potest lex vis centripetae, quae corpus in figurae illius perimetro gyrare faciet. Nimirum computandum est solidum $\dfrac{SP^{\mathrm{q}} \times QT^{\mathrm{q}}}{QR}$ huic vi reciproce proportionale. Ejus rei dabimus exempla in problematis sequentibus.[2]

Prob. 1. Gyrat corpus in circumferentia circuli requiritur lex vis centripetae tendentis ad punctum aliquod in circumferentia.

Esto circuli circumferentia $SPQA$, centrum vis centripetae S, corpus in circumferentia latum P, locus proximus in quem movebitur Q, ad SA diametrum et SP demitte perpendicula PK, QT, et per Q ipsi SP parallelam age LR occurentem circulo in L et tangenti PR in R. Erit RP^{q} (hoc est QRL) ad QT^{q} ut SA^{q} ad SP^{q}. Ergo $\dfrac{QRL \times SP^{\mathrm{q}}}{SA^{\mathrm{q}}} = QT^{\mathrm{q}}$. Ducantur haec aequalia in $\dfrac{SP^{\mathrm{q}}}{QR}$, et punctis P et Q coeuntibus scribatur SP pro RL, sic fiet $\dfrac{SP^{\mathrm{qc}}}{SA^{\mathrm{q}}} = \dfrac{QT^{\mathrm{q}} \times SP^{\mathrm{q}}}{QR}$. Ergo vis centripeta reciproce

---

[1] B reads: proportionalis (vel duplum ejus $SP \times QT$) bis.
[2] Cf. *Principia*, Book I, Prop. VI, 44–5.

est ut $\dfrac{SP^{qc}}{SA^q}$, id est, (ob datum SA^q) ut quadratocubus distantiae SP. Q.E.I.[1]

Schol. Caeterum in hoc casu et similibus concipiendum est, quod postquam corpus pervenerit ad centrum S, id non amplius redibit in orbem, sed abibit in tangente. In spirali quae secat radios omnes in dato angulo, vis centripeta tendens ad spiralis principium est in ratione triplicata distantiae reciproce; sed in principio illo recta nulla positione determinata spiralem tangit.

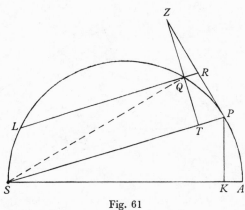

Fig. 61

Prob. 2. Corpus gyrat in Ellipsi veterum; requiritur lex vis centripetae tendentis ad centrum ellipseos.

Sunto CA, CB semiaxes Ellipseos [Fig. 62]; GP, DK diametri conjugatae; PF, QT perpendicula ad diametros; QV ordinatim applicata ad diametrum GP; et $QVPR$ parallelogrammum. His constructis, erit (ex Conicis) PVG ad QV^q ut PC^q ad CD^q, et $\dfrac{QV^q}{QT^q} = \dfrac{PC^q}{PF^q}$, et conjunctis rationibus, $\dfrac{PVG}{QT^q} = \dfrac{PC^q}{CD^q} \times \dfrac{PC^q}{PF^q}$; id est VG ad $\dfrac{QT^q}{PV}$ ut PC^q ad $\dfrac{CD^q \times PF^q}{PC^q}$. Scribe QR pro PV et $BC \times CA$ pro $CD \times PF$, necnon (punctis Q et P coeuntibus) $2PC$ pro VG, et ductis extremis et mediis in se mutuo fiet $\dfrac{QT^q \times PC^q}{QR} = \dfrac{2BC^q \times CA^q}{PC}$, est ergo vis centripeta reciproce ut

[1] B reads: tangenti PR in R. Et coeant TQ, PR in Z. Ob similitudinem triangulorum ZQR, ZTP, SPA erit RP_q (hoc est.... (Cf. *Principia* Book i, Prop. VII, 45–6.)

251

$\dfrac{2BC^q \times CA^q}{PC}$. Id est, ob datum $2BC^q \times CA^q$, ut $\dfrac{1}{PC}$. Hoc est directe ut distantia PC. Q.E.I.[1]

Prob. 3. Corpus gyrat in Ellipsi, requiritur lex vis centripetae tendentis ad umbilicum Ellipseos.

Esto ellipseos superioris umbilicus S, agatur SP secans ellipseos diametrum DK in E [et lineam QV in x, et compleatur parallelogrammum $QxPR$.][2] Patet EP aequalem esse semiaxi majori AC, eo quod, acta ab altero ellipseos umbilico H linea

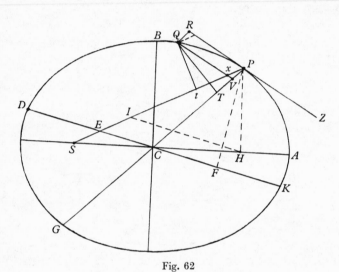

Fig. 62

HI, ipsi EC parallela, ob aequales CS, CH aequentur ES, EI adeo ut EP semisumma sit ipsarum PS, PI, id est (ob parallelas HI, PR et angulos aequales IPR, HPZ) ipsarum PS, PH quae conjunctim totum axem $2AC$ adaequant. Ad SP demittatur perpendicularis Qt, et ellipseos latere recto principali

$$\left(\text{seu } \dfrac{2BC^q}{AC} \right)$$

dicto L, erit $L \times QR$ ad $L \times PV$ ut QR ad PV, id est, ut PE (seu AC) ad PC. Et $L \times PV$ ad $GV \times VP$ ut L ad GV et $GV \times VP$ ad QV^q ut CP^q ad CD^q et QV^q ad Qx^q fiat ut m ad n. Et Qx^q ad

---

[1] B reads T for t and vice versa. C reads: QT perpendicula ad easdem, QV ordinatim.... (Cf. *Principia*, Book I, Prop. X, 48.)

[2] The words in brackets do not occur in B.

Qt^q ut EP^q ad PF^q id est, ut CA^q ad PF^q, sive ut CD^q ad CB^q. Et conjunctis his omnibus rationibus,

$$\frac{L \times QR}{Qt^q} = \frac{AC}{PC} \times \frac{L}{GV} \times \frac{CP^q}{CD^q} \times \frac{m}{n} \times \frac{CD^q}{CB^q},$$

id est ut $\dfrac{AC \times L \,(\text{seu } 2BC^q)}{PC \times GV} \times \dfrac{CP^q}{CB^q} \times \dfrac{m}{n}$ sive ut $\dfrac{2PC}{GV} \times \dfrac{m}{n}$; sed,

punctis Q et P coeuntibus, rationes $\dfrac{2PC}{GV}$ et $\dfrac{m}{n}$ fiunt aequalitatis

ergo $L \times QR$ et Qt^q aequantur. Ducatur pars utraque in $\dfrac{SP^q}{QR}$ et

fiet $L \times SP^q = \dfrac{SP^q \times Qt^q}{QR}$. Ergo vis centripeta reciproce est ut $L \times SP^q$, id est, reciproce in ratione duplicata distantiae. Q.E.I.[1]

Schol. Gyrant ergo Planetae majores in ellipsibus habentibus umbilicum in centro solis; et radiis ad solem ductis, describunt areas temporibus proportionales, omnino ut supposuit Keplerus. Et harum Ellipseon latera recta sunt $\dfrac{Qt^q}{QR}$; punctis P et Q spatio quam minimo et quasi infinite parvo distantibus.

Theor. 4. Posito quod vis centripeta sit reciproce proportionalis quadrato distantiae a centro, quadrata temporum periodicorum in Ellipsibus sunt ut cubi transversorum axium.

Sunto Ellipseos axis transversus AB, axis alter PD latus rectum L, umbilicus alteruter S [Fig. 63]. Centro S intervallo SP describatur circulus PMD. Et eodem tempore describant corpora duo gyrantia arcum ellipticum PR et circularem PM, vi centripeta ad umbilicum S tendente. Ellipsin et circulum tangant PQ et PN in puncto P. Ipsi PS agantur parallelae QR, MN tangentibus occurrentes in Q et N. Sint autem figurae PQR, PMN indefinite parvae, sic ut (per schol. Prob. 3) fiat $[L \times QR = RT^q$, et $2SP \times MN = MV^q$; ob communem a centro $S]^2$ distantiam SP, et inde aequales vires centripetas, sunt MN et QR aequales. Ergo RT^q ad MV^q est ut L ad $2SP$, et RT ad MV ut medium proportionale inter $L \times 2SP$ (seu PD) ad $2SP$.

---

[1] B reads: ...axem totem $2AC$—et QV^q ad Qx^q puta ut m ad n—et m ad n fiunt aequalitatis: Ergo et ex his composita ratio $L \times QR$ ad QT^q. Ducatur pars.... (Cf. *Principia*, Book I, Prop. XI, 50.)

[2] These words in brackets are not found in B.

Hoc est area *SPR* ad aream *SPM* ut area tota Ellipseos ad aream totam circuli. Sed partes arearum singulis momentis genitae sunt ut areae *SPR* et *SPM*, atque adeo ut areae totae, et proinde per numerum momentorum multiplicatae, simul evadent totis aequales. Revolutiones igitur eodem tempore in ellipsibus perficiuntur ac in circulis, quorum diametri sunt axibus transversis Ellipseon aequales. Sed (per Cor. 5, Theor. 2) quadrata temporum periodicorum in circulis sunt ut cubi diametrorum, ergo et in Ellipsibus. Q.E.D.[1]

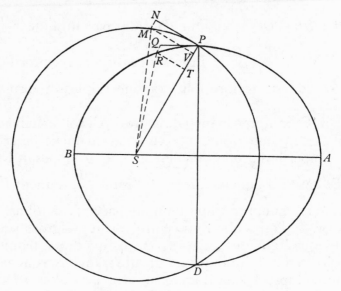

Fig. 63

Schol. Hinc in Systemate coelesti, ex temporibus periodicis Planetarum, innotescunt proportiones transversorum axium Orbitarum. Axem unum licebit assumere. Inde dabuntur caeteri. Datis autem axibus, determinabuntur Orbitae in hunc modum. Sit *S* locus Solis seu umbilicus unus ellipseos, *A*, *B*, *C*, *D*, loca Planetae observatione inventa, et *Q* axis transversus Ellipseos [Fig. 64]. Centro *A* radio *Q*−*AS* describatur circulus *FG* et erit ellipseos umbilicus alter in hujus circumferentia. Centris *B*, *C*, *D*, &c. intervallis *Q*−*BS*, *Q*−*CS*, *Q*−*DS*, &c. describantur itidem alii quotcunque

---

[1] This proposition is Prop. XV, Book i of the *Principia*, 56, but the development of a proof for the relation is quite different there.

circuli, & erit umbilicus ille alter in omnium circumferentiis, atque adeo in omnium intersectione communi *F*. Si intersectiones omnes non coincidunt, sumendum est punctum medium pro umbilico. Praxis hujus commoditas est quod ad unam conclusionem eliciendam adhiberi possint et inter se expedite comparari observationes quam plurimae. Planetae autem loca singula *A, B, C, D,* &c. ex binis observationibus, cognito Telluris orbe magno, invenire docuit Hallaeus. Si orbis ille magnus nondum satis exacte determinatus habetur, ex eo prope cognito determinabitur Planetae alicujus, puta Martis, propius, deinde ex orbita Planetae per eandem

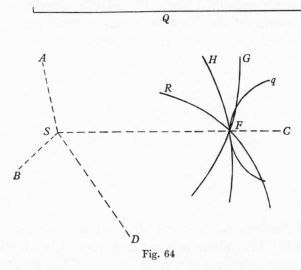

Fig. 64

methodem determinabitur orbita telluris adhuc propius. Tum ex orbita Telluris determinabitur orbita Planetae multo exactius quam prius. Et sic per vices, donec circulorum intersectiones in umbilico orbitae utriusque exacte satis conveniant.

Hac methodo determinare licet orbitas Telluris, Martis, Jovis, et Saturni: Orbitas autem Veneris et Mercurii sic. Observationibus in maxima Planetarum a Sole digressione factis, habentur Orbitarum tangentes [Fig. 65]. Ad ejusmodi tangentem *KL* demittatur a Sole perpendiculum *SL* centroque *L* et intervallo dimidii axis Ellipseos describatur circulus *KM*. Erit centrum Ellipseos in hujus circumferentia; adeoque descriptis hujusmodi pluribus circulis reperietur in omnium intersectione. Cognitis tandem orbitarum dimensionibus, longitudines horum Planetarum postmodum exactius ex transitu suo per discum solis determinabuntur.

Caeterum totum coeli Planetarii Spatium vel quiescit (ut vulgo creditur) vel uniformiter movetur in directum et perinde Planetarum commune centrum gravitatis (per Legem 4) vel quiescit vel una movetur. Utroque in casu motus Planetarum inter se (per Legem 3) eodem modo se habent, et eorum commune centrum gravitatis respectu spatii totius quiescit, atque adeo pro centro immobili Systematis totius Planetarii haberi debet. Inde vero systema Copernicaeum probatur a priori. Nam si in quovis Planetarum situ computetur commune centrum gravitatis hoc vel incidet in corpus Solis vel ei semper proximum erit. Eo Solis a centro gravitatis errore fit ut vis centripeta non semper tendat ad centrum illud immobile,

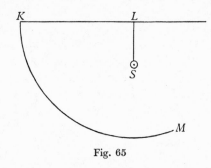

Fig. 65

et inde ut planetae nec moveantur in Ellipsibus exacte neque bis revolvant in eadem orbita. Tot sunt orbitae Planetae cujusque quot revolutiones, ut fit in motu Lunae et pendet orbita unaquaeque ab omnium Planetarum motibus conjunctis, ut taceam eorum omnium actiones in se invicem. Tot autem motuum causas simul considerare et legibus exactis calculum commodum admittentibus motus ipsos definire superat ni fallor vim omnium humani ingenii. Omitte minutias illas et orbita simplex et inter omnes errores mediocris erit Ellipsis de qua jam egi. Si quis hanc Ellipsin ex tribus observationibus per computum trigonometricum (ut solet) determinare tentaverit, hic minus caute rem aggressus fuerit. Participabunt observationes illae de minutiis motuum irregularium hic negligendis adeoque Ellipsin de justa sua magnitudine et positione (quae inter omnes errores mediocris esse debet) aliquantulum deflectere facient, atque tot dabunt

Ellipses ab invicem discrepantes quot adhibentur observationes trinae. Conjungendae sunt igitur et una operatione inter se conferendae observationes quam plurimae, quae se mutuo contemperent et Ellipsin positione et magnitudine mediocrem exhibeant.[1]

Prob. 4. Posito quod vis centripeta sit reciproce proportionalis quadrato distantiae a centro, et cognita vis illius quantitate, requiritur Ellipsis quam corpus describet de loco dato cum data celeritate secundum rectam emissum.

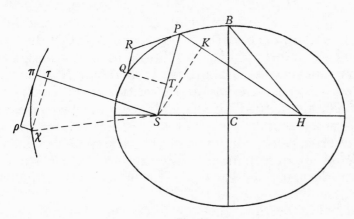

Fig. 66

Vis centripeta tendens ad punctum S ea sit quae corpus π in circulo $\pi\chi$, centro S intervallo quovis descripto, gyrare faciat [Fig. 66]. De loco P secundum lineam PR emittatur corpus P, et mox inde cogente vi centripeta deflectat in Ellipsin PQ. Hanc igitur recta PR tanget in P. Tangat itidem recta $\pi\rho$ circulum in π, sitque PR ad $\pi\rho$ ut prima celeritas corporis emissi P ad uniformem celeritatem corporis π. Ipsis SP et $S\pi$ parallelae agantur RQ et $\rho\chi$, haec circulo in χ, illa Ellipsi in Q occurrens, et a Q et χ ad SP et $S\pi$ demittantur perpendicula QT et $\chi\tau$. Est QR ad $\chi\rho$ ut vis centripeta in P ad vim centripetam in π, id est ut $S\pi^q$ ad SP^q, adeoque datur illa ratio. Datur etiam ratio QT ad RP, et ratio RP ad $\rho\pi$ seu $\chi\tau$ et inde composita ratio QT ad $\chi\tau$. De hac ratione duplicata auferatur

ratio data QR ad $\chi\rho$ et manebit data ratio $\dfrac{QT^q}{QR}$ ad $\dfrac{\chi\tau^q}{\chi\rho}$, id est

[1] This paragraph occurs only in D.

257

(per Schol. Prob. 3) ratio lateris recti Ellipseos ad dimetrum circuli. Datur igitur latus rectum Ellipseos. Sit istud L. Datur praeterea Ellipseos umbilicus S. Anguli RPS complementum ad duos rectos fiat angulus RPH et dabitur positione linea PH, in qua umbilicus alter H locatur. Demisso ad PH perpendiculo SK et erecto semiaxe minore BC, est $SP^q - 2KP \times PH + PH^q$
$$= SH^q = 4CH^q = 4BH^q - 4BC^q = (SP + PH)^{\text{quad}} - L \times (SP + PH)$$
$$= SP^q + 2SP \times PH + PH^q - L \times (SP + PH).\text{ Addantur utrobique}$$
$2KP \times PH + L \times (SP + PH) - SP^q - PH^q$, et fiet $L \times (SP + PH)$
$$= 2SP \times PH + 2KP \times PH;\text{ seu } SP + PH \text{ ad } PH \text{ ut } 2SP + 2KP$$
ad L. Unde datur umbilicus alter H. Datis autem umbilicis una cum axe transverso $SP + PH$ datur ellipsis. Q.E.I.[1]

Haec ita se habent ubi figura Ellipsis est. Fieri enim potest ut corpus moveat in Parabola vel Hyperbola. Nimirum, si tanta est corporis celeritas ut sit latus rectum L aequale $2SP + 2KP$, figura erit Parabola umbilicum habens in puncto S, et diametros omnes parallelas lineae PH. Sin corpus majori adhuc celeritate emittatur, movebitur id in Hyperbola habente umbilicum unum in puncto S, alterum in puncto H sumpto ad contrarias partes puncti P et axem transversum aequalem differentiae linearum PS et PH.[2]

Schol. Jam vero, beneficio hujus Problematis soluti, planetarum orbitas definire concessum est, et inde revolutionum tempora; et ex orbitarum magnitudine, excentricitate, apheliis, inclinationibus ad planum Eclipticae et nodis inter se collatis cognoscere an idem cometa ad nos saepius redeat. Nimirum ex quatuor observationibus locorum cometae, juxta hypothesin quod cometa moveatur uniformiter in linea recta, determinanda est ejus via rectilinea. Sit ea $APBD$, sintque A, P, B, D loca cometae in via illa temporibus observationum, et S locus solis [Fig. 67]. Ea celeritate qua cometa uniformiter percurrit rectam AD, finge ipsum emitti de locorum suorum aliquo P, et vi centripeta mox correptum deflectere a recto tramite et abire in Ellipsi $Pbda$. Haec Ellipsis determinanda est ut in superiori Problemate. In ea sunto a, P, b, d loca Cometae temporibus observationum. Cognoscantur horum locorum e terra longitudines et latitudines. Quanto majores vel minores sunt hae longitudines et latitudines observatae,

---

[1] In B some steps are omitted.
[2] Cf. the more general argument in *Principia*, Book I, Prop. XVII, 58-9.

tanto majores vel minores observatis sumantur longitudines novae. Ex his novis inveniatur denuo via rectilinea cometae, et inde via rectilinea cometae et inde via Elliptica ut prius. Et loca quatuor nova in via Elliptica, prioribus erroribus aucta vel diminuta, jam congruent cum observationibus exacte satis. Aut si forte errores etiamnum sensibiles manserint, potest opus totum repeti. Et ne computa Astronomos

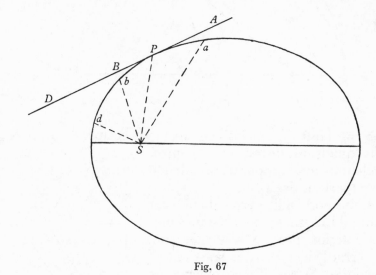

Fig. 67

moleste habeant suffecerit haec omnia per praxin geometricam determinari.

Sed areas aSP, PSb, bSd temporibus proportionales assignare difficile est. Super Ellipseos axe majore EG describatur semicirculus EHG [Fig. 68]. Sumatur angulus ECH tempori proportionalis. Agatur SH eique parallela CK circulo occurrens in K. Jungatur HK, et circuli segmento HKM (per tabulam segmentorum vel secus) aequale fiat triangulum SKN; ad EG demitte perpendiculum NQ, et in eo cape PQ ad NQ, ut Ellipseos axis minor ad axem majorem, et erit punctum P in Ellipsi, atque acta recta PS abscindetur area Ellipseos EPS tempori proportionalis. Namque area $HSNM$ triangulo SNK aucta, et huic aequali segmento HKM diminuta, fit triangulo HSK id est triangulo HSC aequale. Haec aequalia adde areae ESH, fient areae aequales $EHNS$ et EHC.

Cum igitur Sector *EHC* tempori proportionalis sit et area *EPS* areae *EHNS*, erit etiam area *EPS* tempori proportionalis.[1]

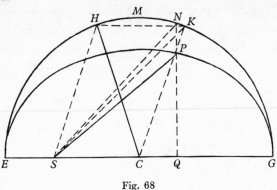

Fig. 68

Prob. 5. Posito quod vis centripeta sit reciproce proportionalis quadrato distantiae a centro, spatia definire quae corpus recta cadendo datis temporibus describit.

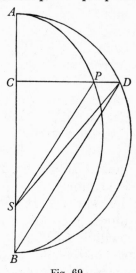

Fig. 69

Si corpus non cadit perpendiculariter describit id Ellipsin puta *APB*, cujus umbilicus inferior puta *S* congruet cum centro terrae [Fig. 69]. Id ex jam demonstratis constat. Super ellipseos axe majore *AB* describatur semicirculus *ADB*, et per corpus decidens transeat recta *DPC* perpendicularis ad axem, actisque *DS*, *PS*, erit area *ASD* areae *ASP* atque adeo etiam tempori proportionalis. Manente axe *AB* minuatur perpetuo latitudo Ellipseos, et semper manebit area *ASD* tempori proportionalis. Minuatur latitudo illa in infinitum, et Orbita *APB* jam coincidente cum axe *AB*, et umbilico *S* cum axis termino *B*, descendet corpus in recta *AC*, et area *ABD* evadet tempori proportionalis. Definietur itaque spatium *AC* quod corpus de loco *A* perpendiculariter cadendo tempore dato describit si modo tempori proportionalis capiatur area *ABD* et a

---

[1] B reads: jam vero...soluti Cometarum orbitas—sufficerit haec omnia per descriptionem linearum determinari.

puncto *D* ad rectam *AB* demittatur perpendicularis *DC*. Q.E.F.[1]

Schol. Hactenus motus corporum in mediis non resistentibus exposui; id adeo ut motus corporum coelestium in aethere determinarem. Aetheris enim puri resistentia quantum sentio vel nulla est vel perquam exigua. Valide resistit argentum vivum, longe minus aqua, aer vero longe adhuc minus. Pro densitate sua quae ponderi fere proportionalis est atque adeo (poene dixerim) pro quantitate materiae suae crassae resistunt haec media. Minuatur igitur aeris materia crassa et in eadem circiter proportione minuetur medii resistentia usque dum ad aetheris tenuitatem perventum sit. Celeri cursu equitantes vehementer aeris resistentiam sentiunt, at navigantes exclusis e mari interiore ventis nihil omnino ex aethere praeter fluente patiuntur. Si aer libere interflueret particulas corporum et sic ageret, non modo in externam totius superficiem, sed etiam in superficies singularum partium, longe major foret ejus resistentia. Interfluit aether liberrime nec tamen resistit sensibiliter. Cometas infra orbitam Saturni descendere jam sentiunt Astronomi saniores quotquot distantias eorum ex orbis magni parallaxi praeterpropter colligere novunt; hi igitur celeritate immensa in omnes coeli nostri partes indifferenter feruntur, nec tamen v l crinem seu vaporem capiti circundatum resistentia aetheris impeditum et abreptum amittunt. Planetae vero jam per annos millenos in motu suo perseverarunt, tantum abest impedimentum sentiant.

Demonstratis igitur legibus reguntur motus in coelis. Sed in aere nostro, si resistentia ejus non consideratur, innotescunt motus projectilium per Prob. 4 et motus gravium perpendiculariter cadentium per Prob. 5 posito nimirum quod gravitas sit reciproce proportionalis quadrato distantiae a centro terrae. Nam virium centripetarum species una est gravitas; et computanti mihi prodiit vis centripeta qua luna nostra detinetur in motu suo menstruo circa terram ad vim gravitatis hic in superficie terrae reciproce ut quadrata distantiarum a centro terrae quamproxime. Ex horologii oscillatorii motu tardiore in cacumine montis praealti quam in valle liquet

---

[1] In B the word 'terrae' is added in the title after 'a centro' and later crossed out; correspondingly it was struck out at the end of the first sentence. 'Gravis' was first written for 'corpus'. (Cf. *Principia*, Book I, Prop. XXXII, 115.)

etiam gravitatem ex aucta nostra a terrae centro distantia diminui, sed qua proportione nondum observatum est.

Caeterum projectilium motus in aere nostro referendi sunt ad immensum et revera immobile coelorum spatium, non ad spatium mobile quod una cum terra et aere nostro convolvitur et a rusticis ut immobile spectatur. Invenienda est Ellipsis quam projectile describit in spatio illo vere immobili et inde motus ejus in spatio mobili determinandus. Hoc pacto colligitur grave, quod de aedeficii sublimis vertice demittitur, inter cadendum deflectere aliquantulum a perpendiculo, ut et quanta sit illa deflexio et quam in partem. Et vicissim ex deflexione experimentis comprobata colligitur motus terrae. Cum ipse olim hanc deflexionem Clarissimo Hookio significarem, is experimento ter facto rem ita se habere confirmavit, deflectente semper gravi a perpendiculo versus orientem et austrum ut in latitudine nostra boreali oportuit.[1]

De motu corporum in mediis resistentibus

Prob. 6. Corporis sola vi insita per medium similare resistens delati motum definire.

Asymptotis rectangulis ADC, CH describatur Hyperbola secans perpendicula AB, DG in B, G [Fig. 70]. Exponatur tum corporis celeritas tum resistentia medii ipso motus initio per lineam AC elapso tempore aliquo per lineam DC et tempus exponi potest per aream $ABGD$ atque spatium eo tempore descriptum per lineam AD. Nam celeritati proportionalis est resistentia medii et resistentiae proportionale est decrementum celeritatis: hoc est, si tempus in partes aequales dividatur, celeritates ipsarum initiis sunt differentiis suis proportionales. Decrescit ergo celeritas in proportione Geometrica dum tempus crescit in Arithmetica. Sed tale est decrementum lineae DC et incrementum areae $ABGD$, ut notum est. Ergo tempus per aream et celeritas per lineam illam recte exponitur. Q.E.D.

Porro celeritati atque adeo decremento celeritatis propor-

---

[1] In place of this long scholium B reads: Schol. Priore Problemate definiuntur motus projectilium in aere nostro hacce motus gravium perpendiculariter cadentium ex Hypothesi quod gravitas reciproce proportionalis sit quadrato distantiae a centro terrae quodque medium aeris nihil resistat. Nam gravitas est species una vis centripetae.

C breaks off in the middle of this short scholium.

tionale est incrementum spatii descripti, sed et decremento lineae *DC* proportionale est incrementum lineae *AD*. Ergo incrementum spatii per incrementum lineae *AD*, atque adeo spatium ipsum per lineam illam recte exponitur. Q.E.D.[1]

Prob. 7. Posita uniformi vi centripeta, motum corporis in medio similari recta ascendentis ac descendentis definire.

Corpore ascendente, exponatur vis centripeta per datum quodvis rectangulum *BC* [Fig. 71] et resistentia medii initio

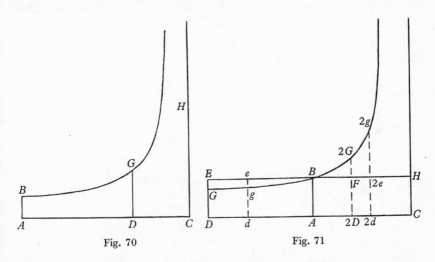

Fig. 70   Fig. 71

ascensus per rectangulum *BD* sumptum ad contrarias partes. Asymptotis rectangulis *AC*, *CH*, per punctum *B* describatur Hyperbola, secans perpendicula *DE*, *de* in *G*, *g*, et corpus ascendendo tempore *DGgd* describet spatium *EGge*, tempore *DGBA* spatium ascensus totius *EGB*, tempore *AB2G2D* spatium descensus *BF2G* atque tempore *2D2G2g2d* spatium descensus *2GF2e2g*, et celeritas corporis resistentiae medii proportionalis erit in horum temporum periodis *ABED*, *ABed*, nulla, *ABF2D*, *AB2e2d*; atque maxima celeritas quam corpus descendendo potest acquirere erit *BC*.

Resolvatur enim rectangulum *AH* in rectangula innumera *Ak*, *Kl*, *Lm*, *Mn*, &c. quae sint ut incrementa celeritatum aequalibus totidem temporibus facta et erunt *Ak*, *Al*, *Am*, *An*, &c. [Fig. 72] ut celeritates totae, et adeo ut resistentiae medii in fine singulorum temporum aequalium. Fiat *AC* ad

1 Cf. *Principia*, Book II, Prop. II, 237.

AK, vel *ABHC* ad *ABkK* ut vis centripeta ad resistentiam in fine temporis primi et erunt *ABHC, KkHC, LlHC, NnHC*, &c. ut vires absolutae quibus corpus urgetur atque adeo ut incrementa celeritatum, id est ut rectangula *Ak, Kl, Lm, Mn*, &c. & proinde in progressione geometrica. Quare si rectae *Kk, Ll, Mm, Nn*, productae occurrant Hyperbolae in *κ, λ, μ, ν*, &c. erunt area *ABκK, KκλL, LλμM, MμνN* &c. aequales, adeoque tum temporibus aequalibus tum viribus centripetis semper aequalibus analogae. Est autem area *ABκK* ad aream *Bκκ*

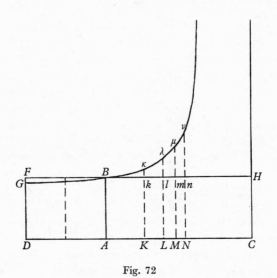

Fig. 72

ut *Kκ* ad ½*kκ* seu *AC* ad ½*AK* hoc est ut vis centripeta ad resistentiam in medio temporis primi. Et simili argumento areae *κKLλ, λLMμ, μMNν* &c sunt ad areas *κklλ, μmnν* &c ut vires centripetae ad resistentias in medio temporis secundi tertii, quarti &c. Proinde cum areae aequales *BAKκ, κKLλ, λLMμ, μMNν*, &c. sint viribus centripetis analogae, erunt areae *Bκκ, κklλ, λlmμ, μmnν* &c resistentiis in medio singulorum temporum, hoc est celeritatibus atque adeo descriptis spatiis analogae. Sumantur analogarum summae et erunt areae *Bκκ, Blλ, Bmμ, Bnν*, &c. spatiis totis descriptis analogae, nec non areae *ABκK, ABλL, ABμM, ABνN*, &c. temporibus. Corpus igitur inter descendendum tempore quovis *ABλL* describit spatium *Blλ* et tempore *LλνN* spatium *λlnν*. Q.E.D.

Et similis est demonstratio motus expositi in ascensu.
Q.E.D.[1]

Schol. Beneficio duorum novissimorum problematum inno-
tescent motus projectilium in aere nostro, ex hypothesi quod
aer iste similaris sit quodque gravitas uniformiter & secundum
lineas parallelas agat. Nam si motus omnis obliquus corporis
projecti distinguatur in duos, unum ascensus vel descensus

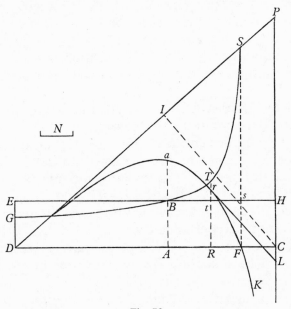

Fig. 73

alterum projectus horizontalis: motus posterior determinabitur per Problema sextum, prior per septimum, ut in hoc
diagrammate [Fig. 73].

Ex loco quovis D ejaculetur corpus secundum lineam
quamvis rectam DP, & per longitudinem DP exponatur
ejusdem celeritas sub initio motus. A puncto P ad lineam
horizontalem DC demittatur perpendiculum PC, ut et ad DP
perpendiculum CI, ad quod sit DA ut resistentia medii ipso
motus initio ad vim gravitatis. Erigatur perpendiculum AB
cujusvis longitudinis et completis parallelogrammis DABE,

---

[1] B reads: ...semper aequalibus analogae. Subducantur rectangula *Ak*, *Kl*, *Lm*, *Mn*,
&c. viribus absolutis analoga et relinquentur areae *Bkκ*, *Kκλl*, *lλμm*, *mμνn*, &c. resistentiis
medii in fine singulorum temporum, hoc est celeritatibus atque adeo descriptis spatiis
analogae. Sumantur analogarum.... (Cf. *Principia*, Book II, Prop. III, 238–40.)

CABH, per punctum *B* asymptotis *DC*, *CP* describatur Hyperbola secans *DE* in *G*. Capiatur linea *N* ad *EG* ut est *DC* ad *CP*, et ad rectae *DC* punctum quodvis *R* erecto perpendiculo *RT* quod occurrat Hyperbolae in *T* et rectae *EH* in *t*, in eo capiatur $Rr = \dfrac{DR \times DE - DRTBG}{N}$ et projectile tempore *DRTBG* perveniet ad punctum *r*, describens curvam lineam *DarFK* quam punctum *r* semper tangit, perveniens autem ad maximam altitudinem *a* in perpendiculo *AB*, deinde incidens lineam horizontalem *DC* ad *F*, ubi areae *DFsE*, *DFSBG* aequantur, et postea semper appropinquans Asymptoto *PCL*. Estque celeritas ejus in puncto quovis *r* ut curvae tangens *rL*.[1]

Si proportio resistentiae aeris ad vim gravitatis nondum innotescit, cognoscantur (ex observatione aliqua) anguli, *ADP*, *AFr*, in quibus curva *DarFK* secat lineam horizontalem *DC*. Super *DF* constituatur rectangulum *DFsE* altitudinis cujusvis, ac describatur Hyperbola rectangula ea lege ut ejus una Asymptotos sit *DF*, ut areae *DFsE*, *DFSBG* aequentur, et ut *sS* sit ad *EG* sicut tangens anguli *AFr* ad tangentem anguli *ADP*. Ab hujus Hyperbolae centro *C* ad rectam *DP* demitte perpendiculum *CI* et a puncto *B* ubi ea secat rectam *Es* ad rectam *DC* perpendiculum *BA*, et habebitur proportio quaesita *DA* ad *CI*, quae est resistentiae medii ipso motus initio ad gravitatem projectilis. Quae omnia ex praedemonstratis facile eruuntur. Sunt et alii modi inveniendi resistentiam aeris quos lubens praetereo. Postquam autem inventa est haec resistentia in uno casu, capienda est ea in aliis quibusvis ut corporis celeritas et superficies sphaerica conjunctim, (nam projectile sphaericum esse passim suppono) vis autem gravitatis innotescit ex pondere. Sic habebitur semper proportio resistentiae ad gravitatem seu lineae *DA* ad lineam *CI*. Hac proportione et angulo *ADP* determinatur specie figura *DarFK*; et capiendo longitudinem *DP* proportionalem celeritati projectilis in loco *D* determinatur eadem magnitudine sic ut altitudo *Aa* inter ascensum et casum projectilis semper sit proportionalis, atque adeo ex longitudine *DF* in agro semel

---

[1] B reads: $Rr = \dfrac{DRtE - DRTBG}{n}$. (Cf. *Principia*, Book II, Prop. IV, 241–5, where a proof is given.)

mensurata semper determinet tum longitudinem illam *DF* tum alias omnes dimensiones figurae *DarFK* quam projectile describit in agro. Sed in colligendis hisce dimensionibus usurpandi sunt logarithmi pro area Hyberbolica *DRTBG*.[1]

Eadem ratione determinatur etiam motus corporum, gravitate vel levitate et vi quacunque semel et simul impressa moventium in aqua.

TRANSLATION

On the Motion of Spherical Bodies in Fluids

Definitions

Def. 1. I call that a *centripetal force* by which a body is attracted or impelled towards some point considered as the centre.

Def. 2. And that the *force of a body*, or the *force innate in a body* by which the body endeavours to persevere in its motion along a straight line.

Def. 3. And *resistance* that which arises from a regularly impeding medium.

Def. 4. The exponents of quantities are any other quantities proportional to the ones being dealt with.

[Hypotheses]

Law 1. A body always goes on uniformly in a straight line by its innate force alone if nothing impedes it.

Law 2. A change of the state of motion or rest is proportional to the impressed force and occurs along the straight line in which the latter is impressed.

Law 3. The relative motions of bodies enclosed in a given space are the same whether that space is at rest or whether it moves perpetually and uniformly in a straight line without circular motion.

Law 4. The common centre of gravity does not alter its state of motion or rest through the mutual actions of bodies. This follows from Law 3.

---

[1] B reads: ...sic ut altitudo *Aa* maximae altitudini projectilis et longitudo *DF* longitudini horizontali inter ascensum et casum. ...

Law 5. The resistance of a medium is as the density of that medium and as the spherical surface of the moving body and its velocity conjointly.[1]

Lemmas

Lemma 1. A body describes by the action of combined forces the diagonal of a parallelogram in the same time as it would describe the sides by the action of separate forces.

If in a given time a body would be carried by the force m alone from A to B [Fig. 56], by the force n alone from A to C, let the parallelogram $ABDC$ be completed and it will be carried by both forces from A to D in the same time. For since the force m acts along the line AC parallel to BD, by Law 2, this force will not at all alter the velocity to-wards that line BD impressed by the other force. Therefore the body reaches the line BD in the same time whether the force AC is impressed or not, and so at the end of that time will be found somewhere on that line BD. By the same reasoning, it will be found on the line CD at the end of the same time, and hence it must be found at D, the meeting-point of the two lines.

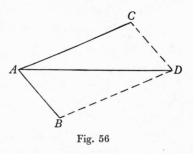

Fig. 56

Lemma 2. The space described by a body urged by any centripetal force at the beginning of its motion, is as the square of the time.

Let the times be represented by the lines AB, AD [Fig. 57]. Take Ab, Ad as proportional to the given line, and the spaces described under a uniform centripetal force are represented by the areas ABF, ADH with perpendiculars BF, DH and bounded by any straight line AFH, as Galileo explained. But let the

[1] B reads:

1. The resistance is zero in the first nine propositions, and in the following ones as the density of the medium and the velocity of the bodies together.

2. Any body progresses uniformly and infinitely along a straight line by its innate force alone, unless something external prevents it.

3. A body is brought in a given time by forces acting together to that point to which it would be brought by the separate forces acting successively in equal times.

4. The space which a body describes at the commencement of its motion, when impelled by any centripetal force, is as the square of the time.

spaces described under a non-uniform centripetal force be represented by the areas *ABC*, *ADE* included under any curve *ACE* to which the straight line *AFH* is tangent at *A*. Draw a straight line *AE*, with parallels *BF*, *bf*, *dh* meeting it in *G*, *g*, *e* and *AFH* produced will meet *bf*, *dh* in *f* and *h*. Since the area *ABC* is greater than the area *ABF*, less than the area *ABG*, and the curvilinear area *ADEC* is greater than the area *ADH*, less than the area *ADEG*, the [ratio of the] area *ABC* to the area *ADEG* will be greater than [that of] the area *ABF* to the area

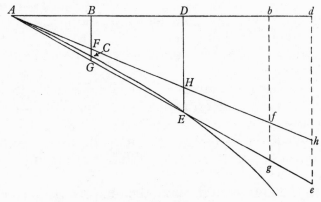

Fig. 57

ADEG, less than [that of] the area *ABG* to the area *ADH*, that is, greater than [that of] the area *Abf* to the area *Ade*, less than that of the area *Abg* to the area *Adh*. Now let the lines *AB*, *AD* decrease in their given ratio until the points *A*, *B*, *D* meet and the line *Ae* coincides with the tangent *Ah*, and so the ultimate ratios of *Abf* to *Ade* and of *Abg* to *Adh* become the same as the ratio of *Abf* to *Adh*. But this ratio is the square of that of *Ab* to *Ad* or *AB* to *AD*; therefore the ratio *ABC* to *ADEC*, intermediate between the ultimate ratios, will be the square of the ratio *AB* to *AD*; that is, the ultimate ratio of the vanishing spaces or the first of the nascent spaces is the square of the ratio of the times.

Lemma 3. Quantities proportional to their differences are continually proportional.

Let *A* be to *A−B* as *B* to *B−C* and *C* to *C−D* etc. and by division, *A* will be to *B* as *B* to *C* and *C* to *D*, etc.

Lemma 4. All parallelograms described about a given ellipse are equal.

This is evident from [the properties of] conic sections.

Theorems

1. All revolving bodies describe areas proportional to the times by means of radius-vectors drawn to the centre.

2. The centripetal forces of bodies revolving uniformly around the circumferences of circles are as the squares of the arcs described in a given time divided by the radii of the circles.

Corollary 1. The centripetal forces are as the squares of the velocities divided by the radii.

2. And reciprocally as the squares of the periodic times divided by the radii.

3. Whence, if the squares of the periodic times are as the radii of the circles, the centripetal forces are equal and vice versa.

4. And if the squares of the periodic times are as the squares of the radii, the centripetal forces are reciprocally as the radii.

5. If the squares of the periodic times are as the cubes of the radii, the centripetal forces are reciprocally as the squares of the radii.

3. That the centripetal force of a body revolving about a point is reciprocally as the solid formed from the square of the distance, multiplied by the square of the perpendicular to the distance from a point on the periphery, divided by the parallel from the same point on the periphery to the tangent.

Corollary. Hence, if any figure is given and a point in it to which the centripetal force is directed, it is possible to determine the law of centripetal force by which a body revolves in the perimeter of that figure, that is, by computing that solid reciprocally proportional to this force. Examples will be given in the following problems.

Problems

1. A body revolves in the circumference of a circle; the law of the centripetal force tending towards any point on the circumference is sought.

2. A body revolves in the ellipse of the ancients; the law of the centripetal force tending to the centre of the ellipse is sought.

3. A body revolves in an ellipse; the law of the centripetal force tending to a focus of the ellipse is sought.

Theorem 4. Suppose that the centripetal force is reciprocally proportional to the square of the distance from the centre: the squares of the periodic times about the ellipses are as the cubes of the transverse axes.

Problem 4. Suppose that the centripetal force is reciprocally proportional to the square of the distance, and the quantity of that force is known; there is required the ellipse which a body describes in moving from a given place, with a given velocity, along a given straight line.

Problem 5. Suppose that the centripetal force is reciprocally proportional to the square of the distance from the centre; to define the distances which a body falling along a straight line describes in given times.

6. To define the motion of a body borne by an innate force through a uniformly resisting medium.

7. Supposing a uniform centripetal force, to define the motion of a body ascending and descending in a straight line in a uniform medium.

By the aid of the last two problems the motions of projectiles in our air may be known, on the hypothesis that air is uniform, and that gravity acts uniformly along parallel lines.

Of the Motion of Bodies in Non-Resisting Mediums

Theorem 1. All revolving bodies describe areas proportional to the times by means of radius-vectors drawn to the centre.

Let the time be divided into equal intervals, and in the first interval let the body by its innate force describe the straight line AB [Fig. 58]. So in the second interval of time, if nothing should impede it, it would go straight on to c, describing the straight line Bc equal to AB, in such a way that when the radii AS, BS, cS are drawn to the centre the equal areas ASB, BSc will be constructed. However, when the body comes to B, let the centripetal force act with a single large impulse and make the body deflect from the straight line *Bc* and proceed along the straight line *BC*. Let *cC* be drawn parallel to *BS* meeting *BC* in *C* and, when the second interval of time is ended, the body will be found at *C*. Join *S* and *C* and the triangle *SBC* will be equal to the triangle *SBc*, because *SB* and *Cc* are parallel, and also to the triangle *SAB*. By a similar argument if the centripetal force acts successively at *C*, *D*, *E* etc. making the

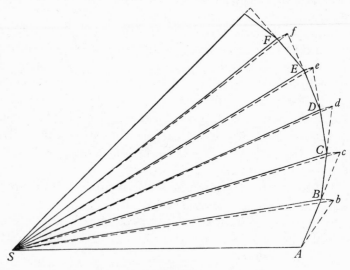

Fig. 58

body describe the straight lines *CD*, *DE*, *EF*, &c. in single moments of time, the triangle *SDC* will be equal to the triangle *SBC*, and *SDE* equal to *SCD*, and *SEF* to *SDE*. In equal times, therefore, equal areas are described. Let there be an infinite number of these triangles, each of them infinitely small, so that each triangle corresponds to a single instant of time, then with the centripetal force acting uninterruptedly the proposition will stand.

Theorem 2. The centripetal forces of bodies revolving uniformly around the circumferences of circles are as the squares of the arcs described in a given time divided by the radii of the circles.

The bodies *B*, *b*, revolving around the circumferences of the circles, *BD*, *bd* describe the arcs *BD*, *bd* in the same time [Fig. 59]. By their innate force alone they would describe the tangents *BC*, *bc* equal to these arcs. Centripetal forces are those which continually draw bodies back from the tangent to the circumference, and so these are to each other as the distances of the differences *CD*, *cd*; that is, with *CD*, *cd* produced to *F* and *f*, the forces are as BC^2/CF to bc^2/cf, or as $BD^2/\frac{1}{2}CF$ to $bd^2/\frac{1}{2}cf$. I speak of the distances *BD*, *bd* as very small, and diminishing to infinity, so that for $\frac{1}{2}CF$, $\frac{1}{2}cf$ may be written the radii of the circles *SB*, *sb*, and when this is done the proposition stands.

Corollary 1. The centripetal forces are as the squares of the velocities divided by the radii.

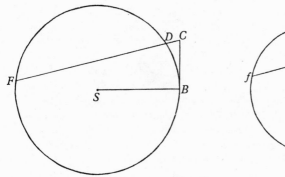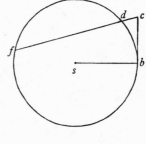

Fig. 59

Corollary 2. And reciprocally as the squares of the periodic times divided by the radii.

Corollary 3. Whence if the squares of the periodic times are as the radii of the circles, the centripetal forces are equal and vice versa.

Corollary 4. If the squares of the periodic times are as the squares of the radii, the centripetal forces are reciprocally as the radii.

Corollary 5. If the squares of the periodic times are as the cubes of the radii, the centripetal forces are reciprocally as the squares of the radii.

Scholium. The case of the fifth corollary obtains in celestial bodies. The squares of the periodic times are as the cubes of the distances from a common centre about which they revolve. That this obtains in the major planets revolving about the Sun

273

and in the minor planets about Jupiter is now accepted by Astronomers.

Theorem 3. If a body P revolving about a centre S describes any curved line APQ [Fig. 60]; and if the straight line PR is tangent to that curve at any point P, and to the tangent from any point Q is drawn QR parallel to SP, and the perpendicular QT is dropped to the line SP; I say that the centripetal force will be reciprocally as the solid $(SP^2 \times QT^2)/QR$, if the magnitude of this expression is taken as the ultimate one which it has when the points P and Q approach coincidence.

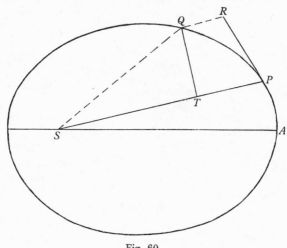

Fig. 60

For in the indefinitely small figure $QRPT$, the little line QR, if the time is given, is as the centripetal force, and if the force is given, as the square of the time, and if neither is given, as the centripetal force and the square of the time conjointly, that is, as the centripetal force taken once and the square of the area SPQ, proportional to the time, or of its double $SP \times QT$. Let each part of this ratio be divided by the little line QR and we have unity proportional to the centripetal force multiplied by $(SP^2 \times QT^2)/QR$, that is the centripetal force is as

$$\frac{QR}{SP^2 \times QT^2} \quad \text{Q.E.D.}$$

Corollary. Hence, if any figure is given, and a point on it to which the centripetal force is directed, the law of centripetal force which causes a body to révolve in the perimeter of that figure can be determined. Of course, the solid $(SP^2 \times QT^2)/QR$, reciprocally proportional to this force, must be computed. Of this we give examples in the following problems.

Problem 1. A body revolves in the circumference of a circle; the law of the centripetal force tending towards any point on the circumference is sought.

Let $SPQA$ [Fig. 61] be the circumference of a circle, S the centre of the centripetal force, P the body borne in the circumference, Q the next place to which it will move, and drop the perpendiculars PK, QT to the diameter SA and to SP; then through Q draw LR parallel to SP, meeting the circle in L and

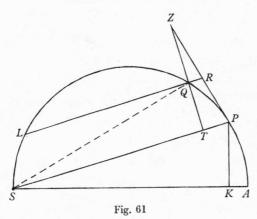

Fig. 61

the tangent PR in R. RP^2 (that is QRL) will be to QT^2 as SA^2 to SP^2. Therefore $[(QRL \times SP^2)/SA^2] = QT^2$. Multiply these equal quantities by SP^2/QR and when the points P and Q coincide we can write SP for RL and so make

$$\frac{SP^5}{SA^2} = \frac{QT^2 \times SP^2}{QR}.$$

Therefore the centripetal force is reciprocally as SP^5/SA^2, that is (since SA^2 is given) [reciprocally] as the fifth power of the distance SP. Q.E.I.

Scholium. Moreover, in this case and similar ones it is to be understood that after the body has reached the centre S, it will no longer return to the orbit but will vanish along the tangent. In a spiral which cuts all radii in a given angle, the centripetal force tending toward the beginning of the spiral is reciprocally as the third power of the distance; but in the beginning no straight line can be a tangent to the spiral in a determined position.

Problem 2. A body revolves in the ellipse of the ancients; the law of the centripetal force tending to the centre of the ellipse is sought.

Assume CA, CB to be the semiaxes of an ellipse [Fig. 62]; GP, DK conjugate diameters; PF, QT perpendicular to the diameters; QV an ordinate to the diameter GP; and $QVPR$ a

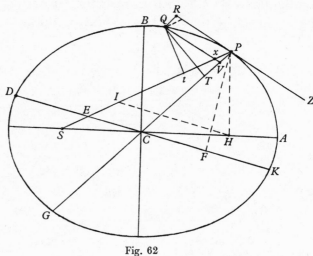

Fig. 62

parallelogram. By this construction it will follow (from the properties of conics) that PVG is to QV^2 as PC^2 to CD^2, and $\dfrac{QV^2}{QT^2} = \dfrac{PC^2}{PF^2}$ and, by joining the ratios, $\dfrac{PVG}{QT^2} = \dfrac{PC^2}{CD^2} \times \dfrac{PC^2}{PF^2}$; that is, VG to $\dfrac{QT^2}{PV}$ is as PC^2 to $\dfrac{CD^2 \times PF^2}{PC^2}$. Write QR for PV and $BC \times CA$ for $CD \times PF$, and also (since Q and P coincide) $2PC$ for VG, and multiplying the ends and middles, one will have $\dfrac{QT^2 \times PC^2}{QR} = \dfrac{2BC^2 \times CA^2}{PC}$; therefore the centripetal force is reciprocally as $\dfrac{2BC^2 \times CA^2}{PC}$. That is, because $2BC^2 \times CA^2$ is a constant, as $\dfrac{1}{PC}$. That is, directly as the line PC. Q.E.I.

Problem 3. A body revolves in an ellipse; the law of the centripetal force tending to a focus of the ellipse is sought.

Construct an ellipse with the principal focus S [Fig. 62], draw SP cutting the diameter of the ellipse DK in E [and the line QV in x, and complete the parallelogram $QxPR$]. It is clear that EP is equal to the major semiaxis AC, because, having drawn from the other focus of the ellipse H the line HI, parallel to EC, since CS, CH are equal, ES, EI will be equal, so that EP will be half the sum of PS, PI; that is (since HI, PR are parallel and IPR, HPZ are equal angles), of PS, PH which are together equal to the whole axis $2AC$. To SP drop the perpendicular Qt, and calling the principal latus rectum of the ellipse L (or $2BC^2/AC$), $L \times QR$ will be to $L \times PV$ as QR to PV; that is, as PE (or AC) to PC. And $L \times PV$ to $GV \times VP$ is as L to GV; and $GV \times VP$ to QV^2 as CP^2 to CD^2; and let QV^2 to Qx^2 be as m to n. And Qx^2 is to Qt^2 as EP^2 to PF^2, that is, as CA^2 to PF^2, or as CD^2 to CB^2. And combining all these ratios,

$$\frac{L \times QR}{Qt^2} = \frac{AC}{PC} \times \frac{L}{GV} \times \frac{CP^2}{CD^2} \times \frac{m}{n} \times \frac{CD^2}{CB^2}$$

that is, as $\dfrac{AC \times L \,(\text{or } 2BC^2)}{PC \times GV} \times \dfrac{CP^2}{CB^2} \times \dfrac{m}{n}$ or $\dfrac{2PC}{GV} \times \dfrac{m}{n}$; but since the points Q and P coincide, the ratios $\dfrac{2PC}{GV}$ and $\dfrac{m}{n}$ are equal and therefore $L \times QR$ and Qt^2 are equal. Multiply both parts by $\dfrac{SP^2}{QR}$, and $L \times SP^2 = \dfrac{SP^2 \times Qt^2}{QR}$. Therefore the centripetal force is reciprocally as $L \times SP^2$, that is, reciprocally as the square of the distance. Q.E.I.

Scholium. Therefore the major planets revolve in ellipses having a focus in the centre of the Sun; and the radius-vectors to the Sun describe areas proportional to the times, exactly as Kepler supposed. And the latera recta of these ellipses are Qt^2/QR, with the points P and Q separated as little as possible, as it were infinitely little.

Theorem 4. Suppose that the centripetal force is reciprocally proportional to the square of the distance from the centre; the squares of the periodic times about the ellipses are as the cubes of the transverse axes.

Assume AB to be the transverse axis of the ellipse, the other axis PD, the latus rectum L, one of the two foci S [Fig. 63]. From S as a centre with radius SP draw the circle PMD. And in the same time two revolving bodies describe the arc of the ellipse PR and of the circle PM, under a centripetal force tending to the focus S. PQ and PN are tangent to the ellipse and the circle in point P. QR, MN are drawn parallel to PS, meeting the tangents in Q and N. Let the figures PQR, PMN be indefinitely small so that (by the scholium to Problem 3)

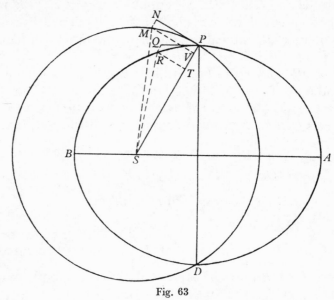

Fig. 63

[$L \times QR = RT^2$ and $2SP \times MN = MV^2$; on account of their common] distance SP from the centre S, and thence the centripetal forces are equal and MN and QR are equal. Therefore RT^2 to MV^2 is as L to $2SP$, and RT to MV as the mean proportional between $L \times 2SP$ (or PD) to $2SP$. That is the area SPR is to the area SPM as the whole area of the ellipse to the whole area of the circle. But the parts of the areas generated in single moments are as the areas SPR and SPM, and so as the whole areas, and hence when multiplied by a number of moments, they become equal to the whole areas. Therefore the revolutions are performed in the same time in ellipses and in circles whose diameters are equal to the transverse axes of the ellipses. But (by corollary 5, Theorem 2) the squares of the periodic times in circles are as the cubes of the diameters, and hence they are in ellipses. Q.E.D.

Scholium. Hence, in the celestial system, from the periodic times of the planets may be known the proportions of the transverse axes of the orbits. Let one axis be assumed. Then the others are known. From given axes, the orbits may be determined in this way. Let S be the position of the Sun or one focus of the ellipse, A, B, C, D the positions of the planet as found out by observation, and Q the transverse axis of the ellipse [Fig. 64]. Let the circle FG be described by the radius

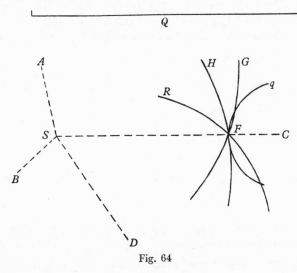

Fig. 64

$Q-AS$ from the centre A, and the other focus of the ellipse will be on its circumference. From the centres B, C, D etc. with radii $Q-BS, Q-CS, Q-DS$ etc. let there be described in the same way any number of other circles, and the other focus will lie on all the circumferences, and therefore in their common intersection F. If all the intersections do not coincide, the middle point is to be taken for the focus. The advantage of this method is that in order to reach a single conclusion many observations can be brought together and readily compared. Halley showed how to discover single positions of a planet $A, B,$ C, D from two observations, if the orbit of the Earth be known. If that orbit has not yet been accurately enough determined, from the approximate orbit that of any planet, say Mars, may be determined more nearly, and thence from the orbit of the planet by the same method may be determined the orbit of the Earth more nearly than before. Then from the orbit of the Earth may be determined the orbit of the planet much more exactly than before. And so turn and turn about until the intersections of the circles meet in the focus of both orbits exactly enough.

By this method the orbits of the Earth, Mars, Jupiter and Saturn may be determined; but the orbits of Venus and Mercury must be determined as follows. From observations made at the maximum digression of the planets from the Sun, the tangents to the orbits are obtained [Fig. 65]. To such a tangent KL is dropped from the Sun a perpendicular SL, and with the centre L and with a radius of half the axis of the ellipse

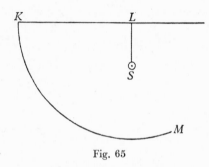

Fig. 65

the circle KM is described. The centre of the ellipse will lie on the circumference of this circle; and when several circles of this kind have been drawn it will be found at the point where all of them intersect. Finally, when the dimensions of the orbits are known, the longitudes of these planets may then be more exactly determined by their transit across the face of the Sun.

Moreover, the whole space of the planetary heavens is either at rest (as is commonly believed) or uniformly moved in a straight line, and similarly the common centre of gravity of the planets (by Law 4) is either at rest or moved at the same time. In either case the motions of the planets among themselves (by Law 3) take place in the same manner and their common centre of gravity is at rest with respect to the whole space, and so it ought to be considered the immobile centre of the whole planetary system. Thence indeed the Copernican system is proved *a priori*. For if a common centre of gravity is computed for any position of the planets, this either lies in the body of the

Sun or will always be very near it. By the displacement of the Sun from the centre of gravity it may happen that the centripetal force does not always tend to that immobile centre, and thence that the planets neither revolve exactly in ellipses nor revolve twice in the same orbit. Each time a planet revolves it traces a fresh orbit, as happens also with the motion of the Moon, and each orbit is dependent upon the combined motions of all the planets, not to mention their actions upon each other. Unless I am much mistaken, it would exceed the force of human wit to consider so many causes of motion at the same time, and to define the motions by exact laws which would allow of an easy calculation. Leaving aside these fine points, the simple orbit that is the mean between all vagaries will be the ellipse that I have discussed already. If any one shall attempt to determine this ellipse by trigonometrical computation from three observations (as is usual) he will be proceeding without due caution. For those observations will be affected by minutiae of irregular motion that I have neglected here, and so they will cause the ellipse to be somewhat displaced from its proper size and position (which should be the mean between all vagaries); as many ellipses, different from each other, may be obtained as there are sets of three observations. Therefore as many observations as possible should be combined and brought together in a single procedure so that they even out, and yield an ellipse of mean position and magnitude.

Problem 4. Suppose that the centripetal force is reciprocally proportional to the square of the distance from the centre, and the quantity of that force is known; there is required the ellipse which a body describes in moving from a given place, with a given velocity, along a given straight line.

Let the centripetal force towards the point S be that which would cause the body π to revolve in the circle $\pi\chi$, described about the centre S with any radius [Fig. 66]. The body P is projected from the place P following the line PR and immediately thereafter is deflected into the ellipse PQ by the action of the centripetal force. Therefore the straight line PR is a tangent to the ellipse at P. In the same way the straight line $\pi\rho$ is a tangent to the circle at π, and let PR to $\pi\rho$ be as the initial velocity of the projected body P to the uniform velocity of the body π. RQ and $\rho\chi$ are drawn parallel to SP and $S\pi$, the former meeting the ellipse in Q, the latter the circle in χ. And perpendiculars QT and $\chi\tau$ are dropped from Q and χ to SP and $S\pi$. QR to $\chi\rho$ is as the centripetal force in P to the centripetal force in π, that is, as $S\pi^2$ to SP^2, and so that ratio is given. The ratio of QT to RP, and the ratio of RP to $\rho\pi$ or $\chi\tau$, are also given, and thence the compound ratio of QT to $\chi\tau$. The square of this ratio is divided by the given ratio QR to $\chi\rho$ and there will remain the given ratio QT^2/QR to $\chi\tau^2/\chi\rho$, that is (by the

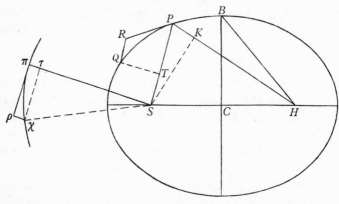

Fig. 66

scholium to problem 3), the ratio of the latus rectum of the ellipse to the diameter of the circle. Therefore the latus rectum of the ellipse is given. Let this be L. Moreover the focus of the ellipse, S, is given. The angle RPS and the angle RPH are together equal to two right angles, and these give the position of the line PH on which is located the other focus H. When the perpendicular SK is dropped to PH and the minor semiaxis BC is drawn,

$$SP^2 - 2KP \cdot PH + PH^2 = SH^2 = 4CH^2 = 4BH^2 - 4BC^2$$
$$= (SP + PH)^2 - L(SP + PH)$$
$$= SP^2 + 2SP \cdot PH + PH^2 - L(SP + PH)$$

Add to each side $2KP \times PH + L(SP + PH) - SP_2 - PH^2$, and it becomes
$$L(SP + PH) = 2SP \cdot PH + 2KP \cdot PH;$$
or
$$SP + PH : PH = 2SP + 2KP : L.$$

Whence the other focus H is given. And when the two foci are given together with the transverse axis $SP+PH$, the ellipse is given. Q.E.I.

It happens thus when the figure is an ellipse. For it may be that the body will move in a parabola or hyperbola. Indeed, if the velocity of the body is such that the latus rectum

$$L = 2SP + 2KP,$$

the figure will be a parabola having its focus in the point S and all diameters parallel to the line PH. But if the body is projected with a still greater velocity it will move in a hyperbola having one focus in the point S, the other in the point H taken on the opposite side of the point P and its transverse axis equal to the difference between PS and PH.

Scholium. Now the advantage of solving this problem is that it enables the planetary orbits to be defined, and thence their times of revolution. Further, from the magnitude, eccentricity, aphelia, inclinations to the plane of the ecliptic and nodes of orbits compared with one another we may know whether the same comet returns time and again. From four observations of the places of the comet, together with the hypothesis that comets move uniformly in straight lines, its rectilinear path is to be determined. Let this be $APBD$, and let A, P, B, D be the places of the comet in its path at the times of the observations, and S the position of the Sun [Fig. 67]. Imagine that the comet moves from some one of its places P, having the same velocity with which it uniformly traverses the straight line AD; and that by the centripetal force it is thereupon drawn inwards so as to diverge from the straight path into the ellipse $Pbda$. This ellipse is determined as in the problem above. In it, a, P, b, d are assumed to be the places of the comet at the times of the observations. The longitudes and latitudes of these places are known from the Earth. The greater or less these observed longitudes and latitudes are, the greater or less the new longitudes or latitudes assumed than the observed ones. From these new ones the rectilinear path of the comet is determined a second time, and thence the rectilinear path of the comet [*sic*] and thence its elliptical path as before. And the four new places in the elliptical path, increased or diminished by the former errors, now agree well enough with the observations.

But if perhaps sensible errors remain, the whole work can be repeated. And that the computation may not be too troublesome for astronomers, it will suffice to determine all these matters by a geometrical method.

But it is difficult to assign areas *aSP*, *PSb*, *bSd* proportional to the times. Upon the major axis *EG* of the ellipse describe

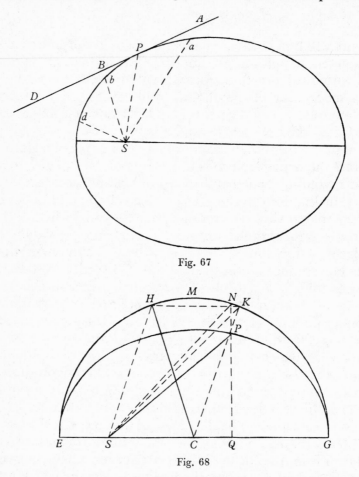

Fig. 67

Fig. 68

the semicircle *EHG* [Fig. 68]. Take the angle *ECH* proportional to the time. Draw *SH*, and *CK* parallel to it, meeting the circle in *K*. Join *HK*, and make the triangle *SKN* equal to the segment of circle *HKM* (by means of a table of segments or otherwise), drop the perpendicular *NQ* to *EG*, and in it take *PQ* to *NQ* as the minor axis of the ellipse to the major axis, and the point *P* will be on the ellipse and when the line *PS* is drawn, it will cut off the area of the ellipse *EPS* proportional to the time. Let the area *HSNM* be increased by the triangle *SNK*, and decreased by the equal segment *HKM*, to make it equal to the triangle *HSK*, that is, equal to the triangle *HSC*. To these equal areas add the area *ESH*, and the areas *EHNS* and *EHC* are equal. Since therefore the sector *EHC* is proportional to the time, and the area *EPS* to the area *EHNS*, the area *EPS* will also be proportional to the time.

Problem 5. Suppose that the centripetal force is reciprocally proportional to the square of the distance from the centre; to define the distances which a body falling along a straight line describes in given times.

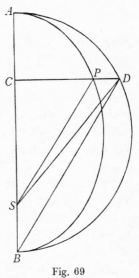

Fig. 69

If a body does not fall perpendicularly it describes an ellipse, suppose *APB*, whose inferior focus (say *S*) coincides with the centre of the Earth [Fig. 69]. That follows from what has been demonstrated already. Upon the major axis of the ellipse *AB* is described the semicircle *ADB*, and the straight line *DPC* perpendicular to the axis passes through the falling body. When *DS* and *PS* have been drawn, the area *ASD* will be proportional to the area *ASP* and so to the time as well. Keeping the axis *AB* the same let the width of the ellipse be continually diminished, and the area *ASD* will always remain proportional to the time. Let the width be infinitely diminished, and since the orbit *APB* now coincides with the axis *AB*, and the focus *S* with the end of the axis *B*, the body descends in the straight line *AC*, and the area *ABD* becomes proportional to the time. Thus the distance *AC* that the body falling perpendicularly from *A* will describe in a given time may be defined if the area *ABD* is taken as proportional to the time, and from the point *D* the perpendicular *DC* is dropped to the straight line *AB*. Q.E.F.

Scholium. Thus far I have explained the motions of bodies in non-resisting mediums, in order that I might determine the motions of the celestial bodies in the aether. For I think that the resistance of pure aether is either non-existent or extremely small. Quicksilver resists strongly, water far less, and air still less.[1] These mediums resist according to their density, which is almost proportional to their weights and hence (I may almost say) according to the quantity of their solid matter. Therefore the solid matter of air may be made less, and the resistance of the medium will be diminished nearly in the same proportion until it reaches the tenuousness of aether. Horsemen riding swiftly feel the resistance of the air strongly, but sailors on the open seas when protected from the winds feel nothing at all of the continuous flow of the aether.[2] If air flowed freely between the particles of bodies and thus acted not only on the external surface of the whole, but also on the surfaces of the single parts, its resistance would be much greater. Aether flows between very freely, and yet does not sensibly resist. All those sounder astronomers think that comets descend below the orb of Saturn, who know how to compute their distances from the parallax of the Earth's orbit, more or less; these therefore are indifferently carried through all parts of our heaven with an immense velocity, and yet they do not lose their tails nor the vapour surrounding their heads, which the resistance of the aether would impede and tear away. Planets will persevere in their motion for thousands of years, so far are they from experiencing any resistance.

Motion in the heavens, therefore, is ruled by the laws demonstrated. But if the resistance of our air is not taken into account, the motions of projectiles in it are known from Problem 4 and the motions of bodies falling perpendicularly from Problem 5, assuming indeed that gravity is reciprocally proportional to the square of the distance from the centre of the Earth. For one kind of centripetal force is gravity, and from my computations it appears that the centripetal force by which our moon is kept in its monthly motion about the Earth is to the force of gravity on the surface of the Earth reciprocally as the squares of the distances from the centre of the Earth, more or less. From the slower motion of pendulum clocks on

[1] Cf. No. 1, Section II, above, 89.
[2] The idea is not very clear. But presumably Newton meant that sailors on a ship moving with the speed of the wind do not feel an aether-draught; how this could be distinguished by sensation from a wind is not stated.

the summits of high mountains than in valleys it is clear also that gravity diminishes with increase of distance from the centre of the Earth, but in what proportion has not yet been observed.

The motions of projectiles in our air, moreover, are to be referred to the immense and indeed motionless space of the heavens, not to the moving space which is revolved along with our Earth and our air, and is naïvely regarded as immobile. The ellipse which the projectile describes in that motionless space is to be found and thence its motion in a moving space is to be determined. With this agreed, it will be gathered that the heavy body which is let fall from the top of a tall building will be deflected a little from the perpendicular in falling, so that the amount of its deflection and the direction thereof [may be determined]. And conversely the motion of the Earth may be gathered from the deflection as established by experiments. When I myself formerly communicated this deflection to the celebrated Hooke, he confirmed that it was so by an experiment three times repeated, the heavy body always deflecting from the perpendicular towards the east and south as in our northern latitude it should.[1]

Of the Motion of Bodies in Resisting Mediums

Problem 6. To define the motion of a body borne by an innate force alone through a uniformly resisting medium.

Let a hyperbola be described with rectangular asymptotes *ADC*, *CH*, cutting the perpendiculars *AB*, *DG* in *B* and *G* [Fig. 70]. Let both the velocity of the body and the resistance of the medium at the beginning of the motion be expressed by the line *AC*, and after the lapse of any period of time by the line *DC*; then the time elapsed can be expressed by the area *ABGD* and the distance passed over in that time by the line *AD*. For the resistance of the medium is proportional to the velocity and the decrease of velocity is proportional to the resistance; that is, if the time is divided into equal parts, the velocities at the beginnings of the times are proportional to their differences.

[1] B reads: The motions of projectiles in our air are defined in the former problem, and in this one the motions of heavy bodies falling perpendicularly on the hypothesis that gravity is reciprocally proportional to the square of the distance from the centre of the Earth and that the airy medium does not resist at all. For gravity is one kind of centripetal force.

Therefore the velocity decreases in geometrical proportion while the time increases in arithmetical proportion. But this is the case with the decrease of the line DC and the increase of the area $ABGD$, as is well known. Therefore the time is correctly expressed by that area and the velocity by that line. Q.E.D.

Further, the increase of the space passed over is proportional to the velocity and to the decrease of the velocity, while the increase of the line AD is proportional to the decrease of the line DC. Therefore the increase of the distance is correctly expressed by the increase of the line AD, and also the distance itself by that line. Q.E.D.

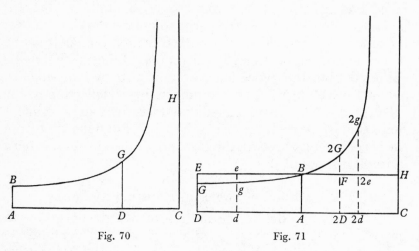

Fig. 70   Fig. 71

Problem 7. Supposing a uniform centripetal force, to define the motion of a body ascending and descending in a straight line in a uniform medium.

When the body ascends, let the centripetal force be expressed by any given rectangle BC [Fig. 71], and the resistance of the medium at the beginning of the ascent by the rectangle BD taken on the opposite side [of AB]. Let a hyperbola with the rectangular asymptotes AC, CH be described through the point B, cutting the perpendiculars DE, de in G and g; and while ascending in the time $DGgd$ the body will pass over the distance $EGge$, in the time $DGBA$ the whole distance of ascent EGB, in the time $AB2G2D$ the distance of descent $BF2G$, and in the time $2D2G2g2d$ the distance of descent $2GF2e2g$, and the velocity of the body (proportional to the resistance of the medium) will be in these periods of time *ABED*, *ABed*, zero, *ABF2D*, *AB2e2d*, and the greatest velocity which the body can acquire during its descent will be *BC*.

For let the rectangle *AH* be resolved into innumerable rectangles *Ak*, *Kl*, *Lm*, *Mn*, etc. [Fig. 72], which will be proportional to the increases of velocity made in so many equal times, and *Ak*, *Al*, *Am*, *An*, etc. will be proportional to the whole velocities, and so proportional to the resistances of the medium

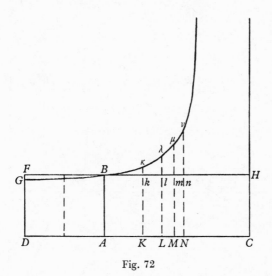

Fig. 72

at the end of each of the equal times. Let *AC* to *AK*, or *ABHC* to *ABkK*, be as the centripetal force to the resistance at the end of the first time, and *ABHC*, *KkHC*, *LlHC*, *NnHC*, etc. will be as the absolute forces by which the body is impelled and therefore as the increases of the velocities, that is, as the rectangles *Ak*, *Kl*, *Lm*, *Mn*, etc. and hence in geometrical progression. For which reason, if the straight lines *Kk*, *Ll*, *Mm*, *Nn* produced meet the hyperbola in κ, λ, μ, ν, etc. the areas *ABκK*, *KκλL*, *LλμM*, *MμνN*, etc. will be equal, and so always analogous both to the equal times and to the equal centripetal forces. But the area *ABκK* is to the area *Bkκ* as *Kκ* to $\frac{1}{2}kκ$ or *AC* to $\frac{1}{2}AK$, that is, as the centripetal force to the resistance in the middle of the first time. By a similar argument the areas $\kappa KL\lambda$, $\lambda LM\mu$, $\mu MN\nu$, etc. are to the areas $\kappa kl\lambda$, $\mu mn\nu$ etc. as the centripetal forces to the resistances in the middle of the second, third, fourth, and so on times. And hence since the equal areas $BAK\kappa, \kappa KL\lambda, \lambda LM\mu, \mu MN\nu$, etc. are analogous to the centripetal forces, the areas $Bk\kappa, \kappa kl\lambda, \lambda lm\mu, \mu mn\nu$, etc. will be analogous to the resistances in the middle of each of the times, that is, to the velocities and therefore to the distances passed over. Add the sums of the analogous quantities, and the areas $Bk\kappa$, $Bl\lambda$, $Bm\mu$, $Bn\nu$, etc. will be analogous to the whole distances passed over,

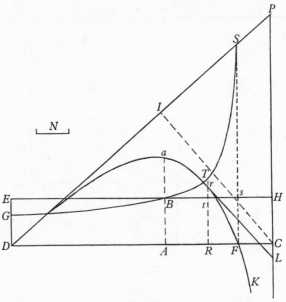

Fig. 73

and also the areas $ABkK$, $AB\lambda L$, $AB\mu M$, $AB\nu N$, etc. to the times. Therefore the body in descending during any time $AB\lambda L$ describes the distance $Bl\lambda$ and in the time $L\lambda\nu N$ the distance $\lambda ln\nu$. Q.E.D. And the demonstration for ascending motion is similar. Q.E.D.

Scholium. By the aid of the last two problems the motions of projectiles in our air may be known, on the hypothesis that this air is uniform and that gravity acts uniformly and along parallel lines. For if the oblique motion of any projected body may be analysed into two components, one of ascent or descent and the other of horizontal projection, the latter motion will be determined by Problem 6, and the former by Problem 7, as in this figure [Fig. 73].

Let the body be hurled from any place D along any straight line DP, and let its velocity at the beginning of motion be represented by the length DP. Drop the perpendicular PC from the point P to the horizontal line DC, and draw CI perpendicular to DP, and let DA to CI be as the resistance of the medium at the beginning of the motion to the force of gravity. Erect the perpendicular AB of any length and completing the parallelograms $DABE$, $CABH$, let a hyperbola with asymptotes DC, CP, be described through the point B, cutting DE in G. Take the line N to EG as DC is to CP, and at any point R on the straight line DC erect the perpendicular RT which meets the hyperbola in T and the straight line EH in t. In RT take $Rr = (DR . DE - DRTBG)/N$ and in the time $DRTBG$ the projectile arrives at the point r, describing the curved line $DarFK$ on which the point r always lies. It reaches a maximum altitude at a on the perpendicular AB, then crosses the horizontal line DC at F, where the areas $DFsE$, $DFSBG$ are equal, and afterwards continually approaches closer to the asymptote PCL. And its velocity at any point r is proportional to the tangent to the curve rL.

If the ratio between the resistance of air and the force of gravity is not yet known, the angles ADP, AFr in which the curve $DarFK$ cuts the horizontal line DC may be learnt from any observation. Let the rectangle $DFsE$ be constructed upon DF, having any altitude, and let a rectangular hyperbola be described in such a way that one of its asymptotes is DF, so that the areas $DFsE$, $DFSBG$ are equal and so that sS may be to EG as the tangent of the angle AFr is to the tangent of the angle ADP. From the centre C of this hyperbola drop the perpendicular CI to the straight line DP and also the perpendicular BA from B where the hyperbola cuts the straight line Es to the straight line DC, and the desired ratio between DA and CI, which is that of the resistance of the medium at the beginning of the motion to the gravity of the projectile, will be obtained. All of which is easily established from what has been demonstrated before. There are also other methods of discovering the resistance of the air which I gladly set aside. However, when this resistance has been found in one case, it is to be taken in any others as proportional to the velocity of the body and its spherical surface conjointly (for I suppose throughout that the projectile is spherical) while the force of gravity may be known from its weight. Thus the ratio of the resistance to the gravity, or of the line *DA* to the line *CI* will always be obtained. From this ratio and the angle *ADP*, the form of the curve *DarFK* will be determined; and by taking the length *DP* proportional to the velocity of the projectile at the place *D*, its size; so that the altitude *Aa* between the rise and fall of the projectile is always in proportion, and so from the length *DF* once measured on the ground, both that length *DF* and all other dimensions of the figure *DarFK* that the projectile describes over the ground may always be determined. But in working out these dimensions logarithms should be used in place of the hyperbolic area *DRTBG*.

By the same reasoning, the motions of bodies in water may be determined, whether they are moved by gravity or levity or any other force acting once and at the same time.

2

ON MOTION IN ELLIPSES

MS. Add. 3965, fols. 1–4

John Locke was an early reader of the *Principia*, apparently when he was still in exile in Holland, if the story that he obtained an assurance of the accuracy of its geometry from Huygens is to be trusted. He returned to England with Princess Mary in February 1689, and seems to have become acquainted with Newton within the next twelve months. Newton wrote for Locke a simpler set of demonstrations of the properties of elliptical motion, which was printed by Lord King from the copy found among Locke's papers.[1] No. 2 printed here is Newton's original holograph copy, which differs somewhat from Locke's copy; it was incompletely published by Rouse Ball.[2]

Locke's copy was endorsed 'Mr. Newton, March 1689', that is, presumably, March 1690 N.S., since it is very unlikely that Locke would have received this paper from Newton, to whom he was up to this point unknown, within a month of his return from exile.

Hypoth. 1. Bodies move uniformly in straight lines unless so far as they are retarded by the resistence of ye Medium or disturbed by some other force.

Hyp. 2. The alteration of motion is ever proportional to ye force by which it is altered.

Hyp. 3. Motions imprest in two different lines, if those lines be taken in proportion to the motions & completed into a parallelogram, compose a motion whereby the diagonal of ye Parallelogram shall be described in the same time in wch ye sides thereof would have been described by those compounding motions apart. The motions AB & AC compound the motion AD [Fig. 74].

Prop. 1

If a body move in vacuo & be continually attracted toward an immovable center, it shall constantly move in one & the same plane, & in that plane describe equal areas in equall times.

---

[1] *Life of Locke*, 2nd ed., London, 1830, I, 389–400. Prop. 2 of the version that follows is missing from this copy, and the subsequent propositions are renumbered.
[2] Rouse Ball, 116–20.

Let *A* be ye center towards wch ye body is attracted [Fig. 75], & suppose ye attraction acts not continually but by discontinued impressions made at equal intervalls of time wch intervalls we will consider as physical moments. Let *BC* be ye right line in wch it begins to move from *B* & wch it describes wth uniform motion in the first physical moment before ye attraction make its first impression upon it. At *C* let it be attracted towards ye center *A* by one impuls or impression of force, & let *CD* be ye line in wch it shall move after that impuls. Produce *BC* to *I* so that *CI* be equall to *BC* & draw *ID* parallel to *CA* & the point *D* in wch it cuts *CD* shall be ye

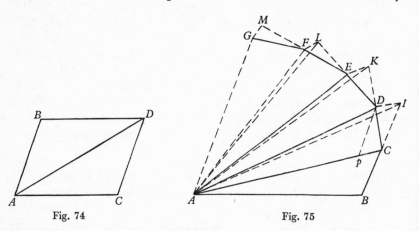

Fig. 74   Fig. 75

place of ye body at the end of ye second moment. And because the bases *BC CI* of the triangles *ABC*, *ACI* are equal those two triangles shall be equal. Also because the triangles *ACI*, *ACD* stand upon the same base *AC* & between two parallels they shall be equall. And therefore, the triangle *ACD* described in the second moment shall be equal to ye triangle *ABC* described in the first moment. And by the same reason if the body at ye end of the 2d, 3d, 4th, 5t & following moments be attracted by single impulses in *D*, *E*, *F*, *G* &c describing the line *DE* in ye 3d moment, *EF* in the 4th, *FG* in ye 5t &c: the triangle *AED* shall be equall to the triangle *ADC* & all the following triangles *AFE*, *AGF* &c to the preceding ones & to one another. And by consequence the areas compounded of these equal triangles (as *ABE*, *AEG*, *ABG* &c) are to one another as the times in wch they are described. Suppose now that the moments of time be diminished in length & encreased in number in infinitum, so yt the impulses or impressions of ye attraction may become continuall & that ye line *BCDEFG* by ye infinite number & infinite littleness of its sides *BC*, *CD*, *DE* &c may become a curve one: & the body by the continual attraction shall describe areas of this Curve *ABE*, *AEG*, *ABG* &c proportionall to the times in wch they are described. W. W. to be Dem.

Prop. 2

If a body be attracted towards either focus of an Ellipsis & the quantity of the attraction be such as suffices to make ye

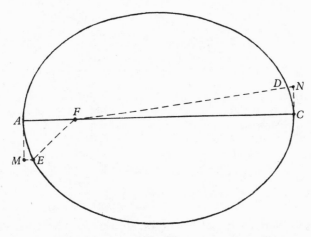

Fig. 76

body revolve in the circumference of the Ellipsis: the attraction at ye two ends of the Ellipsis shall be reciprocally as the squares of the body in those ends from that focus.

Let *AECD* be the Ellipsis [Fig. 76], *A*, *C* its two ends or vertices, *F* that focus towards wch the body is attracted, & *AFE*, *CFD* areas wch the body with a ray drawn from that focus to its center describes at both ends in equal times: & those areas by the foregoing Proposition must be equal because proportionall to the times: that is the rectangles $1/2AF \times AE$ & $1/2FC \times DC$ must be equal supposing the arches *AE* & *CD* to be so very short that they may be taken for right lines & therefore *AE* is to *CD* as *FC* to *FA*. Suppose now that *AM* & *CN*

295

are tangents to the Ellipsis at its two ends A & C & that EM & DN are perpendiculars let fall from the points E & D upon those tangents: & because the Ellipsis is alike crooked at both ends those perpendiculars EM & DN will be to one another as the squares of the arches AE & CD, & therefore EM is to DN as FC^q to FA^q. Now in the times that the body by means of the attraction moves in the arches AE & CD from A to E & from C to D it would without attraction move in the tangents from A to M & from C to N. Tis by ye force of the attractions that

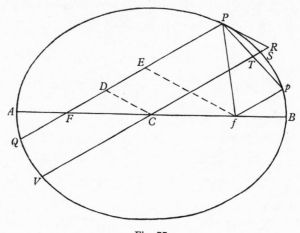

Fig. 77

the bodies are drawn out of the tangents from M to E & from N to D & therefore the attractions are as those distances ME & ND, that is the attraction at the end of the Ellipsis A is to the attraction at ye other end of ye Ellipsis C as ME to ND & by consequence as FC^q to FA^q. W.w. to be dem.

Lemma 1.

If a right line touch an Ellipsis in any point thereof & parallel to that tangent be drawn another right line from the center of the Ellipsis wch shall intersect a third right line drawn from ye touch point through either focus of the Ellipsis: the segment of the last named right line lying between ye point of intersection & ye point of contact shall be equal to half ye long axis of ye Ellipsis.

Let $APBQ$ be the Ellipsis [Fig. 77]; AB its long axis; C its center; F, f its Foci; P the point of contact; PR the tangent; CD the line parallel to the tangent, & PD the segment of the line FP. I say that this segment shall be equal to AC.

For joyn Pf & draw fE parallel to CD & because Ff is bisected in C, FE shall be bisected in D & therefore $2PD$ shall be equal to the summ of PF & PE that is to the summ of PF & Pf, that is to AB & therefore PD shall be equal to AC. W.W. to be Dem.

Lemma 2.

Every line drawn through either Focus of any Ellipsis & terminated at both ends by the Ellipsis is to that diameter of the Ellipsis wch is parallel to this line as the same Diameter is to the long Axis of the Ellipsis.

Let $APBQ$ be ye Ellipsis, AB its long Axis, F, f its foci, C its center, PQ ye line drawn through its focus F, & VCS its diameter parallel to PQ & PQ will be to VS as VS to AB.

For draw fp parallel to QFP & cutting the Ellipsis in p. Joyn Pp cutting VS in T & draw PR wch shall touch the Ellipsis in P & cut the diameter VS produced in R & CT will be to CS as CS to CR, as has been shewed by all those who treat of ye Conic sections. But CT is ye semisumm of FP & fp that is of FP & FQ & therefore $2CT$ is equal to PQ. Also $2CS$ is equal to VS & (by ye foregoing Lemma) $2CR$ is equal to AB. Wherefore PQ is to VS as VS to AB. W.W. to be Dem.

Corol. $AB \times PQ = VS^q = 4CS^q$.

Lem. 3.

If from either focus of any Ellipsis unto any point in the perimeter of the Ellipsis be drawn a right line & another right line doth touch ye Ellipsis in that point & the angle of contact be subtended by any third right line drawn parallel to the first line: the rectangle wch that subtense conteins wth the same subtense produced to the other side of the Ellipsis is to the rectangle wch the long Axis of the Ellipsis conteins wth ye first line produced to the other side of the Ellipsis as the square of the distance between the subtense & the first line is to the square of the short Axis of the Ellipsis.

Let $AKBL$ be the Ellipsis [Fig. 78], AB its long Axis, KL it's short Axis, C its center, F, f its foci, P ye point of the perimeter,

PF ye first line *PQ* that line produced to the other side of the Ellipsis *PX* the tangent, *XY* ye subtense produced to ye other side of the Ellipsis & *YZ* the distance between this subtense & the first line. I say that the rectangle *YXI* is to the rectangle *AB* × *PQ* as *YZ*ꟼ to *KL*ꟼ.

For let *VS* be the diameter of the Ellipsis parallel to the first line *PF* & *GH* another diameter parallel to ye tangent *PX*, & the rectangle *YXI* will be to the square of the tangent *PX*ꟼ as the rectangle *SCV* to ye rectangle *GCH* that is as *SV*ꟼ to *GH*ꟼ.

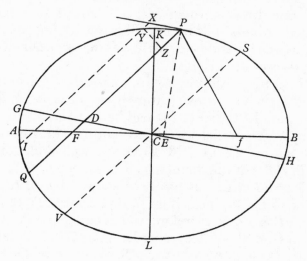

Fig. 78

This a property of the Ellipsis demonstrated by all that write of the conic sections. And they have also demonstrated that all the Parallelogramms circumscribed about an Ellipsis are equall. Whence the rectangle 2*PE* × *GH* is equal to ye rectangle *AB* × *KL* & consequently *GH* is to *KL* as *AB* that is (by Lem. 1) 2*PD* to 2*PE* & in the same proportion is *PX* to *YZ*. Whence *PX* is to *GH* as *YZ* to *KL* & *PX*ꟼ to *GH*ꟼ as *YZ*ꟼ to *KL*ꟼ. But *YXI* was to *PX*ꟼ as *SV*ꟼ that is (by Cor. Lem. 2) *AB* × *PQ* to *GH*ꟼ, Whence invertedly *YXI* is to *AB* × *PQ* as *PX*ꟼ to *GH*ꟼ & by consequence as *YZ*ꟼ to *KL*ꟼ. W.w. to be Dem.

Prop. III

If a body be attracted towards either focus of any Ellipsis &
by that attraction be made to revolve in the Perimeter of ye
Ellipsis: The attraction shall be reciprocally as the square of
the distance of the body from that focus of the Ellipsis.

Let P be the place of the body in the Ellipsis at any moment
of time [Fig. 79] & PX the tangent in wch the body would
move uniformly were it not attracted & X ye place in that
tangent at wch it would arrive in any given part of time & Y
the place in the perimeter of the Ellipsis at wch the body doth
arrive in the same time by means of the attraction. Let us
suppose the time to be divided into equal parts & that those

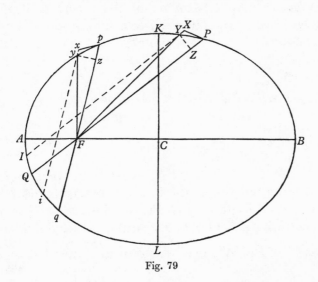

Fig. 79

parts are very little ones so yt they may be considered as
physical moments & yt ye attraction acts not continually but
by intervalls once in the beginning of every physical moment &
let ye first action be upon ye body in P, the next upon it in Y &
so on perpetually, so yt ye body may move from P to Y in the
chord of ye arch PY & from Y to its next place in ye Ellipsis in
the chord of ye next arch & so on for ever. And because the
attraction in P is made towards F & diverts the body from ye
tangent PX into ye chord PY so that in the end of the first
physical moment it be not found in the place X where it would

have been without ye attraction but in Y being by ye force of ye attraction in P translated from X to Y: the line XY generated by the force of ye attraction in P must be proportional to that force & parallel to its direction that is parallel to PF. Produce XY & PF till they cut the Ellipsis in I & Q. Joyn FY & upon FP let fall the perpendicular YZ & let AB be the long Axis & KL ye short Axis of ye Ellipsis. And by the third Lemma YXI will be to $AB \times PQ$ as YZ^q to KL^q & by consequence YX will be equall to $\dfrac{AB \times PQ \times YZ^q}{XI \times KL^q}$.

And in like manner if py be the chord of another Arch py wch the revolving body describes in a physical moment of time & px be the tangent of the Ellipsis at p & xy the subtense of the angle of contact drawn parallel to pF, & if pF & xy produced cut ye Ellipsis in q & i & from y upon pF be let fall the perpendicular yz: the subtense yx shall be equal to $\dfrac{AB \times pq \times yz^{\text{quad}}}{xi \times KL^{\text{quad}}}$. And therefore YX shall be to yx as

$$\frac{AB \times PQ \times YZ^q}{XI \times KL^q} \text{ to } \frac{AB \times pq \times yz^{\text{quad}}}{xi \times KL^q},$$

that is as $\dfrac{PQ}{XI} YZ^q$ to $\dfrac{pq}{xi} yz^{\text{quad}}$.

And because the lines $PY py$ are by the revolving body described in equal times, the areas of the triangles $PYF py F$ must be equal by the first Proposition; & therefore the rectangles $PF \times YZ$ & $pF \times yz$ are equal, & by consequence YZ is to yz as pF to PF. Whence $\dfrac{PQ}{XI} YZ^q$ is to $\dfrac{pq}{xi} yz^{\text{quad}}$ as $\dfrac{PQ}{XI} pF^{\text{quad}}$ to $\dfrac{pq}{xi} PF^{\text{quad}}$. And therefore YX is to yx as $\dfrac{PQ}{XI} pF^{\text{quad}}$ to $\dfrac{pq}{xi} PF^{\text{quad}}$.

And as we told you that XY was the line generated in a physical moment of time by ye force of the attraction in P, so for the same reason is xy the line generated in the same quantity of time by the force of the attraction in p. And therefore the attraction in P is to the attraction in p as the line XY to the line xy, that is as $\dfrac{PQ}{XI} pF^{\text{quad}}$ to $\dfrac{pq}{xi} PF^{\text{quad}}$.

Suppose now that the equal times in wch the revolving body describes the lines PY & py become infinitely little, so that the attraction may become continual & the body by this attraction revolve in the perimeter of the Ellipsis: & the lines PQ, XI as also pq, xi becoming coincident & by consequence equal, the quantities $\dfrac{PQ}{XI} pF^{\text{quad}}$ & $\dfrac{pq}{xi} PF^{\text{quad}}$ will become pF^{quad} & PF^{quad}. And therefore the attraction in P will be to the attraction in p as pF^{q} to PF^{q}, that is reciprocally as the squares of the distances of the revolving bodies from the focus of the Ellipsis. W.W. to be Dem.

3

PARTIAL DRAFT OF THE PREFACE

MS. Add. 3965, fol. 620

This paper is a holograph draft, incomplete but commencing with sentences included in the final printed Preface of the first edition of the *Principia*. It was presumably written in the spring of 1687.[1] The chief interest of this draft lies in the extended discussion of the particulate theory of matter, which Newton may have inserted here because he had already decided to omit his *Conclusion* (No. 7, below). Again, however, he changed his mind and followed the cautious policy of silence and suppression. In the printed version of the Preface the sentence ending '...vel in se mutuo impelluntur & cohaerent, vel ab invicem fugantur & recedunt' is followed simply by the remark, 'These forces being unknown, philosophers have hitherto attempted the search of Nature in vain...' and the passages relating to these forces were struck out. Newton was not yet prepared to reveal to the world what he conjectured about these forces.

The content of this draft is discussed in the Introduction to Section III. The empirical facts are identical with those of the *Conclusio* (No. 7, below) where their similarity to those later used in the *Quaeries* is noted.

This is a very much interlineated and corrected draft. There is a fair copy by an amanuensis of a part of the Preface in MS. Add. 3963, fol. 183, which differs very slightly from this.

Et huc spectant Propositiones generales in Libro primo et secundo. In Libro autem tertio exemplum hujus rei proposui in explicatione Systematis mundani. Ibi enim, ex phaenomenis caelestibus, per Propositiones in Libris prioribus mathematice demonstratas, derivantur vires gravitatis quibus corpora ad Solem & Planetas singulos tendunt ut et virium illarum quantitates & leges: deinde ex his viribus per Propositiones etiam Mathematicas deducuntur motus Plane-

[1] It is impossible to believe that Newton composed the *final* draft of this Preface as early as 8 May 1686, the date appearing beneath it in the *second* edition of the *Principia* thirty years later. At that time neither Book II nor Book III were complete in their ultimate form, and the allusions to them in the second half of the Preface could not have been made. Newton may have prepared a draft preface for Book I (the *Principia* as first contemplated) by May 1686, shortly after the book as it then was had been laid before the Royal Society, but he must have revised and enlarged it considerably at a later stage.

tarum, Cometarum, Lunae & Maris. Utinam caetera naturae Phaenomena ex Principijs Mechanicis eodem argumentandi genere derivare liceret. Nam suspicor ea omnia ex viribus quibusdam pendere quibus corporum particulae per causas nondum cognitas vel in se mutuo impelluntur & cohaerent, vel ab invicem fugantur & recedunt. Menstrua, sales, spiritus & corpora per hujusmodi vires agere in se mutuo vel non agere; promptius vel tardius congredi vel non misceri facilius vel difficilius ab invicem separari vel non cohaerere; pro vi et more congressus et cohaesionis particularum formare corpora dura, mollia, fluida, elastica, malleabilia, densa, rara, volatilia, fixa; lucem emittere, refringere, reflectare, sistere; in dissolutionibus cum impetu congredi, ex concussione particularum congredientium incalescere & particulas per caloris motum vibratorium excutere, quae si crassiores sint in vapores, exhalationes, aerem abeant, si omnium minimae in lucem; eorum particulas conjunctas ita cohaerere ut simul distillentur, sublimentur vel fixae maneant, per novos tamen coitus se mutuo saepe deserere & corpus derelictum praecipitari: Aeris, exhalationum, vaporum, Salium in aqua dissolutorum et tincturarum particulas distantes se mutuo fugere, at contiguas factas cohaerere denuo et cohaerendo in corpora compacta reverti. Corporum omnium particulas contiguas cohaerere, distantes sese saepius refugere. Ideo muscas ambulando super aquam pedes non madefacere, nec pulveres coalescere et cohaerendo corpus unum formare nisi per aquam affusam quacum componant corpus continuum, tunc autem exhalante aqua ad contactum adduci & cohaerendo corpus durum formare. Mercurium quoque in tubo Torricelliano per cohaesionem partium ad altitudinem digitorum 40 50 60 & amplius suspendi & sustineri. Denique per hujusmodi vires fieri ut particulae corporum non coacerventur pro more lapidum sed instar nivium et salium secundum figuras regulares coalescant, & ex minimis majores ex majoribus maximae per texturas retiformes formentur ex maximis autem corpora sensibilia quae et luci omnifariam pateant, et gradibus densitatis valde different inter se sicut aqua sit novendecim vicibus rarior quam aurum. nam et talium corporum particulae motu vibratorio facillime agitari possint & agitationem diutissime conservare (ut natura caloris requirit) & per talem agitationem si lenta

sit et continua situm inter se paulatim mutare & vi qua
cohaerent arctius congregari ut fit in fermentatione et vege-
tatione qua rarior aquae substantia in densiores animalium,
vegetabilium salium & lapidum substantias paulatim convertitur; sin agitatio illa satis vehemens sit per copiam excussae
lucis candere et ignem fieri. Nam carbones candentes et exhalationes candentes sunt ign[is] & flamma, et candere nihil
aliud est quam per vehementem calorem lucere ut in metallis
et lapidibus candentibus videre licet. Flamma autem non
aliter differt a carbone candente quam ignes fatui qui sunt
vapores putridi a lignis putridis lucentibus. Hi per putre-
factionis agitationem lucent, illi per agitationem caloris. Tales
operationes imaginari licet per vim duplicem quarum altera
particulas vicinas in se mutuo impellat sitque fortior sed recedendo a particulis celerius decrescat, altera imbecilior sit
sed tardius decrescat adeoque in majoribus distantiis vim
priorem superet & particulas abigat a se mutuo. Certum est
quod parti[culae] ferri nunc impelluntur in magnetem nunc
vero ab eodem fugantur idque eo fortius quo propius ad magnetem accedunt quodque corporum omnium particulae vi
gravitatis impelluntur in terram ut et leviora corpuscula vi
electrica in vitrum & succinum. Inquirendum igitur pro-
pono annon plures extent hujusmodi vires nondum animad-
versae per quas corporum particulae se mutuo agitent et in
varias texturas coalescant. Nam si Natura simplex sit et sibi
ipsi satis consona, idem erit causarum tenor in phaenomenis
universis ut quemadmodum motus majorum corporum a
majori illa vi gravitatis regantur sic etiam motus minorum a
[viri]bus quibusdam minoribus pendeant. Superest igitur ut
per experimenta commoda quaeramus an extent ejusmodi
vires in rerum natura[e] dein quaenam sint earum proprietates
quantitates et effectus. Nam si motus omnes naturales tam
minorum quam majorum corporum per ejusmodi vires explicari possint, nihil amplius restabit quam ut causas gravitatis, attractionis magneticae et aliarum virium quaeramus.

In his edendis Vir acutissimus et in omni literarum genere
eruditissimus Edmundus Halleus operam navavit, nec solum
Typothetarum sphalmata correxit & schemata incidi curavit,
sed etiam author fuit ut horum editionem aggrederer, qui cum
figuram Orbium coelestium olim a me inventam impetrare[t]

PLATE V

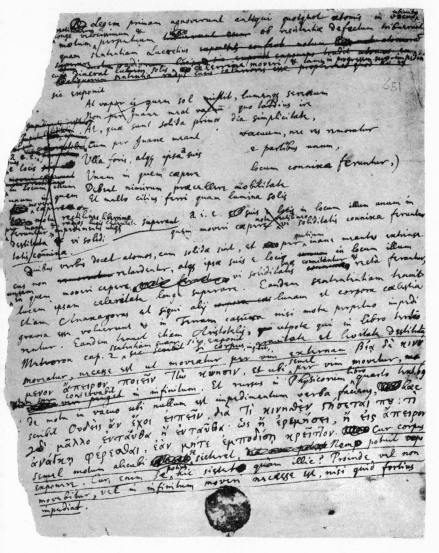

A manuscript scrap, reproduced in Section IV (no. 4). Newton has written around the hole made by tearing the seal on a letter, of which this sheet originally formed a part. The appearance of the leaf is typical of his rough drafts.

communicare coepi deque eadem in lucem edenda cogitare adhortante etiam Societate Regali: at postquam motuum Lunarium inaequalitates agressus essem, deinde etiam alia tentare caepissem quae ad leges et mensuras gravitatis et aliarum virium ad figuras a corporibus secundum datas quascunque leges attractis describendas, ad motus corporum plurium inter se, ad motus corporum in Medijs resistentibus, ad vires, densitates et motus Mediorum, ad Orbes Cometarum et similia spectant, aeditionem in aliud tempus differendam esse putavi, ut caetera rimarer et una in lucem emitterem. [In his scribendis methodum analyticam nonnunquam secutus sum idque vel quod aptior videretur (ut in Prop. VII & sequentibus quibusdam Libri primi) vel quod nonnunquam brevior esset et satis tamen perspicua.] Quae ad motus Lunares spectant, (imperfecta cum sint,) in Corollarijs Propositionis LXVI simul complexus sum, ne singula methodo prolixiore quam pro rei dignitate proponere, & sigillatim demonstrare tenerer, & seriem reliquarum Propositionum interrumpere. Nonnulla sero inventa locis minus idoneis inserere malui quam numerum Propositionum et citationes mutare. [Quaedam inserui quae cum sint pure Geometrica ad libri scopum minus spectare videantur. Talia sunt Problemata in sectione IV & V Lib. 1. Horum aliqua quidem composui ob eorum usum in describendis orbibus Planetarum et Cometarum, alia vero ob analogiam cum caeteris. Si quid in his et similibus peccatum sit, id meum esse agnosco. Quae vero recte se habent clarissimo Societatis Regalis nomini jure non minore ascribi debebunt quam olim apud Ægyptios omnia omnium inventa Mercurio uni in columnis inscribi et huic soli attribui solebant][1]

TRANSLATION

...For I suspect that all these things depend upon certain forces by which the particles of bodies, through causes still unknown, either are impelled towards one another and cohere, or repel each other and fly apart. Through forces of this kind solvents, salts, spirits and bodies either act upon one another

---

[1] The above paragraph is similar to the printed preface with the exception of the sentences in square brackets, which Newton crossed out.

or not; come together either more swiftly or more slowly; either do not mix easily or are separated from each other with difficulty or do not cohere. Depending on the force and manner of the coming together and cohering of the particles, they form bodies which are hard, soft, fluid, elastic, malleable, dense, rare, volatile, fixt; [capable of] emitting, refracting, reflecting or stopping light. In solutions, when particles rush together violently they grow hot by concussion, and by the vibratory motions of heat particles are driven off; if these particles are coarser they go off as air, vapours and exhalations; if least of all as light. The particles when joined so cohere that they can be distilled, or sublimed together or remain fixed, yet often in new associations they abandon one another and the deserted body is precipitated. When the particles of air, exhalations, vapours, salts dissolved in water, and particles of dyes are at a distance they repel each other; but when they are brought to touch each other the particles at last cohere and by cohering are turned into solid bodies again. The contiguous particles of all bodies cohere, and their distant ones frequently repel one another. And so flies walking on water do not wet their feet, and powders do not coalesce nor cohere to form a single body unless they are made as it were a continuous body by pouring water on them, when by evaporating the water they are brought into contact and form a hard body by cohesion. Mercury, also, is suspended and sustained in the Torricellian tube at a height of 40, 50, 60 or more inches by the cohesion of the parts. Lastly, such forces account for the fact that the particles of bodies do not collect together like a heap of stones, but like snow and salts coalesce into regular figures. From the very smallest particles bigger ones are formed, and from these the largest ones, all in a lattice structure. From the largest particles, then, sensible bodies are formed which allow light to pass through them in all directions, and differ markedly among themselves in density, just as water may be 19 times rarer than gold. The particles of such bodies can very easily be agitated by a vibrating motion and this agitation may last a long time (as the nature of heat requires); through such agitation, if it is slow and continuous, they little by little alter their relative arrangement, and by the force by which they cohere they are more strongly united, as happens in fermentation and the growth of plants, by which the rarer substance of water is gradually converted into the denser substance of animals, vegetables, salts and stones. If however that agitation is vehement enough, they will glow through the abundance of the emitted light and become fire. For glowing coals and glowing exhalations are fire and flame, and to glow is nothing else than to shine by vehement heat, as one may see in glowing metals and stones. Flame does not differ, however, from glowing coal otherwise than the *ignes fatui* (which are putrid vapours) do from shining putrid wood. The latter shine from the agitation due to putrefaction, the former from the agitation due to heat. Such operations may be supposed to be effected by double forces: one of these impels adjacent particles towards one another; this is the stronger but decreases more quickly with distance from the particle. The other force is weaker but decreases more slowly and so at greater distances exceeds the former force; this drives the particles away from each other. It is certain that particles of iron are sometimes impelled towards the magnet, sometimes again repelled from it, and that the more strongly the more closely they approach the magnet; and that the particles of all bodies are impelled towards the Earth by the force of gravity just as lighter bodies are impelled towards glass and amber by the electric force. I therefore propose the inquiry whether or not there be many forces of this kind, never yet perceived, by which the particles of bodies agitate one another and coalesce into various structures. For if Nature be simple and pretty conformable to herself, causes will operate in the same kind of way in all phenomena, so that the motions of smaller bodies depend upon certain smaller forces just as the motions of larger bodies are ruled by the greater force of gravity. It remains therefore that we inquire by means of fitting experiments whether there are forces of this kind in nature, then what are their properties, quantities and effects. For if all natural motions of great or small bodies can be explained through such forces, nothing more will remain than to inquire the causes of gravity, magnetic attraction and the other forces.

In publishing these things the most acute Edmond Halley, very learned in all forms of literature, has assisted the work greatly. He not only has corrected the printers' errors and taken care of the cutting of the blocks, but has been responsible for my publishing these things. For when he asked for the shape of the celestial orbits formerly discovered by me, I began to communicate and to think of publishing it; the Royal Society also offered its encouragement. But when I began upon the inequalities of the motions of the moon, and when also I began to attempt some other things that relate to the laws and measures of gravity and other forces, to the figures described by bodies attracted according to any given laws, to the motions of many bodies among themselves, to the motions of bodies in resisting mediums, to the forces, densities and motions of mediums, to the orbits of comets and the like, I thought of putting off the publication to another time, so that I might investigate the remaining matters and publish all at once.

[In writing of these matters I have sometimes adopted the analytical method, either because it seemed more suitable (as in Prop. VII and certain succeeding ones in the first Book) or because it was sometimes briefer and yet clear enough.] What pertains to the Lunar motions (imperfect as it is) I have put together in the Corollaries to Prop. LXVI, rather than be obliged to propose and demonstrate [it] in every particular by a special method more prolix than the business merits, and interrupt the series of the rest of the propositions. I have chosen to insert some things discovered late in less suitable places, rather than change the numbers of the propositions and references to them. [Certain ones I have introduced although they are purely geometrical and seem less relevant. Such are the problems in Book I, Sections IV and V. Some of these indeed I have composed on account of their usefulness in describing the orbits of planets and comets, others for their analogy with the rest. If there is any error in these propositions or similar ones, I acknowledge that the fault is mine. But if these things are correct, they ought by right to be assigned to the illustrious name of the Royal Society, just as once all discoveries of any kind used among the Egyptians to be ascribed to Mercury alone, and inscribed upon columns dedicated to him.]

4

FRAGMENT ON THE LAW OF INERTIA

MS. Add. 3970, fol. 652a

This is written on a sheet of paper blank except for an address to Newton as 'Mathematic Professor at Cambridge'.

Although he refers to Pappus in the first line of the Preface to the *Principia*, Newton did not often make historical allusions to the ideas of the ancients in his published scientific writings. It is clear that this was not from ignorance, since he was well read in the classical authorities. This passage is of special interest as indicating that Newton was prepared to find antecedents for the First Law of Motion not merely in the moderns, Galileo and Descartes, but in the ancients, Lucretius and Aristotle—an historical impulse which he later overcame.

This fragment seems to be connected with the *Principia* because of its reference to the 'legem primam': the 'laws of motion' were not so denominated by Newton before the summer of 1684.[1]

Legem primam agnoverunt antiqui quotquot atomis in vacuo infinito motum rectilineum longe velocissimum & perpetuum ob resistentiae defectum tribuerunt quam sententiam Lucretius cum dixerat lucem solis celerrime moveri & tamen in progressu suo impediri sic exponit

> At vapor is quem sol mittit, lumenque serenum
> Non per Inane meat vacuum quo tardius ire
> At, quae sunt solida primordia simplicitate,
> (Cum per Inane meant vacuum, nec res remoratur
> Ulla foris, atque ipsa[a] suis e partibus unum [una],
> Unum in quem caepere locum connixa feruntur,)
> Debent nimirum praecellere mobilitate
> Et multo citius ferri quam lumina solis

(a). i.e. suis e locis in locum illum unum in quem moveri caepere motu rectilineo vi soliditatis connixa feruntur.[2]

Quibus verbis docet atomos, cum solidae sint, et per spatium inane meantes extrinsecus non retardentur, atque

---

[1] See above, p. 243.
[2] Newton crossed out this note after incorporating its substance in the succeeding paragraph.

ipsae suis e locis in locum illum unum in quem moveri caepere, vi soliditatis suae connitantur & recta ferantur, lucem ipsam celeritate longe superare. Eandem sententiam tenuit etiam Anaxagoras et siqui alij lunam et corpora coelestia gravia esse voluerunt & in Terram casura nisi motu perpetuo impedirentur. Eandem tenuit etiam Aristotelis, utpote qui in libro tertio Meteoron cap. 2 sententiam suam sic exponit *Si corpus*, inquit, *gravitate et levitate destitutum moveatur, necesse est ut moveatur per vim externam* βίᾳ δὲ κινούμενον ἄπειρον ποεῖν τὴν κίνησιν *et ubi semel per vim movetur, motum suum conservabit in infinitum.* Et rursus in Physicorum libro quarto text. 69 de motu in vacuo ubi nullum est impedimentum verba faciens, haec scribit Οὐδεὶς ἄν ἔχοι εἰπεῖν, διὰ τὶ κινηθὲν στήσεταί που: τί γὰρ μᾶλλον ἐνταῦθα ἢ ἐνταῦθα. ὡς ἢ ἐρεμήσει, ἢ εἰς ἄπειρον ἀνάγκη φέρεσθαι, ἐὰν μήτε ἐμποδίσῃ κρεῖττον. *Cur corpus semel motum alicubi se sisteret, Nemo potuit exponere. Cur enim se potius hic sistet quam illic? Proinde vel non movebitur, vel in infinitum moveri necesse est, nisi quid fortius impediat.*

TRANSLATION

All those ancients knew the first law who attributed to atoms in an infinite vacuum a motion which was rectilinear, extremely swift and perpetual because of the lack of resistance. This is the opinion that Lucretius expresses thus, when he has said that the light of the sun moves most quickly and yet is resisted in its motion:

But that heat and that serene light which the Sun sends, does not pass through empty void; for this reason it is forced to go more slowly....[1]

But the first-beginnings, which are solid in their simplicity, when they pass through the empty void, are not hindered by anything exterior, and being units[a] though made of parts, when they are carried each to that one point towards which their first efforts tend, they must most certainly excel in swiftness, and be borne much more quickly than the light of the sun....[2]

[1] Lucretius, *De rerum natura*, II, 150–1. Lucretius thought that the light had to beat its way through waves of air.

[2] *Ibid.* 157–62. Our translation is based on that of Cyril Bailey, *Titi Lucretii de Rerum Natura* (Oxford, 1947), I, 245.

(a) i.e. they are carried from their places towards that one place towards which their first efforts tend with a rectilinear motion arising from the force of solidity.

In which words he teaches that atoms, as they are solid, and not retarded externally in passing through the empty vacuum, and strive and are borne straightly by the force of their solidity from their places towards that place to which their first efforts tend, far exceed light itself in velocity. Anaxagoras held the same opinion too, and all those others who thought that the Moon and celestial bodies were heavy and would fall towards the Earth, unless they were prevented by their perpetual motion. Aristotle was of the same mind, since he expresses his opinion thus in the third book of the *Meteors*, chapter 2: *If a body,* he says, *destitute of gravity and levity, be moved, it is necessary that it be moved by an external force* [Moreover if there is to be a moving body which is neither light nor heavy, its motion must be enforced and it must perform this enforced motion to infinity];[1] *And when it is once moved by a force, it will conserve its motion indefinitely.* And again in Book IV of the *Physics*, text 69, speaking of motion in the void where there is no impediment he writes [Nor if it did move could a reason be assigned why the projectile should ever stop—for why here rather than there? It must therefore either not move at all, or continue its movement without limit, unless some stronger force impedes it]. *Why a body once moved should come to rest anywhere no one can say. For why should it rest here rather than there? Hence either it will not be moved, or it must be moved indefinitely, unless something stronger impedes it.*[2]

---

[1] Newton's reference is wrong: the quotation is from *On the Heavens*, Book III, chapter II, 301 b. The version in brackets is that of W. K. C. Guthrie, from the Loeb Library edition; Newton's would seem more precise.

[2] *Physics*, Book IV, chapter VIII, 215 a 19. We have printed in brackets the translation of P. H. Wickstead and F. M. Cornford, from the Loeb Library edition.

DRAFT ADDITION TO THE *PRINCIPIA*

Book III, Proposition VI, Corollaries 4 and 5

MS. Add. 4005, fols. 28–9

This draft clearly has no connexion with the *Conclusion* of the *Principia* (No. 7) nor with the passages on comets in Book III. Yet it belongs to the *Principia* and seems intended to form the fourth and fifth corollaries to some proposition therein. It would appropriately follow the third corollary to Book III, Prop. VI where (in the first edition) Newton discussed the vacuity of the celestial spaces. In the second edition he added to the third corollary (which corresponds to the fourth corollary of the first edition) the sentence: 'And if the quantity of matter in a given space can, by any rarefaction, be diminished, what should hinder a diminution to infinity?' This is the very question discussed at length in this draft.

Clearly it was written after the first publication of the *Principia*; and the mention of Fatio de Duillier suggests a date not long after the publication. Fatio's hypothesis for the explanation of gravity (cf. p. 205) was first examined by Newton and Halley at Gresham College on 19 March 1690,[1] and even three years later Newton was still impressed with his work. He proposed to make Fatio an allowance to enable him to reside in Cambridge, probably to facilitate the preparation of a second edition of the *Principia* as Fatio then thought of doing, with Newton's approval.[2] This draft may therefore be assigned to the period of the 1690's when Newton held so high an opinion of Fatio's ability, and especially of his explanation of gravity.

pervolent non agent in materiam centralem eodem impetu quo in superficialem feruntur. Errant igitur qui corporum particulas minimas ad modum particularum arenae aut lapidum coacervatorum confertim jungunt. Si particulae aliquae tam dense constipentur, causa gravitans minus aget in interiores quam in exteriores et sic gravitas desinet esse proportionalis materiae. Excogitandae sunt aliae particularum

[1] Cf. Bernard Gagnebin, 'De la Cause de la Pesanteur. Mémoire de Nicolas Fatio de Duillier Présenté à la Royal Society le 26 Février 1690', *Notes and Records of the Roya Society*, **6**, 1949, 117.

[2] Rouse Ball, 125.

texturae quibus interstitia earum reddantur amplissima. Et hae sunt necessariae conditiones Hypotheseos per quam gravitas explicetur mechanice. Hujus autem generis Hypothesis est unica per quam gravitas explicari potest, eamque Geometra ingeniosissimus D. N. Fatio primus excogitavit. Et vacuum ad ejus constitutionem requiritur cum particulae tenuiores motibus et rectilineis et longe rapidissimis et uniformiter continuatis in omnes partes ferri debeant neque resistentiam minime sentire nisi ubi in particulas crassiores impingunt. Vacuum igitur cum ex hac Hypothesi aeque ac ex tertia et quarta sequatur omnino dabitur.

Idem concluditur ex motu projectilium. Nam si verbi gratia materia aliqua subtilis in particulas quam minimas divisa uniformiter spargeretur per caelorum spatia vacua et eorum partem tantum millesimam implerent; corpora autem Planetarum aut Cometarum aut aliorum globorum solida essent et poris omnino destituta: haec (per Prop [XL] Lib II et ejus Corol 2)[1] millesimam partem motus sui prius amitterent quam longitudinem trium semidiametrorum describerent. Et rationi consentaneum est quod in coelis plenioribus majorem partem motus sui amitterent in ratione plenitudinis, atque adhuc majorem si non essent solida corpora. Unde Cometae qui per caelos Planetarum (ut posthac ostendetur) in omnes partes trajiciuntur motum suum cito amitterent, et globi plumbei e tormentis bellicis explosi citissime sisterentur nisi spatia caelorum et aeris prope vacua existerent.

Sophisma illud vulgare quod ex natura corporum in extensione posita contra vacuum adduci solet, nil moror; cum corpora non tam extensio sint quam extensa, et ab extensione per soliditatem suam et mobilitatem & vim resistendi ac duritiem omnino distinguantur. Nam dura sunt omnia corpora prima seu minima ex quibus reliqua componuntur.

Corol. 5. Corpora longe rariora esse quam credi solet jam ante insinuavimus et ex hac etiam Propositione consequitur. Nam cum aqua novendecim vicibus levior sit quam aurum sibi magnitudine aequale (ut notum est) haec erit etiam rarior in eadem ratione, adeoque si aurum omnino solidum esset, aqua in massam solidam condensata evaderet novendecim

---

[1] Newton neglected to fill in the number of the Proposition.

vicibus minor quam prius adeoque octodecim haberet partes vacui ad unam materiae. Aurum vero solidum non est sed poris abundat. Ab ingredientibus enim aquis acidis dissolvitur, et argentum vivum per poros ejus ad partes omnes interiores facile penetrat easque ad centrum usque dealbat. Sed et aqua pluvialis in sphaera vel aurea vel plumbea vehementer compressa permeat poros metalli et per sphaeram adhuc integram et fissuris omnino vacantem percolatur et undique exudat, ut ex testibus oculatis et fide dignis audivi. Quinimo tanta est pororum copia in auro ut actio magnetis in ferrum per aurum interpositum liberrime et absque diminutione sensibili propagatur. Aurum vero ex particulis majoribus et has ex minoribus constare credendum est et liquores jam dictos interstitia majorum permeare et magnetis effluvia per ea quoque minorum transire. Si proportio particularum utriusque generis ad ipsarum interstitia assumatur eadem quae est arenae ad ejus loca vacua hoc est quae est numeri 7 ad numerum 6; vacuum in auro erit ad ejus materiam ut quinque ad duo circiter et vacuum in aqua ad ejus materiam ut 65 ad 1. Et tamen partes aquae ita contexuntur inter se ut per compressionem condensari nequeant. Porro cum aer in nostris regionibus sit nongentis vicibus levior quam aqua, ejus materia solida vix implebit partem quinquagesimam millesimam spatij per quod dilatatur. In regionibus autem superioribus erit adhuc rarior in infinitum ut ex Prop XXII Lib II manifestum est. Haec est corporum raritas minima quam fingere licet, at maximam definire perdifficile est. Siquis enim Hypothesin excogitaverit qua corpus aquae tam rarum esse possit ut partibus suis solidis vix impleat partem sexagesimam magnitudinis suae, et tamen compressioni minime cedat: hic facile intelliget quomodo aurum aeque rarum esse possit atque aquam esse probavimus et quomodo raritas aquae et aeris augeatur in eadem ratione; sed et raritatem auri et corporum omnium pro lubitu augebit. Certe raritas aquae per vulgares Hypotheses explicari nequit. Confugiendum est ad mirabilem aliquam et valde artificiosam particularum texturam qua corpora omnia more retium effluvijs magneticis et radijs lucis quaqua versum pateant et transitum liberrimum praebeant: et per talem hypothesin potest raritas corporum pro lubitu augeri. Salia inter congelandum figuras inducunt regulares et eorum aliqua ut nitrum et Sal Ammoniacum in ramos semper abeunt. Quidni prima rerum omnium semina vi naturae in figuras retiformes Geometrice coeanti

(ends)

TRANSLATION

. . . they do not act towards the central matter with the same force as that with which they are borne towards the surface. They are mistaken therefore who join the least particles of bodies together in a compact mass like grains of sand or a heap of stones. If any particles were pressed together so densely, the gravitating cause would act less towards the interior ones than towards the exterior ones and thus gravity would cease to be proportional to the [quantity of] matter. Other textures of the particles must be devised by which their interstices are rendered more ample. And these are the necessary conditions of an Hypothesis by which gravity is to be explained mechanically. The unique hypothesis by which gravity can be explained is however of this kind, and was first devised by the most ingenious geometer Mr. N. Fatio. And a vacuum is required for its operation since the more tenuous particles must be borne in all directions by motions which are rectilinear and very rapid and uniformly continued and these particles must experience no resistance unless they impinge upon denser particles. Therefore the vacuum is to be completely taken for granted since it follows from this hypothesis and equally from the third and fourth.

The same conclusion follows from the motion of projectiles. For if (for example) a certain subtle matter divided into least particles were uniformly scattered through the empty spaces of the heavens and filled as much as their thousandth part; and if the bodies of Planets or Comets or other globes were solid and destitute of all pores: these (by Prop. [XL] Book II and Corol. 2) would lose a thousandth part of their motion before they would describe the length of three [of their] semidiameters.[1] And it is reasonable that in a heaven more filled with matter they would lose a larger part of their

[1] Cf. the different formulation of *Opticks*, *Quaery* 28, 368.

motion, in proportion to the density [of the celestial matter], and even more if they were not solid bodies. Whence Comets, which pass through the planetary heavens (as may be shown later) in all directions would soon lose their motions, and balls of lead shot from guns would very quickly stop unless the spaces of the heaven and of air were nearly vacuous.

I am not at all disturbed by that vulgar sophism by which inferences opposed to the concept of the vacuum are drawn from the nature of bodies as extension; since bodies are not so much extension as extended, and they are utterly distinguished from extension by their solidity, mobility, force of resistance and hardness. For all first or least bodies, from which other bodies are made, are hard.[1]

Corol. 5. That bodies are more rare by far than is usually believed we have previously suggested and the same follows from this proposition as well. For as water is 19 times lighter than gold of the same volume (as has been remarked), it will also be more rare in the same proportion, and so if gold were wholly solid, water condensed in a solid mass would prove to be 19 times less than before and thus it would have 18 parts of vacuum to one of matter.[2] Yet gold is not solid; it abounds in pores. For it is dissolved by penetrating acid waters, and quicksilver entering through its pores easily pierces its interior parts and whitens them to the very centre. And rain-water enclosed in a sphere of gold or lead and strongly compressed permeates the pores of the metal and percolates the sphere even though it is whole and quite free from cracks, and exudes everywhere, as I have heard from credible eye-witnesses. Truly, so great is the abundance of pores in gold that the action of a magnet on iron is propagated through intervening gold very freely and without sensible diminution. Gold indeed must be believed to consist of larger particles and these of smaller ones; and the aforesaid liquids must be supposed to permeate the interstices between the larger ones, and magnetic effluvia to pass through the interstices between the

[1] Cf. *Opticks*, *Quaery* 31 (400) and *Principia*, second edition, Book III, Regula III (358).

[2] Newton seems to have believed that all the ultimate particles of matter—whether in gold or water—were of the same density. (Cf. *Principia*, second edition, Prop. VI, Corol. IV, 368.) Hence if water is 19 times less dense than gold, for equal volumes the mass of the least particles in water is 19 times less than in gold, and their volume less in the same proportion. Hence $\frac{18}{19}$th of water must be vacuum, if gold be completely solid matter.

smaller ones. If the proportion of the particles of both kinds to their respective interstices is assumed to be as that between [the particles of] sand and the vacant spaces between them, which is as 7 to 6, the vacuum in gold will be to its matter as 5 to 2 roughly and the vacuum in water to its matter as 65 to 1. And yet the parts of water are so arranged among themselves that they cannot be condensed by compression. Moreover, since air in our regions is 900 times lighter than water, its solid matter will scarcely fill the 50,000th part of the space through which it is spread. In higher regions it will be more rare still *ad infinitum*, as is manifest from Prop. XXII, Book II. This is the least rarity of bodies that may be postulated; to define the greatest rarity is extremely difficult.[1] For if anyone shall devise an Hypothesis by which the body of water can be so rare that its solid parts scarcely fill the sixtieth part of its own bulk, and yet does not at all yield to compression, then he may easily understand how gold can be as rare as we proved water to be, and how the rarity of water and air may be increased in the same proportion; but the rarity of gold and of all bodies he will increase at will. Certainly the rarity of water cannot be explained on common hypotheses. One must have recourse to a certain wonderful and exceedingly artificial texture of the particles of bodies by which all bodies, like networks, allow magnetic effluvia and rays of light to pass through them in all directions and offer them a very free passage: and by such an hypothesis the rarity of bodies may be increased at will. Salts form regular figures in congelation and certain of these such as nitre and sal ammoniac are always transformed into branches. Why should it not be that the first seeds of all things coming together geometrically in net-like figures by the force of nature . . . .

[1] Cf. *Opticks*, *Quaery* 28, 366–7.

6

[ATMOSPHAERA SOLIS]

MS. Add. 4005, fol. 69

This unfinished paragraph was drafted upon a begging letter (one of several thus accidentally preserved). The letter is as follows:

Address: Sir Isaac Newton at Leicester Fields

Honoured Sr
    I hope your goodness will pardon this my bouldness and to lett your good honour know that I am poor James Newton that is latly come out of sickness and humbly begs for godallmity sake as your goodness has saved my life heither too for to signe this note and Sr John Newton will give me some Relief and If Ever I trouble againe I will suffer death
        and in so Doing your poor petitioner in duty bound
        will Ever pray
                                James Newton

Newton moved to Leicester Fields in 1710; the scrap may therefore be associated with the preparation of the second edition of the *Principia*. From the final broken sentence it appears to have been intended for Book III, 505, where the words, 'Ad ascensum vaporum conducit etiam...' appear, with a reference to the immobility or slow rotation of the Sun's atmosphere.

Solem vero ingente Atmosphere cingi patet ex solis eclypsibus, in quibus Luna ubi solem totum teget apparet ut circulus niger: corona splendente, Halonis ad instar, cinctus. Coronae limbus interior ubi circulum lunae contingit lucidissimus est. Minus quidem splendet quam Sol ipse at nubes splendidissimas luce superat. Et quo remotior est haec lux a circulo Lunae eo minus splendet perviendo distantiam graduum plusquam duorum vel trium a centro Lunae antequam cessat. Cessat vero gradatim sic ut terminus ejus exterior definiri nequeat et per totum Lunae circuitum ejusdem apparet latitudinis coloris et splendoris. Totum vero caelum luce sua non minus illuminat quam solet aurora paulo ante ortum Solis ubi stellae fixae magnitudinis secundae jam modo videri desinunt.

Concipe Atmosphaeram Solis non ibi cessare ubi videri desinit, sed aura tenuiore ad orbem usque Mercurii et longe ultra ascendere. Ad ascensum vaporum conducit etiam

TRANSLATION

That the Sun is indeed surrounded by a huge Atmosphere appears from eclipses of the Sun, in which the Moon where it covers the whole Sun appears as a black circle, surrounded by a shining corona like a halo. The interior limb of the corona where it touches the circle of the Moon is most brilliant. It shines less than the Sun itself, but its light exceeds that of the brightest clouds. And the more distant this light is from the circle of the moon, the less brilliant it is, reaching a distance of more than two or three degrees from the centre of the Moon before it stops. It stops gradually so that its exterior limit cannot be defined and around the whole circle of the Moon it appears of the same breadth, colour and brightness. Indeed, it no less illuminates the whole sky by its light than does the dawn a little before the rising of the Sun when fixed stars of the second magnitude are just becoming invisible. Imagine that the atmosphere of the Sun does not end where it ceases to be visible but that it extends as far as the orb of Mercury and far beyond as a more tenuous medium. It is also conducive to the ascent of vapours. . . .

7

CONCLUSIO

MS. Add. 4005, fols. 25–8, 30–7

The suppressed *Conclusio*, originally intended for publication in the first edition of the *Principia*, is one of the most fascinating of Newton's manuscripts. There are two versions: the first (fols. 25–8) consists of five sheets in Newton's hand; the second (fols. 30–7) is a fresh copy in the hand of an amanuensis but with corrections and additions by Newton himself.

The holograph sheets are much written over, amended and re-drafted as was usual with Newton at work. Fragments are written in all directions wherever a blank space remained on the paper. The corrections in the holograph are faithfully followed in the clean copy, but there are some passages in the holograph not found in the copy, and others in the copy not found in the holograph. We have given everything of substance here from both versions, but have not attempted to note every small variation in phrase between them, following the copy in general.

Obviously both versions belong to the period when Newton was writing the *Principia*, presumably to its latest stages in the spring of 1687. Newton had his ideas and examples to hand—had had them for many years—and it could not have taken him long to assemble them in this form (cf. above, Introduction to Section III). The copy version ends with an incomplete draft of the first twelve lines of the Preface to the first edition, almost identical with the printed version, struck out with a single stroke of the pen. (These we have printed but not translated.) Moreover, the holograph also contains passages written later than the main body of the *Conclusio* (for they are written in blank spaces around it) which were later printed in the *Principia*: these passages belong to the latter part of Book III, and are to be found on pp. 504–7 of the first edition. (As differences in phraseology from the printed text are insignificant, we have not reprinted them here.) The inference must be that Newton drafted his *Conclusio* before he had finished composing the final version of Book III. The complete copy of Book III was dispatched to Halley in London before the end of March 1687;[1] it must have been completed at great speed, since Halley was not told that there was to be a third book until the beginning of the month.[2]

[1] Halley to Newton, 5 April 1687: Rouse Ball, 173.
[2] Halley to Newton, 7 March 1687: *ibid.* 172. Cf. above p. 235.

However, the *Conclusio* itself never reached Halley or the printer, and apparently it was never completed, though Newton thought enough of it to take the trouble of having a fair copy made of as much as he had written. Both the holograph and the copy end inconclusively, yet nothing seems to have been lost from them; presumably for some reason Newton abandoned them just short of completion. There is no sign, indeed, that Halley ever heard of the *Conclusio*. Newton did mention it to Roger Cotes, in a letter quoted above, p. 186. Not until the final paragraph was added to the General Scholium in the second edition was there any public hint of the note on which the *Principia* was originally to have ended.

We have discussed the content of the *Conclusio* in the Introduction to Section III.

Hactenus explicui Systema hujus Mundi aspectabilis quoad motus majores qui facile sentiri possunt. Sunt autem alij motus locales innumeri qui ob parvitatem corpusculorum moventium, sentiri nequeunt, uti motus particularum in corporibus calidis, in fermentantibus, in putrescentibus, in vegetantibus, in organis sensuum et similibus. Hos omnes siquis feliciter aperuerit, naturam prope dixerim totam corpoream quoad rerum causas mechanicas detexerit. Philosophiam hac in parte excolere minime suscepi. Dicam tamen breviter quod natura valde simplex est et sibi consona. Quam rationem tenet in majoribus motibus, eandem in minoribus tenere debebit. Illi a majoribus majorum corporum viribus attractivis pendent, hos a minoribus particularum insensibilium viribus nondum animadversis pendere suspicor. Nam varia esse virium naturalium genera ex viribus gravitantibus, magneticis, et electricis manifestum est, et adhuc plura esse posse non est timere [*sic*] negandum. Viribus istis majora corpora in se mutuo agere notissimum est, et cur minora viribus similibus se invicem non agitent, plane non video. Si Spiritus Vitrioli (qui ex aqua communi et acido spiritu constat) commisceatur cum Sale Alcali vel cum idoneo aliquo pulvere metallico oritur statim commotio et ebullitio vehemens. Sed et in ejusmodi operationibus calor ingens plerumque generatur. Motus iste et calor inde excitatus arguit vehementem esse congressum particularum acidarum cum particulis alijs seu metallicis seu Salis Alcali, et congressus particularum cum impetu fieri nequit nisi particulae prius incipiant ad invicem accedere quam se mutuo tangunt. Vis autem omnimodo quo corpuscula distantia in se invicem ruunt, sermone populari attractio appellari solet. Nam cum vulgo loquor vim omnem qua corpuscula distantia vel impelluntur in se mutuo vel quomodocunque coeunt et cohaerent attractionem vocans. Porro si metallum in spiritu Vitrioli solutum distilletur, spiritus iste qui solus calore aquae ebullientis ascenderet, non prius ascendit quam materia fere ad candorem incalescit, attractione scilicet Metalli retentus. Sic et spiritus nitri (qui ex aqua ibidem et spiritu quodam acido constat) cum Sale Tartari violenter congreditur, deinde quamvis seorsim in balneo leviter calefacto destillari potest, tamen a Sale Tartari non nisi per vehementem ignem abstrahitur. Et similis est ratio acidorum omnium quae satis fortia sunt, et corporum quae in ipsis dissolvuntur. Ad eundem modum si oleum Vitrioli cum aqua communi misceatur, oritur calor vehemens ex impetuoso particularum congressu, dein particulae conjunctae ita cohaerent ut aqua, quamvis seorsim summe volatilis, non prius tota destillari possit quam olei pars bene magna calore aquae ebullientis vel paulo majori cogatur simul ascendere. Et simili cohaesione aqua flegmatica et spiritus acidus, ex quibus aqua fortis constat, se mutuo retinent et simul destillantur. Et spiritus acidus in butyro Antimonij quia copiosus est et seorsim summe volatilis, cohaerendo cum particulis metallicis easdem in destillatione secum rapit, sic ut ambo simul calore non admodum violento in forma salis fusibilis ascendant. Eodem modo spiritus urinae et salium cohaerendo componunt Salem Armoniacum; et sal iste cohaerendo cum alijs quibusdam corporibus attollit earum particulas in sublimatione. [Et sal alcali vaporem ex aere attrahendo liquescit & liquorem attractum ita retinet ut is aegerrime per destillationem separari possit.

Et quia particulae omnes compositae majores sunt quam particulae componentes et particulae majores difficilius agitantur, ideo particulae Salis armoniaci sunt minus volatiles quam minores spirituum particulae ex quibus componebantur. Sic aurum quod corpus omnium fixissimum est, ex particulis compositis constare videtur, quae totae ob molem suam per agitationem caloris attolli nequeunt, et quarum partes componentes arctius inter se cohaerent quam ut possint per solam illam agitationem ab invicem separari. Particulas autem aquae et spirituum esse omnium subtillissimas et minimas et ea de causa maxime volatiles, rationi consentaneum est. Et salium spiritus acidus in mercurij sublimatione subtiliatus Antimonium et metalla in sublimati sublimatione nova magis dividit et dividendo volitizat quam spiritus idem per se ex salibus ijsdem destillatus.][1]

Porro ubi corpora in acidis soluta praecipitantur per Salem tartari, praecipitationem fieri verisimile est per attractionem fortiorem qua Sal Tartari spiritus illos acidos a corporibus solutis ad se allicit. Nam si spiritus ad utrumque retinendum non sufficit cohaerebit cum attrahente fortiore. Sic etiam spiritus acidus in mercurio sublimato agendo in metalla mercurium deserit. Iste in butyro Antimonij cum aqua affusa coalescit et Antimonium a se derelictum praecipitari sinit. Et spiritus acidus qui in aqua forti vel spiritu vitrioli cum aqua communi conjungitur agendo in metalla in menstruo illo dissoluta, aquam deserit et permittit ut aqua illa quae prius a spiritu per destillationem separari nequibat jam sola calore levissimo ascendat. Et spiritus vitrioli congrediendo cum Salis Nitri particulis fixis, laxat spiritum nitri qui prius cum particulis illis fixis conjungebatur, sic ut Spiritus ille facilius destillari possit quam antea.

Attractione etiam mutua particulas corporum homogeniorum tam solidorum quam fluidorum cohaerere probabile est, et inde fieri ut argentum vivum in Tubo Torricelliano, quoties ejus particulae et sibi ipsis et tubo ubique contiguae sunt, ad altitudinem digitorum 40, 50, 60 et amplius in vertice suspendi et sustineri possit. Corpora autem perinde ut earum particulae cohaerentes, pro varietate magnitudinum figurarum et virium suarum facilius vel difficilius inter se labendo, vel ab invicem utcunque recedendo moveri possint fluida esse vel firma, mollia vel dura, ductilia vel elastica, et ad ignem facilius vel difficilius liquescere rationi consentaneum est.[2]

[1] In original draft as: Tali etiam attractione particulas corporum tum fluidorum tum durorum cohaerere suspicor et inde fieri ut argentum vivum quoties ejus particulae et sibi ipsis et Tubo vitreo Torricelliana ubique contiguae sunt, in Tubo illo pleno ad altitudinem digitorum 40, 50 vel 60 et amplius suspendi et sustineri possit.

[2] This last paragraph is placed in a different position in the original draft, and ends as follows:
...difficilius liquescere [fluida autem humectantia esse vel non humectantia, mollia tenacia esse, vel non tenacia, ac dura facilius frangi vel difficilius] rationi consentaneum est.

Et quemadmodum corpora magnetica se mutuo non solum attrahunt sed etiam fugiunt, sic etiam corporum particulae viribus quibusdam se mutuo fugere possunt. Aqua cum oleo et plumbum cum ferro vel cupro ob fugam particularum non miscentur. Quod mercurius, aurum, argentum, stannum et plumbum penetrat, at lignum & vesicas et sales & saxa non penetrat; aqua vero e contra lignum, vesicas, sales et saxa quaedam penetrat, at non penetrat metalla, non accedit ex majori vel minori subtilitate partium aquae vel mercurij, sed mutuae particularum cognatarum attractioni, minus cognatarum fugae ascribendum esse suspicor.[1] Quemadmodum vero Magnes vi duplici praedita est, altera gravitatis & altera magnetica; sic earundem particularum variae possunt esse vires ex varijs causis oriundae. Alia debet esse vis particularum olei qua se mutuo attrahunt, et alia qua fugiunt particulas aquae.[2] Particulae metallorum attrahunt particulas

[1] In the original this paragraph begins with the following sentence:
Sed et omnia colorum phaenomena ex diversis particularum magnitudinibus oriri per experimenta quaedam probari potest. Quemadmodum vero Magnes...

[2] The original interpolates the following addition:
Vires autem quibus particulae se mutuo attrahunt, si recedatur a particulis celerrime decrescere colligi potest ex parva quantitate menstrui solventis quam particulae corporis soluti attrahere & retinere possunt. [Nam si metallum in aqua forti solvatur deinde humore superfluo ad ignem lenem exhalante in crystallos vertatur et crystalli ad solem arescant & albescant, materia crystallorum ad tres partes quartas metallica erit et pars una quarta ex acido menstruo desumetur, secundum mensuram ponderis, ideoque secundum mensuram magnitudinis pars una quarta materiae totius crystallorum metallica erit et partes plus minus tres ex acido desumentur. Nam acidum quasi octuplo vel decuplo rarius est quam metallum. Igitur particulae unius metallicae, quae ob fixitatem crassior est, ideoque menstrui acidi particulas plures quae volatiles sunt et eo nomine subtiliores ad se undique attrahet et circum se retinebit, vis attractiva ad brevissimam ab eadem distantiam extenditur et cito terminatur et propterea recedendo a particula celeriter decrescit. Nam diameter compositi totius in quo virtus attractiva fere clauditur duplo major non erit quam diameter particulae illius metallicae quae in centro est et materiam acidam attrahit in circuitu.] At particularum vis expulsiva sese longius exerere videtur & minor quidem esse in earum confiniis at in majoribus distantiis ubi vis attractiva decrevit effectus suos edere. Nam sales ex metallis extracti si in aqua solvantur dilatant sese et per aquam totam licet seipsis millecuplo majorem et amplius uniformiter dimanant. Specifice graviores sunt hae acido-metallicae particulae quam aqua et eo nomine fundum petere deberent sed impedit vis qua se mutuo fugiunt et ab invicem recedendo per aquam omnem diffunduntur. Particulae quoque vaporum et aeris quamdiu sunt in aqua et corporibus solidis et nondum formam vaporis vel aeris induerunt, cohaerent cum particulis contiguis et quamprimum per motum caloris vel fermentationis separentur et ultra certam distantiam tam a corporibus quam ab invicem recesserint fugiunt se mutuo (uti diximus) et ab invicem recedendo Medium componunt quod sese expandere vehementer conatur. Nam et in angustarum fistularum vitrearum cavitatibus cylindricis vitrum fugiunt, adeo ut aer ibi rarior sit quam in spatiis liberis (ut quidam propriam secutus hypothesin ingeniose notavit), et aqua, quae vitrum non fugit, quamprimum fistulae orificium inferius in eam immergitur celeriter ascendat (ut is expertus est) et cavitatem impleat qua aer recedere conatur. Et per similem liquorum ascensum in parvis spon-

acidorum solventium, at particulae ex utrisque compositae in aqua natando se mutuo fugiunt. Nam quamvis specifice graviores sint quam aqua et eo nomine fundum petere deberent, tamen sese fugiendo dilatantur uniformiter per aquam totam cui innatant, quamvis seipsis millecuplo majorem et amplius. Concipe particulas metallicas ob fixitatem majores esse particulis acidis et vi duplici pollere, alteram attractivam quae fortior sit sed si recedatur a particulis celeriter decrescat, et alteram fugatricem quae tardius decrescat, et propterea latius extendatur, et quoniam limes quidam extabit intra quem vis attractiva fortior erit quam vis fugatrix, et extra quem vis fugatrix praevalebit; attrahet particula unaquaeque particulas acidas quotquot intra limitem illum consistere possunt, easque circum se retinebit, et una cum eis componet particulam salis, caeterasque quae extra limitem sunt tam acidas particulas quam metallicas fugabit. Et quamvis per evaporationem aquae fieri possit ut particulae ad invicem accedant ac tandem (ob angustiam spatij in quo continentur) in crystallos coeant, idque regulariter ob regularem situm inter se quem antea in aqua separatim natantes habebant; in crystallis autem per vim attractivam particularum acidarum se mutuo contingentium cohaereant: si tamen crystalli in aqua fontana satis copiosa mergantur; et vires quibus particulae attrahuntur in se mutuo non multo majores sint quam vires quibus attrahuntur in particulas aquae; particulae illae motu levi agitatae facile excutientur a crystallis, et tum denuo extra vires mutuas attractivas positae fugient se mutuo, et per fugam illam dilatabuntur per liquorem totum. Sic et particulae liquorum et corporum solidorum dum sunt contiguae attrahunt se mutuo, at per calorem vel fermentationem separatae sese fugant ab invicem, et vapores exhalationes et aerem constituunt. Et rursus per frigus vel fermentationem concurrentes cohaerent et in liquores et corpora compacta revertuntur. Nam aerem verum per fermentationem et oriri ex corporibus densis et in eadem redire Vir nobilis et insignis Philosophus D. Boylius nos docuit. Id quod de particulis

giarum & aliorum corporum meatibus intelligitur quod particulae aeris a corporibus universis recedere conantur. Unde nec aer vesicae inclusus et compressus per ejus poros egreditur quos tamen aqua facillime permeat. Eodem spectat quod corporum durorum partes divulsae aegerrime adduci.... (The holograph breaks off at this point, which the copy continues.)

ramosis vi elastica sese vehementer expandentibus (quae vulgaris est opinio) difficulter intelligetur. Quinetiam particulae aeris in angustis fistularum vitrearum cavitatibus vitrum fugiunt adeo ut aer ibi rarior esse videatur quam in spatijs liberis et aqua stagnans si fistularum orificium inferius immergatur, celeriter ascendat (ut quidam propriam secutus hypothesin ingeniose notavit) et cavitatem impleat a qua aer recedere conatur. Et per similem liquorum ascensum in parvis spongiarum et aliorum corporum meatibus intelligitur quod particulae aeris a corporibus universis recedere conantur. Unde nec aer vesicae inclusus et compressus per ejus poros egreditur quos tamen aqua facillime permeat. Eodem spectat quod corporum durorum partes divulsae aegerrime adduci possunt ut se mutuo penitus contingant. Unde nec cohaerent uti prius. Supra vitrum objectivum convexo-convexum Telescopij pedum quinquaginta longitudinis collocavi vitrum planum molis minoris et in medio vitrorum circa punctum illud ubi sese quamproxime tangebant, apparebant circuli quidam colorum. Dein leviter premendo vitrum superius ut id propius accederet ad inferius emergebant circuli novi colorati in medio circulorum priorum, ac tandem augendo pressionem emergebat in medio circuli albi quasi macula quaedam caeteris coloribus nigrior. Haec macula index erat contactus vitrorum. Nam superficies vitrorum quae alibi luce a se reflexa satis splendebant, in macula illa lucem pene nullam reflectebant sed luci transitum integrum et minime impeditum praebebant perinde ac si vitra ibi continua essent, et per fusionem unita superficiebus intermedijs destituerentur. Igitur vitrum superius vi ponderis sui adduci non potuit ut vitrum inferius contingeret. Opus erat vi majori manuum comprimentium. Quinimo vi illa omni qua vitra comprimere potui ea se mutuo vix aut ne vix quidem tetigere. Nam sublatis manibus minime cohaerebant. Res plenius intelligetur per hoc experimentum. Bullam satis amplam qualem pueri ludendo conflare solent, tegi vase vitreo, ne aer externus colores in bulla apparentes, motu suo turbaret. Cingebatur autem vertex bullae circulis concentricis colorum diversorum et circuli illi dilatabantur et versus latera et inferiores partes bullae serpebant circulis novis interea in vertice surgentibus. Tandem albor in vertice apparuit et mox in ejus medio macula caerulei suboscuri coloris in cujus medio macula nigra deinceps apparuit et post in ejus medio macula nigrior quae lucem fere nullam reflectebat et tum demum bulla disrumpi solebat. Ejusdem generis et originis sunt hi colores cum coloribus inter vitra. Nam aquea bullae cuticula hic inter aerem externum et internum idem praestat in coloribus exhibendis atque aeris cuticula illic inter duo vitra. Perinde ut aeris cuticula per compressionem vitrorum tenuior redditur oriuntur colores novi donec macula nigra appareat, et perinde ut aquea bullae cuticula in vertice suo per perpetuum de partibus superioribus ad partes inferiores defluxum tenuior evadit, colores novi successive apparent et ultimo macula nigra conspicitur. Et quemadmodum aquea bullae cuticula in vertice tenuissima fit ubi macula nigra ibidem apparet sed omni tamen crassitudine non vacat, sic ex macula nigra in medio vitrorum concludi non potest quod cuticula aeris intermedij juxta maculam illam crassitudine omni destituitur. Igitur corpora disjuncta contactum mutuum vehementer fugiunt. Et inde fit ut musca super aquam ambulet nec tamen madefaciat pedes, utque particulae pulveris aridi nulla compressionis vi adduci possint ut cohaereant ad modum partium corporis unius. At si aqua affundatur quacum componant corpus continuum, dein aqua exhalet; adducuntur ad contactum et per contactum illum cohaerendo formant corpus unum, quod ubi ad ignem siccatur, ut fluidum quo mollescit, exhalet, in corpus firmum ac durum convertitur.

Haec breviter exposui non ut vires particularum attractivas et expulsivas extare temere affirmem sed ut ansam darem experimenta plura excogitandi per quae tandem certius constet utrum extent necne. Nam si veras esse consisterit, reliquum erit ut earum causas et proprietates diligenter investigemus, tanquam vera principia a quibus omnes particularum minimarum secretiores motus secundum rationes Geometricas non minus oriantur quam motus majorum corporum ex legibus Gravitatis in praecedentibus derivari vidimus. Per vires enim quibus particulae contiguae cohaereant et distantes conentur ab invicem discedere,[1] particulae illae in compositione

---

[1] The beginning of this sentence appears in the original draft as:

Nam particulae corporum contiguae per vires quibus cohaereant, et distantes conentur ab invicem recedere, particulae illae....

corporum naturalium minime coacervabuntur ad modum lapidum, sed secundum texturas artificiales et summe regulares coalescent, ut fit in formatione nivis et salium. In virgas utique praelongas et elasticas et per connexionem virgarum in particulas retiformes et per harum compositionem in particulas majores ac tandem in corpora sensibilia secundum leges Geometricas formari possunt. Ac talia corpora luci per se transienti omnifariam patebunt et inter se quoad gradus densitatis valde differre possunt, sic ut aurum sit novendecim vicibus densius quam aqua nec tamen meatibus destituatur per quos mercurius ingredi, et lux, si aurum in Aqua Regia solvatur et in crystallos redigatur, liberrime transire possit. Hujusmodi corpora motum quoque caloris per vibrationes liberrimas particularum elasticarum facile concipient ac diutissime conservabunt per motum illum si lentus sit ac diuturnus, particulae eorum novis modis inter se coalescent et per contiguarum vires attractivas densius congredientur: qua ratione rarior illa substantia aquae per fermentationem continuam in densiores animalium, vegetabilium, salium, lapidum ac terrarum diversarum substantias abire potest, ac tandem per operationis diuturnitatem coagulari in substantias minerales et metallicas. Nam rerum omnium una et eadem est materia, quae per operationes naturales in species innumeras transmutatur, et corpora subtiliora & rariora per fermentationem et vegetationem incrassari et condensari solent. Eodem fermentationis motu corpora particulas quasdam excutere possunt, quae mox per vires expulsivas ab invicem cum impetu resiliant, et si crassiores sunt, constituant vapores, exhalationes et aerem, sin omnium sunt minimae in lucem abeant. Hae utique, cum corpora non nisi per calorem vehementem luceant, fortius adhaerebunt, et postquam separantur, majori cum impetu a corporibus resilient, deinde vero transeundo per alia corpora, nunc in eadem attrahi possunt, nunc vero fugari ab iisdem; et per attractionem quidem refringi, et quandoque reflecti, ut supra explicui; per fugam vero semper reflecti. Corpora vero, quae lucem omnem incidentem fugant, splendebunt ob copiam lucis reflexae ut mercurius et alba metalla, quaeque particulis translucidis satis magnis per materiam uniformem diversae densitatis disseminatis abundant, pro magnitudine particularum radios hujus vel alterius generis copiosius reflectent, et perinde colorati apparebunt. Nam varios corporum colores a varijs particularum reflectentium magnitudinibus oriri, patet ex bulla aquae in praedicto experimento, cujus bullae cuticula pellucida pro varia sua crassitudine colores varios reflexit.

Corpora autem, quae per calorem vehementem particulas lucis excutiunt, sunt ignis. Nam corpora ignita ab non ignitis per calorem et lucem solummodo distingi puto; suntque duplicia, vapores et corpora solida. Nam qua ratione ignes fatui, qui sunt vapores frigidi lucentes, differunt a lignis putridis lucentibus, eadem ratione flammae, quae sunt vapores et exhalationes calefacti et per caloris agitationem vehementem lucent[es], distinguuntur a carbonibus et alijs corporibus densis, seu firmis seu fluidis, quae per calorem vehementem similiter candent. Et qua ratione metallum candens differt a non candente, eadem carbo candens et vapor candens a non candente distinguitur. Flammam enim nihil aliud esse quam vaporem candentem ex eo colligere videor, quod corpora flammam non alunt nisi quae vaporem sulphureum copiose emittunt, quodque lampas admota talem vaporem accendit, et per propagationem fermentationis totam vaporis materiam in flammam statim convertit uti in vapore spiritus vini et vapore candelae recens extinctae experiri licet. Jam vero candentia a non candentibus per motum solum partium internarum distingui videntur. Nam corpora per motum incalescunt et intense calida semper lucent. Sic enim ferrum sub malleo incalescit, et ita malleo agitari potest ut tandem candeat. Axes curruum ex motu rotarum saepe flagrant, Particulae chalybis silice abrasae et excussae (ut quidam ingeniose notavit) ex ictu candent et faenum congestum ex motu vaporum tepet et nonnunquam tependo ignescit. In carbonibus autem et vaporibus ignitis calor excitari et conservari videtur per actionem spiritus sulphuris. Nam ignis absque materia pingui et sulphurea aegre accenditur et conservatur, admixto autem sulphure intendi solet. Sulphuris enim fumus spiritu acido abundat, qui pungit oculos et sub campana condensatus decurrit in liquorem corrosivum ejusdem generis cum spiritu [nitri] et oleo vitrioli: quae non differunt nisi per flegma in spiritu, & cum alijs corporibus seu aridis seu fluidis commixta calorem excitare solent, eumque non raro vehementem. Congreditur igitur spiritus sulphuris cum particulis carbonum et fumorum, easque calefacit ad candorem usque. Nam congressus calidorum vehementior est. Et quoniam aer spiritibus sulphureis abundat, hic etiam materiam ignitam calefacit et propter subtilitatem spirituum ad ignis conservationem requiritur. Unde suspicor quod calor solis per ipsius Atmosphaeram sulphuream conservatur. Interea partes volatiles corporis igniti per calorem exhalant, et partes fixae manent in forma cinerum. Spiritus autem iste sulphureus congrediendo cum salium vegetabilium partibus fixis laxat eorum spiritus et partibus illis fixis unitus componit salem illum fixum qui in cineribus reperiri solet & sal Alcali dicitur quique virtute spiritus sulphurei per deliquium currit. Nam hae sunt proprietates Olei et spiritus Vitrioli. Humorem attrahunt ex aere et congrediuntur cum partibus fixis salis communis et salis nitri (qui duo sunt sales vegetabiles) et spiritum eorum laxant et expellunt. Sed et nitrum in aere libero cum sulphure vel cum corpore quovis sulphureo in tigillo deflagrando, in Salem Alcali convertitur, at si Nitrum solum cum argilla commistum in vitro clauso destilletur, Sal Alcali ex capite mortuo minime extrahitur. Confirmantur etiam jam dicta ex phaenomeno pulveris tormentarij. Hic ex nitro sulphure et carbonibus componitur. Pulvis carbonum ignem facillime concipit et pulverem sulphuris accendit. Hujus fumus acidus nitrum invadit, congressu calefacit, & spiritum ejus laxat; qui spiritus per calorem expulsus et in vaporem conversus vehementer se expandit. Quod si sal alcali in debita proportione admisceatur, et pulvis ad ignem exiccetur, ut humor omnis, quem sal iste ad se attrahere solet, exhalet, pulvis jam cum explosione vehementiore deflagabit ad instar auri fulminantis. Nam spiritus sulphureus cum Sale Alcali lubentius et vehementius congreditur, quam cum sale nitri, et congressu illo materiam totam magis agitat et intensius calefacit.[1] Jam vero per ignis causas materiales quae in his compositionibus vehementius agunt, ignem culinarem in quo causae illae minus abundant lente et paulatim cieri et conservari credendum est. At metalla ignita, quoniam nec partibus volatilibus abundant

---

[1] The original draft omits all that follows from this point. There are however a number of additional paragraphs in the original. These are added here at the end.

quae in fumos per calorem vertantur, nec meatibus satis amplis spiritus sulphureos in partes internas admittunt; ideo nec flammam nutriunt, nec calorem ignitionis per se conservant, nec in cineres facile vertuntur. Praeterquam quod metallum illud imperfectum et volatile quod Zinetum dicitur emittendo fumum copiosum flammam alit. Et D. Boylius idem probavit in cupro ubi per mercurium sublimatum ita dividitur et subtiliatur ut fumum emittere possit. Sic ut lapides candendo non uruntur nec ignem per se conservant praeter bituminosos illos qui vaporem sulphureum et copiosum emittunt.

Cum Philosophi antiqui (uti Author est Pappus) mechanicam in rerum naturalium investigatione maximi fecerint, et recentiores, missis formis substantialibus et qualitatibus occultis phaenomena naturae ad leges Mathematicas revocare aggressi sint: visum est hoc Tractatu Mathesin excolere quatenus ea ad Philosophum spectat. Mechanicam vero duplicem veteres constituerunt: rationalem quae per Demonstrationes accurate procedit, et practicam. Ad practicam spectant Artes omnes manuales, a quibus utique Mechanica nomen mutuata est. Cum autem Artifices parum accurate operari soleant, fit ut mechanica omnis a Geometria ita distinguatur ut quicquid accuratum sit ad Geometriam referatur, quicquid minus accuratum ad mechanicam. Attamen errores non sunt Artis sed Artificum. Qui minus accurate operatur imperfectior est mechanicus, & siquis accuratissime operari posset, hic foret me.

Additional passages found in the original only:

Sic omnia fere naturae phaenomena a particularum viribus pendebunt si modo hujusmodi vires extare probari potest. Et quamquam attractionum et virium fugatricum nomina plerisque displiceant si tamen considerent partes corporum certe cohaerere et particulas distantes iisdem de causis ad invicem impelli posse quibus cohaerent, me vero modum attractionum non definire sed cum vulgo loquentem vires omnes nominare attractivas quibus corpora ad invicem per causas quascunque impelluntur coeunt & cohaerent: quae hactenus de his viribus diximus, a ratione minus aliena videbuntur. Vera quidem esse minime affirmo, et valde imperfecta esse agnosco, simplicia tamen sunt et conceptu facilia, et ejusdem generis cum philosophia Naturali systematis cosmici a maiorum corporum viribus attractivis pendente. Cum interea Philosophia vulgaris nec verae systemati analoga sit nec commode exponiat quomodo corpora luci lineas rectas affectanti satis pervia sint aut aqua novendecim vicibus rarior esse possit quam aurum aut corporum partes (quas volunt ad modum lapidum coacervari) inter se facile agitari possint et motus impressos diutissime conservare aut quomodo solvantur per menstrua & praecipitentur & similia. Ob analogiam itaque quam vires illae habent cum viribus majorum corporum, easdem hic leviter attingere volui ut aliis ansa daretur hoc philosophandi genus plenius excolendi at quoniam incertae sunt...[1]

Et quemadmodum in aere motus iste vibratorius in quo sonus consistit per vires particularum non contiguarum celerrime propagatur: et in ligno praelongo motus vibratorius: sic motus vibratorius in solidis per particularum etiam non contiguarum vires celerrime propagari potest. Qua ratione calor corpus totum cito invadet & corpus calidum calorem suum cum aqua cui immergitur cito communicabit. Sed et motus isti quibus tunica retiformica a luce aures a sonis palatum a salibus nervi olfactorii ab odoribus & nervi reliquii ab objectis tangibilibus agitari solent; per solida & continua nervorum capillamenta in sensorium propagari possunt. Et contra per similes motus a sensorio per solida nervorum capillamenta propagatos potest substantia aliqua in musculis agitari et per agitationem illam dilatari ad musculos contrahendos in motu membrorum. Nam spiritus animales (quos fingunt) per compactam nervorum substantiam ad musculos implendos haud satis facile celeriter et copiose propagari possunt. Et qui onera gravia sustinent etiam absque motu membrorum ad sudorem incalescunt. Oritur calor iste non ab incursu spirituum animalium, in musculos qui jam ante pleni erant, sed a partium corporis agitatione aliqua qua musculi distenti retinentur, et qua cessante flaccescunt. Porro per caloris motum lentum & continuum particulae corporum situs inter se paulatim mutare possunt et novis modis inter se coalescere et per particularum

---

[1] This passage breaks off here. The next passage is written sideways across the same page (fol. 27).

contiguarum vires attractivas (quae expulsivis fortiores sunt) densius congredi.[1]

Conclusion

Hitherto I have explained the System of this visible world, as far as concerns the greater motions which can easily be detected. There are however innumerable other local motions which on account of the minuteness of the moving particles cannot be detected, such as the motions of the particles in hot bodies, in fermenting bodies, in putrescent bodies, in growing bodies, in the organs of sensation and so forth. If any one shall have the good fortune to discover all these, I might almost say that he will have laid bare the whole nature of bodies so far as the mechanical causes of things are concerned. I have least of all undertaken the improvement of this part of philosophy. I may say briefly, however, that nature is exceedingly simple and conformable to herself.[2] Whatever reasoning holds for greater motions, should hold for lesser ones as well. The former depend upon the greater attractive forces of larger bodies, and I suspect that the latter depend upon the lesser forces, as yet unobserved, of insensible particles. For, from the forces of gravity, of magnetism and of electricity it is manifest that there are various kinds of natural forces, and that there may be still more kinds is not to be rashly denied. It is very well known that greater bodies act mutually upon each other by those forces, and I do not clearly see why lesser ones should not act on one another by similar forces. If spirit of vitriol (which consists of common water and an acid spirit) be mixed with Sal Alkali[3] or with some suitable metallic powder, at

---

[1] End of this passage. Written sideways in margin of 26 verso:

et in ligno praelongo motus vibratorius percussione excitatus de extremo ad extremum celerrime propagatur Et quemadmodum ille vibratorius in quo sonus consistit propagatur tam per lignum et alia soli[da] corpora praelonga mediantibus partibus contiguis quam per aerem mediantibus viribus particularum non contiguarum; sic motus vibratorius in quo calor forte consistit propagari potest tam per vires partium non contiguarum quam per impulsus contiguarum. Nam corpora omnia ubi satis indurescunt elastica sunt; indeque si fluida ut mihi videtur ex particulis duris constant hae etiam erunt elasticae

(ends)

[2] Principia, Book III, Hypotheses: 'Natura enim simplex est & rerum causis superfluis non luxuriat' (p. 402).

Opticks, Quaery 31, 'For Nature is very consonant and conformable to her self' (p. 376).

[3] Presumably potash, but we preserve Newton's term throughout.

once commotion and violent ebullition occur. And a great heat is often generated in such operations. That motion and the heat thence produced argue that there is a vehement rushing together of the acid particles and the other particles, whether metallic or of Sal Alkali; and the rushing together of the particles with violence could not happen unless the particles begin to approach one another before they touch one another.[1] The force of whatever kind by which distant particles rush towards one another is usually, in popular speech, called an attraction. For I speak loosely when I call every force by which distant particles are impelled mutually towards one another, or come together by any means and cohere, an attraction.[2] Moreover, if the solution of a metal in spirit of vitriol be distilled, that spirit which by itself will ascend at the heat of boiling water does not ascend before the material is heated almost to incandescence, being held down by the attraction of the metal.[3] So also spirit of nitre (which is composed of water and an acid Spirit) violently unites with salt of tartar; then, although the spirit by itself can be distilled in a gently heated bath, nevertheless it cannot be separated from the salt of tartar except by a vehement fire. And similar reasoning applies to all sufficiently strong acids and the bodies dissolved in them. In the same way if oil of vitriol be mixed with common water, a vehement heat arises from the impetuous rushing together of the particles, and then the united particles so cohere that water, although by itself very volatile, cannot be wholly distilled off before a large part of the oil is driven to ascend by the heat of the water at boiling point, or rather more. And by a similar cohesion the phlegmatic water and the acid spirit, of which aqua fortis is composed, hold on to each other and distill over together. And the acid spirit in butter of antimony, because it is copious and by itself ex-

---

[1] *Quaery* 31, 377. 'And does not this Motion argue, that the Parts of the two Liquors in mixing coalesce with Violence, and by consequence rush towards one another with an accelerated Motion?'

[2] *Quaery* 31, 376. 'What I call Attraction may be perform'd by impulse, or by some other means unknown to me. I use that Word here to signify only in general any Force by which Bodies tend towards one another, whatsoever be the Cause.'

[3] *Quaery* 31, 377–8. 'And when the acid Particles, which alone would distill with an easy Heat, will not separate from the Particles of the Metal without a very violent Heat, does not this confirm the Attraction between them?' The examples in our text are very similar to those in this Quaery.

tremely volatile, cohering with the metallic particles takes them with it in distillation, so that both ascend together by a not very violent heat in the form of a fusible salt. In the same way the spirits of urine and salt by cohering compose Sal Ammoniac; and that salt by cohering with other bodies carries up the particles of these in sublimation. [And Sal Alkali liquefies by attracting vapour from the air, and retains the attracted liquor so firmly that the two can scarcely be separated by distillation.

And because all the particles of a compound are greater than the component particles, and larger particles are agitated with greater difficulty, so the particles of sal ammoniac are less volatile than the smaller particles of the spirits of which they are composed. So gold, which is the most fixed of all bodies, seems to consist of compound particles, not all of which, on account of their massiveness, can be carried up by the agitation of heat, and whose component parts cohere with one another too strongly to be separated by that agitation alone.[1] That the particles of water and spirits, however, are most subtle and small of all, and for that reason exceedingly volatile, is consonant with reason. And the acid spirit of salts rendered more subtle in mercury sublimate divides Antimony and metals already sublimed in a fresh sublimation, and in dividing volatilizes, better than the same spirit by itself distilled from these same salts.][2]

Moreover, when bodies dissolved in acids are precipitated by salt of tartar, the precipitation is probably caused by the stronger attraction by which the salt of tartar draws those acid spirits from the dissolved bodies to itself. For if the spirit does not suffice to retain them both, it will cohere with that which attracts more strongly.[3] Thus also the acid spirit in mercury sublimate, acting on metals, leaves the mercury. That spirit in butter of antimony coalesces with water poured on to it, and allows the antimony abandoned by it to be

[1] Cf. *De natura acidorum*.

[2] *Original*: I suspect that by such an attraction the particles of both fluid and solid bodies cohere, and thus it is that whenever the particles of quicksilver are everywhere contiguous to each other and to the glass Torricellian tube, the quicksilver can be suspended and sustained in that tube to a height of 40, 50, or 60 inches or more.

[3] Newton apparently means that if there is not a sufficient quantity of acid to satisfy the attractions of both the solute and the alkali, then the acid particles will join with the alkaline ones. Cf. *Opticks*, *Quaery* 31, 380.

precipitated. And the acid spirit, joined with common water in aqua fortis and spirit of vitriol, by acting on metals dissolved in those menstruums, leaves the water and allows the water to ascend by itself with a merely gentle heat, whereas before it could not be separated from the spirit by distillation. And spirit of vitriol, meeting with the fixed particles of salt of nitre, looses the spirit of nitre which was formerly joined to those fixed particles, so that that latter spirit can be more easily distilled than before.[1]

It is probable that the particles of homogeneous bodies, whether solid or fluid, cohere by a mutual attraction, and thence it is that whenever the particles of quicksilver in a Torricellian Tube are everywhere contiguous to each other and to the tube, the quicksilver can be suspended and sustained to a vertical height of 40, 50, 60 inches and more.[2] It is agreeable to reason, however, that bodies, like their cohering particles (which according to the variation in their magnitudes, shapes and forces move either by gliding between one another more or less easily, or by receding from each other in some way) should be fluid or solid, soft or hard, ductile or elastic and more or less apt to be liquefied by fire.[3]

And just as magnetic bodies repel as well as attract each other, so also the particles of bodies can recede from each other by certain forces. Water does not mix with oil, or lead with iron or copper, on account of the repulsion between the particles.[4] Mercury penetrates gold, silver, tin and lead, but it does not penetrate wood and bladders and salts and rock; water on the contrary penetrates wood, bladders, salts and rock but does not penetrate metals: the cause of this is not the greater or lesser subtlety of the parts of water or mercury, but is, I suspect, to be attributed to the mutual attraction of particles of a like kind, less to the repulsion of likes.[5] Just as the magnet is endowed with a double force, the one of gravity and the other magnetic, so there are various forces of the same particles arising from various causes. There must be one force

[1] Cf. *Opticks, Quaery* 31, 378.
[2] *Ibid.* 390.
[3] *Original adds*: '...by fire, if fluid; wetting or non-wetting, if soft, sticky or not sticky, if hard, either brittle or not.'
[4] Cf. *Opticks, Quaery* 31, 383.
[5] *Original*: 'But that all the phenomena of colours arise from the different sizes of the particles can be proved by certain experiments. Just as the magnet....'

whereby particles of oil attract each other, and another whereby they repel the particles of water.[1] The particles of metals attract the particles of the acids which dissolve them, and the particles composed of them both swimming in the water flee from each other.[2] For although their specific gravity is greater than that of water, and for that reason they ought to fall to the bottom, nevertheless by fleeing from each other they spread uniformly throughout the water in which they swim, even though its volume is over a thousand times

[1] The following addition is interpolated in the original: 'That the forces by which particles attract one another, however, decrease rapidly with distance from the particles may be gathered from the small quantity of a dissolving menstruum which the particles of a dissolved body can attract and retain. [For if a metal is dissolved in aqua fortis, and then by exhaling the superfluous humour on a slow fire is turned into crystals, and the crystals dry in the sun and grow white, the matter of the crystals will be three-fourths parts metallic by weight and one-fourth part taken from the acid menstruum, and similarly one-fourth part of all the matter of the crystals by volume will be metallic and three parts more or less taken from the acid. For acid is about eight or ten times more rare than metal. Therefore one single metallic particle, which on account of its fixity is more dense, attracts to itself many of the particles of the acid menstruum which are volatile and for that reason more subtle, and will retain them around itself. The attractive force extends to a very short distance from it and is soon terminated, and therefore in receding from the particle quickly decreases. For the diameter of the whole compound [particle] to which the attractive force is practically confined, will not be twice the diameter of the metallic particle which is in the centre and attracts the acid matter about itself.] But the expulsive force of particles seems to be exercised at greater distances, and even to be less in their immediate neighbourhood; but at greater distances, where the attractive force has decreased [the repulsive] makes its effects felt.[3] For if salts extracted from metals are dissolved in water they expand themselves and spread uniformly throughout a volume of water say a thousand times greater than their own or more. The specific gravity of these acido-metallic particles is greater than that of water, and for that reason they ought to fall to the bottom, but their force of repulsion prevents this and in receding from each other they diffuse throughout the water.[4] Also the particles of vapours and of air, so long as they are in water and solid bodies and not yet assuming the form of vapour or air, cohere with contiguous particles. But as soon as they are separated by the motion of heat or fermentation and recede beyond a certain distance from the bodies and from each other, they fly apart (as we said) and in receding from each other compose a Medium which has a strong tendency toward expansion. For in the cylindrical cavities of narrow glass tubes they flee from the glass, so that air is more rare there than in free spaces (as a certain person [Boyle] has ingeniously remarked, following an hypothesis of his own) and water, which does not flee from glass, as soon as the lower orifice of the tube is immersed in it, quickly rises inside it (as he found by experiment) and fills the cavity from which the air tries to recede. And from the similar ascent of liquids in the small passages of sponges and other bodies it is learnt that the particles of air endeavour to recede from all bodies universally. Whence it is that air when confined in a bladder does not escape through its pores even when compressed, although water easily permeates them. And in the same manner we see why the separated parts of hard bodies can only with great difficulty be brought....

[2] Newton seems to mean that as soon as the metallic and acid particles have satiated their mutual attraction by joining together, the lesser repulsive forces of each kind of particle combine in the compound particle, thus rendering the compound particles mutually repulsive.

[3] Cf. *Opticks*, *Quaery* 31, 395.   [4] *Ibid.* 387.

greater than theirs. Conceive the metallic particles on account of their fixity to be larger than the particles of acid and let them have a double force. The first force is an attractive one and is the stronger, but it quickly decreases with distance from the particle; the second is a repulsive force which decreases more slowly and on that account extends more widely. Since there will be a certain limit within which the attractive force will be stronger than the repulsive, beyond this limit the repulsive force will prevail. Each and every particle [of metal] attracts so many particles of acid as can remain within that limit and will retain them about itself, and together with them composes a particle of salt, and it will repel the other particles, whether acid or metallic, outside that limit. And although it is possible to make the particles come together by evaporation of the water and finally join together into crystals (because of the narrowness of the space in which they are contained) and that regularly on account of the regular arrangement between [the particles] which they had before when swimming separately in the water; yet they cohere in such crystals by the attractive force of the acid particles, now in contact with each other.[1] If, however, the crystals are immersed in sufficient spring-water, and if the forces by which the [crystalline] particles are attracted to each other are not much greater than the forces by which they are attracted towards the particles of water, the former particles are easily driven off from the crystals when agitated by a gentle motion; and then, being beyond [the limit of] the forces of mutual attraction, they flee from each other, and by that repulsion are diffused throughout the liquid. So also the particles of liquid and solid bodies attract each other when contiguous, but when separated by heat or fermentation they flee from each other and constitute vapours, exhalations, and air. And on the other hand coming together by the action of cold or fermentation they cohere and revert into liquids and compact bodies. For that remarkable Philosopher, the Honourable Mr Boyle, teaches us that true air arises from the fermentation of dense bodies and returns into them. And this, in connexion with the violent expansion of branched particles by an elastic force (as is commonly supposed), is difficult to understand.[2]

[1] Cf. *Opticks, Quaery* 31, 388.   [2] Cf. *Opticks, Quaery* 31, 395–6.

Indeed, particles of air in the narrow cavities of glass tubes flee from the glass, so that air seems to be more rare there than in free spaces and if the lower orifice of the tube is immersed in stagnating water it ascends quickly (as a certain person has ingeniously remarked following an hypothesis of his own) and fills the cavity from which air tries to recede. And from the similar ascent of liquids in the small passages of sponges and other bodies it is learnt that the particles of air endeavour to recede from all bodies universally.[1] Whence it is that air when confined in a bladder does not escape through its pores even when compressed although water very easily permeates them. And in the same manner we see why the separated parts of hard bodies can only with great difficulty be brought together so that they touch each other completely. Whence they do not cohere as they did before. I placed on a bi-convex objective lens from a fifty-foot telescope a plane glass of less weight, and in the middle of the glasses about the point where they almost touched there appeared certain coloured circles.[2] Then lightly pressing upon the upper glass, so that it came closer to the lower one, there emerged new coloured circles in the middle of the former circles; and then increasing the pressure there emerged in the middle of the white circle a sort of spot darker than the other colours. This spot was the indication of the contact of the pieces of glass. For the surfaces of the glasses which elsewhere reflected light from themselves, reflected almost no light at all at that spot, but offered very little hindrance to the unimpaired passage of light just as if the glasses were continuous there, and being joined together by fusion were deprived of intermediate surfaces. Therefore the upper glass could not be brought by its weight into contact with the lower one. That was effected by the greater force of the hands pressing down. Indeed, with all the force with which I compressed those glasses I could scarcely make them touch at all. For on removing my hands they did not cohere in the least. The matter may be better understood from this experiment. I covered a fairly large [soap-]bubble, such as boys blow in play, with a glass vessel so that the external air should not by its motion disturb the colours appearing in the

[1] *Ibid.* pp. 391–2.
[2] Cf. *Opticks, Quaery* 29, 372.

bubble.[1] The top of the bubble was ringed round with concentric circles of different colours; these circles [as the bubble became thinner] widened out; they crept down the sides towards the lower part of the bubble while new circles arose towards the top. At length a whiteness appeared towards the top and soon in the midst of it a spot of dark blue colour in whose centre a black spot then appeared and afterwards in its centre a blacker spot which hardly reflected the light at all, and then at last the bubble usually burst. These colours are of the same kind and have the same origin as the colours between glasses. For the watery skin of the bubble between the internal and external air serves in this case to exhibit colours just as the layer of air between the two glasses does. Just as when the layer of air was made thinner by compression of the glasses new colours arose until the black spot appeared, so when the skin of the soap-bubble becomes thinner at the top by the continual flowing down of the upper parts to the lower part, new colours appear successively and ultimately the black spot is observed. And as the skin of the soap-bubble is thinnest at the top where the black spot appears but notwithstanding this is not wholly lacking in thickness, so from the black spot in the middle of the glasses it cannot be concluded that the layer of intermediate air near that spot is destitute of all thickness. Therefore separate and distinct bodies vehemently avoid mutual contact. And thence it happens that a fly walks upon water without wetting its feet, and that particles of dry powder cannot be brought together by a compressive force so that they cohere as the parts of a solid body do. But if water is poured on so that [the particles] compose a continuous body, and then the water is evaporated, they are brought into contact and through that contact they form a body by cohesion; when it is dried by fire so that the fluid softening it is driven off, it is converted into a firm and hard body.[2]

I have briefly set these matters out, not in order to make a rash assertion that there are attractive and repulsive forces in bodies, but so that I can give an opportunity to imagine further experiments by which it can be ascertained more

---

[1] This experiment is described in the *Hypothesis of Light* (1675): Birch, *History*, III, 281–4; *Papers & Letters*, 211–14.   [2] *Opticks*, *Quaery* 31, 396–7.

certainly whether they exist or not. For if it shall be settled that they are true [forces] it will remain for us to investigate their causes and properties diligently, as being the true principles from which, according to geometrical reasoning, all the more secret motions of the least particles are no less brought into being than are the motions of greater bodies which as we saw in the foregoing [books] derived from the laws of gravity. For through the forces by which contiguous bodies cohere, and distant ones seek to separate from one another, these particles will not collect together in the composition of natural bodies like a heap of stones, but they coalesce into the form of highly regular structures almost like those made by art, as happens in the formation of snow and salts. Undoubtedly, following the laws of geometry they can be formed into very long and elastic rods, and by the connexion of the rods into retiform particles, and by the composition of these into greater particles, and so at length into perceptible bodies. And such bodies suffer light to pass through them every way, and there can be great differences of density between them. Thus gold is nineteen times denser than water yet it is not destitute of pores through which mercury can penetrate, and light (if the gold be dissolved in aqua regia and reduced into crystals) very freely be transmitted. Bodies of this kind easily receive the motion of heat by means of the free vibrations of the elastic particles and they will conserve [heat] a long time through that motion if it is slow and long-lasting. Their particles coalesce in new ways and by means of the attractive forces of contiguous ones they come together more densely: for which reason that rare substance water can be transformed by continued fermentation into the more dense substances of animals, vegetables, salts, stones and various earths. And finally by the very long duration of the operation be coagulated into mineral and metallic substances.[1] For the matter of all things is one and the same, which is transmuted into countless forms by the operations of nature, and more subtle and rare bodies are by fermentation and the processes of growth commonly made thicker and more condensed. By the same motion of fermentation bodies can expel certain particles, which thereupon by their repulsive forces are caused to recede from each other

[1] Cf. *Principia*, 506.

violently; if they are denser, they constitute vapours, exhalations and air; if on the other hand they are very small they are transformed into light. These last undoubtedly adhere more strongly, since bodies do not shine save by a vehement heat. After they are separated they recede from bodies more violently, then in passing through other bodies sometimes they are attracted towards them, sometimes repelled; and by attraction they are certainly refracted, and sometimes reflected as I explained above;[1] by repulsion they are always reflected. Bodies which reflect all incident light will shine on account of the abundance of reflected light, as do mercury and white metals. Others which abound in rather large translucent particles disseminated through uniform material of a different density, according to the size of the particles, more copiously reflect rays of one kind or of some other kind, whence they will appear coloured. For that the various colours of bodies arise from various sizes of reflecting particles appears from the preceding soap-bubble experiment, where the pellucid skin of the bubble reflected various colours according to variations in its thickness.

Bodies which expel particles of light because of vehement heat are fire.[2] For I think that burning bodies are distinguished from non-burning ones only by heat and light; and they are of two kinds, vapours and solid bodies. For just as *ignes fatui*, which are cold shining vapours, differ from shining rotten wood, so flames, which are vapours and exhalations made hot and shining by the vehement agitation of heat, are distinguished from coals and other dense bodies whether solid or fluid which likewise glow because of vehement heat. And just as glowing metal differs from non-glowing, so glowing coal and glowing vapour are distinguished from non-glowing. For I seem to understand that flame is nothing else but a glowing vapour from the fact that bodies do not feed flame unless they copiously emit a sulphureous vapour, and that if a [lighted] lamp is brought close to such a vapour it ignites it, and by the propagation of the fermentation it immediately converts all the material of the vapour into flame, as may be found with

[1] *Principia*, Prop. XCVI, 230–1.
[2] *Opticks, Quaery* 9, 341: 'Is not Fire a Body heated so hot as to emit Light copiously?' Compare also the whole of this paragraph with *Quaery* 8, 340–1, which gives many of the same examples.

the vapour of spirit of wine and with the vapour of a recently extinguished candle.[1] Accordingly, glowing bodies seem to be distinguished from non-glowing ones solely by the motion of their internal parts. For bodies grow hot by motion and always shine by intense heat. Thus iron grows warm under the hammer, and can be so agitated by hammering that at length it glows. The axles of waggons often burst into flame from the motion of the wheels. Particles of steel when rubbed or struck with flint (as someone [Boyle] has ingeniously noted) glow from the stroke; and closely packed hay grows warm from the motion of the vapours, and sometimes takes fire from the warmth. In coals and ignited vapours, however, heat seems to be excited and conserved by the action of a sulphureous spirit. For fire can hardly burn and be supported without fatty and sulphureous matter; with the addition of sulphur it generally becomes intense. For the fume of sulphur abounds in an acid spirit, which makes the eyes smart, and when condensed under the bell runs down as a corrosive liquid of the same kind as spirit of nitre and oil of vitriol. These only differ by the phlegm in the spirit, and when mixed with other bodies whether dry or fluid excite heat in them, and not infrequently vehement heat. Therefore spirit of sulphur meeting with the particles of coals and fumes heats them till they glow; for the encounter of hot bodies is the more vehement. And inasmuch as air abounds in sulphureous spirits, it also makes ignited matter grow hot and because of the subtlety of the spirit is required for the maintenance of fire. Whence I suspect that the heat of the Sun may be conserved by its own sulphureous atmosphere. However the volatile parts of ignited bodies are thrown off by their heat, while the fixed parts remain in the form of ashes. That sulphureous spirit meeting with the fixed parts of vegetable salts looses their spirits, however, and united with those fixed parts forms that fixed salt which is usually found in ashes and is called Sal Alkali, and by virtue of that sulphureous spirit runs *per deliquium*. For such are the properties of oil and spirit of vitriol. They attract a

[1] Cf. *Opticks, Quaery* 10, 341–2: 'Is not Flame a Vapour, Fume or Exhalation heated red hot, that is, so hot as to shine? For Bodies do not flame without emitting a copious Fume, and this Fume burns in the Flame. The *Ignis Fatuus* is a Vapour shining without heat, and is there not the same difference between this Vapour and Flame, as between rotten Wood shining without heat and burning Coals of Fire?'

humour from the air and meeting with the fixed particles of common salt and salt of nitre (which are both vegetable salts), they loose their spirits and expel them. But nitre deflagrated in the open air with sulphur or with any sulphureous body is converted into Sal Alkali, whereas if nitre alone is mixed with clay and distilled in a closed glass, no Sal Alkali is extracted from the *caput mortuum*. What has been said before is confirmed by gunpowder, which is composed of nitre, sulphur, and charcoal. The powdered charcoal readily takes fire and ignites the powdered sulphur. The acid fume of the sulphur invades the nitre and by this encounter makes it hot, and looses its spirit, which spirit driven out by heat and converted into vapour expands itself violently. If Sal Alkali be mixed in due proportion with it, and the powder dried by fire to drive off all the humour that this salt commonly attracts to itself, the powder now deflagrates with a more violent explosion after the manner of fulminating gold. For the sulphureous spirit meets more willingly and violently with Sal Alkali than with salt of nitre, and by that meeting agitates all the material more and makes it grow more intensely hot.[1] Indeed, it is to be believed that the culinary fire is excited and maintained at a slow and gradual rate by the material causes of fire which act more vigorously in these compositions, for those causes abound less in it. But ignited metals, because they do not abound with volatile parts which are turned into fumes by heat, nor do they have passages wide enough to admit sulphureous spirit into their internal parts, so they do not nourish flame, nor do they by themselves conserve the heat of ignition, nor are they easily turned into ashes; except that zinc, as that imperfect and volatile metal is called, by emitting a copious fume does nourish a flame,[2] and Mr. Boyle has proved the same for copper divided and made more subtle by mercury sublimate which could emit a fume. So also stones are not consumed by burning, nor (except those bituminous ones which emit a copious sulphureous vapour) do they conserve fire.

[1] Cf. *Opticks*, *Quaery* 10, 342–3.
[2] *Ibid*. 342. 'Metals in fusion do not flame for want of a copious Fume, except Spelter, which fumes copiously and thereby flames.'

As noted above (p. 320), the following additional paragraphs are found only in the original:

Thus almost all the phenomena of nature will depend on the forces of particles, if only it be possible to prove that forces of this kind do exist. And although the names of attractive and repulsive forces will displease many, yet what we have thus far said about these forces will appear less contrary to reason if one considers that the parts of bodies certainly do cohere, and that distant particles can be impelled towards one another by the same causes by which they cohere, and that I do not define the manner of attraction, but speaking in ordinary terms call all forces attractive by which bodies are impelled towards each other, come together and cohere, whatever the causes be.[1] I am far from affirming that my views are correct, and I acknowledge their great imperfection, nevertheless they are simple and easy to conceive, and of the same kind as the natural philosophy of the cosmic system which depends on the attractive forces of greater bodies. Since sometimes the philosophy of the common sort[2] is not consonant with the true system, nor does it conveniently expound the way in which bodies are pretty pervious to light pursuing straight lines, nor how water can be nineteen times more rare than gold, nor how the parts of bodies (which they wish to imagine as amassed together like a heap of stones) can be easily agitated among themselves, and conserve impressed motions for a very long time, nor how they are dissolved by menstruums and precipitated and the like. And thus because of the analogy that these forces have with the forces of greater bodies, I wished here to touch lightly upon them in order to give an opportunity to others for cultivating this kind of philosophizing more fully, and since these things are uncertain...

And in the same way that in air that vibratory motion of which sound consists is very quickly propagated by the forces of non-contiguous particles; and in a very long piece of wood a vibratory motion...[3] so a vibratory motion can be very quickly propagated in solids by the forces of even non-

---

[1] Cf. *Principia*, 4–5, 191; also *Opticks*, *Quaery* 31, 376.
[2] Newton means the natural philosophy of the Cartesians.
[3] See further below.

contiguous particles. For which reason heat soon spreads through the whole of a body, and a hot body quickly communicates its heat to water in which it is immersed. But also those motions by which the Tunica Retiformica [retina of the eye] is agitated by light, the ears by sounds, the palate by salts, the olfactory nerves by odours and the remaining nerves by tangible objects, can be propagated to the sensorium [of the brain] through the solid and continuous capillamenta [fibres] of the nerves.[1] And on the other hand, by a similar motion propagated from the sensorium through the solid capillamenta of the nerves, a certain substance in the muscles can be agitated and by that agitation dilated so as to contract the muscles and move the limbs. For the animal spirits (which they feign) can hardly be propagated easily, swiftly and copiously enough through the compact substance of the nerves to swell the muscles. And those who hold up heavy weights are heated to sweating even without moving their limbs. That heat does not arise from the inrush of animal spirits into muscles which were already full before, but from a certain agitation of the parts of the body by which the muscles are kept distended, and when this ceases they grow flaccid. Furthermore through the slow and continued motion of heat the particles of bodies can gradually change their arrangement and coalesce in new ways and by the attractive forces of contiguous particles (which are stronger than expulsive ones) come together more densely.

(end of this passage.)

The following passage clearly belongs to the first sentence of the last paragraph of the preceding:

and in a very long piece of wood a vibratory motion caused by percussion is very rapidly propagated from end to end. And just as that vibratory motion of which sound consists is as well propagated through wood and other long solid bodies by transmission through their contiguous parts as through air by transmission through the forces of non-contiguous particles; so the vibratory motion of which heat possibly consists can be

[1] Cf. *Opticks, Quaery* 12, 345, *Quaery* 23, 353, and *Quaery* 24, 353–4.

propagated as well through the forces of non-contiguous parts as through the impulses of contiguous ones.[1] For all bodies when they have become sufficiently hard are elastic and hence, as it seems to me, if fluids are composed of hard particles these also will be elastic.

[1] Cf. *De Aere & Aethere*, Section III, No. 1, p. 214, where this is discussed at greater length.

8

SCHOLIUM GENERALE

MS. Add. 3965, fols. 357–65

There are five holograph drafts of the *Scholium Generale*:

A. MS. Add. 3965 fols. 357–8
B. MS. Add. 3965 fols. 359–60
C. MS. Add. 3965 fols. 361–2
D. MS. Add. 3965 fols. 363–4
E. MS. Add. 3965 fol. 365

Of these, manuscripts A and C offer matter of interest and are printed below. Manuscript B corresponds very closely to the printed text, save that it includes this sentence:

Et haec de Deo, de quo utique ex Phaenomenis disserere, ad Philosophiam experimentalem pertinet. Ex Phaenomenis prodeunt proximae rerum causae: ex his causae superiores donec ad causam summam perveniatur.[1]

And in the final paragraph there is a slightly different, but familiar and characteristic phrase:

...cujus vi & actionibus particulae corporum ad parvas distantias se mutuo attrahunt; & corpora electrica...[2]

Manuscript D lies between C and B, omitting part of C and offers nothing of scientific interest, while Manuscript E is a copy of D.

Manuscript A appears to be the earliest draft of all; Newton rearranged its paragraphs into a fresh order followed in the later drafts and printed text. The earlier part agrees generally with the printed version, though it shows many differences of expression; then it develops the theory of attraction in a manner that Newton ultimately decided to abandon (as he had abandoned the *Conclusio*).

Manuscript C is a revision of Manuscript A. Here the earlier part is so close to the printed text that we have omitted much of the draft, merely printing passages that differ markedly from the published

---

[1] 'And thus much concerning God, to discourse of whom from the phaenomena undoubtedly pertains to experimental philosophy. The intermediate causes of things appear from the phaenomena, and from these the more profound causes, until one arrives at the highest cause.'
[2] '...by whose force and actions the particles of bodies mutually attract each other at near distances; & electric bodies...'

version in the second edition of the *Principia*. The latter part consists of rather incoherent paragraphs, written in different ways on the paper, and sometimes left unfinished or at least incomplete. Here ideas of attraction and the structure of matter are developed even more fully than in Manuscript A. This version throws a very revealing light on the final paragraph of the General Scholium as actually printed in 1713.

The General Scholium never formed a part of the first edition. It was written in January 1712/13 and despatched to Roger Cotes on 2 March following.[1]

MS. A[2]

(2) Projectilia in aere nostro resistentiam parvam sentiunt. In vacuo Boyliano nullam ut in corporibus cadentibus colligitur. In coelis supra atmosphaeram Terrae cessat etiam aeris resistentia & corpora liberrime moveri debent & motus suos diutissime conservare ideoque legibus gravitatis obequentur [*sic*] & orbes in quibus incipiunt moveri perpetuo describent. Nam et ex motibus cometarum valde excentricis demonstratur corpora coelestia in omnes partes liberrime moveri. Et corpora omnia coelestia seu Planetarum et Lunarum seu Cometarum motus suos semel inceptos diutissime conservabu[n]t. At motus illi sub initio ex causis mechanicis oriri non [possunt].

(3) Revolvuntur Planetae sex principales circum Solem in circulis Soli concentricis eodem ordine in eodem plano. Revolvuntur decem Lunae circum Terram Jovem et Saturnum in circulis concentricis, eodem ordine in planis orbium Terrae Jovis et Saturni. Et hi motus regulares ex causis mechanicis non sunt orti, siquidem Cometae in Orbibus valde excentricis in omnes caelorum partes libere ferantur. Ex consilio & dominio solo Entis intelligentis & potentis oriri potuit elegantissima haecce Solis et Planetarum compages. Et si stellae fixae sunt centra similium systematum, subsunt haec omnia unius dominio: Hic omnia regit non ut anima mundi sed ut universorum Dominus. Omnipraesens est et in ipso continentur & moventur universa idque sine resistentia cum sit Ens non corporeus neque corpore resistiatur.

---

[1] Edleston, 147, Letter LXXV.
[2] The numbers in parentheses were inserted by Newton to indicate the revised paragraph order.

(4) Caeterum causam gravitatis nondum exposui neque exponendam suscepi siquidem ex phaenomenis colligere nondum potui. Enim non oritur ex vi centrifuga vorticis alicujus siquidem tendit non ad axem vorticis sed ad centrum Planetae.

(1) Hypothesis vorticum multis premitur difficultatibus. Ut Planetarum tempora periodica sint in proportione sesquialtera distantiarum a Sole, tempora periodica partium Vorticis deberent esse in eadem proportione distantiarum. Ut Planeta unusquisque radio ad solem ducto areas describat tempori proportionales, tempora periodica partium vorticis deberent esse proportione reciproca distantiarum a Sole. Ut vortices Lunarum conserventur et tranquille natent in vorticibus primariis, tempora periodica partium vorticis primarii deberent esse aequalia, Revolutiones corporum Solis et Planetarum et Cometarum, cum his omnibus hypothesibus discrepant. Motus Cometarum sunt summe regulares, & easdem leges cum Planetarum motibus observant, at cum vorticum motibus omnino discrepant & iisdem saepe contrarii sunt.

(5) Oritur ex causa aliqua quae penetrat ad usque centra solis et Planetarum sine virtutis diminutione, quaeque agit non in solas superficies particularum sed in omnem materiam usque ad centrum siquidem actio ejus quantitati materiae in corporibus universis proportionalis est, Oritur ex causa qua singulae corporum particulae agunt ad immensas distantias virtute decrescente in duplicata ratione distantiarum reciproce. Nam vis solis componitur ex viribus omnium particularum ejus & vires particularum omnium hac lege propagantur per omnes Planetarum orbes ut ex quiete Apheliorum Planetarum colligitur. Ex Phaenomenis certissimum est gravitatem dari et in omnia corpora secundum leges in superioribus descriptas pro ratione distantiarum agere, et ad motus omnes Planetarum et Cometarum sufficere, adeoque legem esse naturae quamvis causam legis hujus ex phaenomenis nondum colligere licuit. Nam hypotheses seu metaphysicas seu physicas seu mechanicas seu qualitatum occultarum fugio. Praejudicia sunt et scientiam non pariunt.

Quemadmodum Systema Solis Planetarum & Cometarum viribus gravitatis agitatur & partes ejus in motibus suis persevera[n]t, sic etiam minora corporum systemata viribus aliis agitari videntur & eorum particulae inter se diversimode moveri, & maxime vi electrica. Nam particulae corporum plurimorum vi electrica praedita [agitari] videntur & in se mutuo ad parvas distantias agere etiam absque frictione, et quae maxime electrica sunt, spiritum quendam per frictionem ad magnas distantias emitunt quo festucas & corpora levia nunc attrahunt nunc fugant nunc agitant diversimode.

Si vitra duo plana & polita & quam proxime contigua superficiebus parallelis in aquam stagnantem immergantur; aqua inter vitra ascendet supra superficiem aquae stagnantis & altitudo ascensus erit reciproce ut distantia vitrorum. Et hoc experimentum succeedit [*sic*] in vacuo Boyliano ideoque a gravitate atmosphaerae incumbentis non pendet. Partes vitri ad superficiem aquae ascendentis attrahi[un]t aquam ipsis proximam & inferiorem & ascendere faci[un]t. Attractio eadem est in variis distantiis vitrorum & idem pondus aquae attolit, ideoque aquam eo altius ascendere facit quo minor est distantia vitrorum Et simili de causa aqua ascendit in tubulis tenuibus vitreis idque eo altius quo tenuiores sunt tubulae, et liquores omnes ascendunt in substantiis spongiosis.

Vitra duo plana et polita longitudine viginti digitorum latitudine [   ] parabantur. Horum alterum horizonti parallelum jacebat, & ad unum ejus terminum gutta erat olei malorum citriorum. Alterum priori sic imponebatur ut vitra ad alterum eorum extremum se mutuo contingerent, ad alterum vero ubi gutta jacebat, a se invicem distarent intervallo quasi decimae sextae partis digiti, & vitrum superius contigeret guttam. Quo facto gutta statim incipiebat moveri versus concursum vitrorum. Et quo propius accedebat ad concursum vitrorum eo velocius movebatur. Succesit etiam hoc experimentum in vacuo. Et ortus est hic motus ab attractione vitrorum.

Si vitra ad concursum suum paululum attollerentur ut vitrum inferius inclinaretur ad horizontem gutta ascenderet, & vitrum superius positionem suam ad vitrum inferius servaret: gutta ascendendo tardius movebitur quam prius & quo major esset vitri inferioris inclinatio eo tardior erat motus guttae donec gutta quiesceret, pondere ejus attractionem vitrorum aequante. Sic ex inclinatione vitri inferioris dabatur pondus guttae et ex pondere guttae dabatur attractio vitrorum. Inclinationes autem vitri inferioris quibus gutta stabat in aequilibrio et distantiae guttae a concursu vitrorum exhibentur in Tabula sequente.

TRANSLATION

Projectiles suffer little resistance in our air. In the Boylian vacuum there is none, as may be gathered from bodies falling [in it]. The resistance of the air ceases in the heavens above the Earth's atmosphere and there bodies should move very freely and conserve their motion for a very long time. And thus they will obey the laws of gravity and perpetually describe the orbits in which they begin to move. For from the markedly eccentric motions of Comets it is demonstrated that celestial bodies can move in all directions with complete freedom. And all celestial bodies whether Planets or Moons or Comets conserve for a very long time the motions that they have once begun. But these motions cannot at their beginning arise from mechanical causes.

The six principal Planets revolve around the Sun in circles concentric with the Sun in the same order and in the same plane. Ten Moons revolve about the Earth, Jupiter and Saturn in concentric circles, in the same order and in the planes of the orbits of the Earth, Jupiter and Saturn. And these regular motions have not sprung from mechanical causes, since Comets are freely borne to all parts of the heavens in highly eccentric orbits. This most elegant structure of Sun and Planets could arise only from the wisdom and dominion of an intelligent and powerful Being. And if the fixed stars are the centres of similar systems, all these are under the same one dominion: This Being rules all things not as the soul of the world but as Lord of the Universe. He is omnipresent and in him all things are contained and move, and that without resistance since this Being is not corporeal and is not resisted by body.

Moreover I have not yet disclosed the cause of gravity nor have I undertaken to explain it since I could not understand it from the phenomena. For it does not arise from the centrifugal force of any vortex, since it does not tend to the axis of a vortex but to the centre of a Planet.

The hypothesis of vortices is pressed with many difficulties. So that the periodic times of the planets may be as the 3/2 power of their distances from the Sun, the periodic times of the parts of the vortex would have to be in the same ratio of the distances. So that each Planet by a radius drawn from the Sun may describe areas proportional to the times, the periodic times of the parts of the vortex would have to be in reciprocal proportion to the distances from the Sun.[1] So that the vortices of the Moons may be conserved and swim tranquilly in the primary vortices, the periodic times of the parts of the primary vortices ought to be equal; the Revolutions of the Sun and the Planets and the Comets do not agree with all these hypotheses. The motions of the Comets are exceedingly regular, and they observe the same laws as the motions of the Planets, but they differ from the motions of vortices in every particular, and are often contrary to them.

[Gravity] proceeds from some cause that penetrates to the very centres of the Sun and Planets without any diminution of its virtue, and which acts not on the surfaces of particles alone, but on all matter to the very centre since its action is proportional to the quantity of matter in all bodies. It proceeds from a cause by which the single particles of bodies act at immense distances with a virtue decreasing in the duplicate ratio of the distances reciprocally.[2] For the force of the Sun is composed of the forces of all its particles and the forces of all the particles are propagated according to this law through all the orbs of the Planets, as may be gathered from the quiescence of the aphelia of the Planets. From the phenomena it is very certain that gravity is given and acts on all bodies according to the laws described above in proportion to the distances, and suffices for all the motions of Planets and Comets, and thus it is a law of nature although it has not yet been possible to understand the cause of this law from the phenomena. For I avoid hypotheses, whether metaphysical or physical or mechanical or of occult qualities. They are harmful and do not engender science.

Just as the System of the Sun, Planets and Comets is put in motion by the forces of gravity and its parts persist in their

---

[1] The second edition has *duplicata ratione*.
[2] 'Reciprocally' is a mistake.

motions, so the smaller systems of bodies also seem to be set in motion by other forces, and their particles to be variously moved in relation to each other, and especially by the electric force. For the particles of most bodies seem to be set in motion by the electric force which they possess and to act upon each other at small distances even without being rubbed, and those which are most electric, when rubbed emit a spirit to great distances, by which straws and light objects are now attracted, now repelled and now moved in various ways.

If two polished plane pieces of glass, with their parallel surfaces as contiguous as possible, are immersed in still water, the water ascends between the pieces of glass above the surface of the water, and the height of ascent will be inversely as the distance between the glasses.[1] This experiment succeeds in the Boylian vacuum and so does not depend on the weight of the incumbent atmosphere. The parts of the glass near to the rising water attract the water close to and below themselves, and make it rise. The attraction is the same when the pieces of glass are at various distances apart and draws up the same weight of water, so that it causes the water to ascend higher, the closer the glasses are together. And for a similar reason water ascends in narrow glass tubes, and the higher when the tubes are narrower; and all liquids ascend in spongy substances.

Two polished plane pieces of glass were prepared, 20 inches long and [   ] wide.[2] One lay parallel to the horizon, and upon one end of it was a drop of oil of oranges. The other was so placed upon the first that they touched one another at the other end; and at that end where the drop lay they were about $\frac{1}{16}$th of an inch apart, and the upper glass touched the drop. As soon as this happened the drop began to move towards the meeting-point of the glasses. And the closer it was to the meeting-point, the faster it moved. This experiment also succeeds *in vacuo*. And the origin of this motion lies in the attraction of the glasses.

If the glasses are raised slightly at their meeting-point so that the lower glass is inclined to the horizontal, the drop will

---

[1] Cf. *Quaery* 31, 390–1.
[2] Cf. *Quaery* 31, 392, for an account of the same experiment, with slightly different figures.

ascend, and the upper glass will maintain its position with respect to the lower one. The ascending drop moves more slowly than at first and the greater the inclination of the glass, the slower the motion of the drop, until it will come to rest, its weight being equal to the attraction of the glass. Thus from the inclination of the lower glass the weight of the drop is given, and from the weight of the drop the attraction of the glasses is given. The inclinations of the lower glass by which the drop was maintained in equilibrium, and the distances of the drop from the meeting-point of the glasses, are shown in the following table [not given].

MS. C

Para. 1.: Hypothesis vorticum... (*practically as printed; sentences 2 and 3 are reversed*).[1]

Para. 2.: Projectilia in aere...

Para. 3.: Planetae sex...Entis intelligentis & potentis oriri potuit. Et si stellae fixae sint centra similium Systematum, subsunt haec omnia unius dominio. Hic omnia regit non ut anima mundi (nam corpus non habet) sed ut...dici solet. Æternus est et infinitus id est. Semper durat & ubique adest: nam quod nunquam nusquam est nihil est. An Deus erit nusquam cum momento temporis sit ubique? Certe. Omnipraesens est non per virtutem...subsistere non potest, et quae fingitur sine substantia subsistere jam fingisus [? fingitur] esse substantia. In ipso...omnipraesentia Dei. Eadem necessitate idem est semper et ubique. Totus est sui similis. Totus oculus, totus auris, totus cerebrum, totus manus, totus vis sentiendi intelligendi & agendi sed more minime humano, more incorporeo more nobis prorsus incognito. Vivit sine corde et sanguine praesens praesentia sentit et intelligit sine organis sensuum et sine cerebro agit sine manibus, et corpore minime vestitus videri non potest sed Deus est prorsus invisibilis. Iamvero si Deus Systema Solis et Planetarum in ordinem redegit, causae finales in Philosophia naturali locum habebunt, et quem in finem conditus est mundus, quos in fines membra animalium formata sunt quo consilio situm elegantem inter se habent licebit inquirere.

[1] Where the wording coincides very nearly with that of the *Principia* we have omitted the text of the draft, except for a few words to permit identification.

Hactenus phenomena...si modo Aphelia illa quiescant, et corpora impellit non in axes vorticum sed in centra Solis et Planetarum, et undique ad aequales a centro distantias aequaliter agit. Causam vero harum proprietatum ejus ex phaenomenis nondum potui invenire. Nam hypotheses seu mechanicas seu qualitatum occultorum fugio. Praejudicia sunt et scientiam non pariunt. Satis est quod gravitas revera detur, & agat secundum leges a nobis expositas & ad corporum coelestium et maris nostri motus omnes sufficiat.

Substantias rerum non cognoscimus. Nullas habemus earum ideas. Ex phaenomenis colligimus earum proprietates solas & ex proprietatibus quod sint substantiae. Corpora se mutuo non penetrare colligimus ex solis phaenomenis: substantias diversi generis se mutuo non penetrare ex phaenomenis minime constat. Et quod ex phaenomenis minime colligitur temere affirmari non debet.

Ex phaenomenis cognoscimus proprietates rerum & ex proprietatibus colligimus res ipsas extare easque vocamus substantias sed ideas substantiarum non magis habemus quam caecus ideas colorum. Ex phaenomenis solis colligimus corpora se mutuo non penetrare; substantias diversi generis se mutuo non penetre [*sic*] ne quidem ex phaenomenis constat. Sed Deum summum necessario existere in confessio est et eadem necessitate est semper et ubique. Unde etiam totus est sui similis, totus oculus, totus auris, totus cerebrum, totus brachium, totus vis   [ends]

Ex phaenomenis cognoscimus rerum proprietates, ex proprietatibus colligimus res ipsas existere easque vocamus substantias sed ideas substantiarum non habemus. Videmus tantum corporum figuras et colores, audimus tantum sonos, tangimus tantum superficies externas, olfacimus odores & gustamus sapores: substantias vel essentias ipsas nullo sensu, nulla actione reflexa cognoscimus, proindeque ideas earum non magis habemus quam caecus ideas habet colorum. Et ubi dicitur nos habere ideam Dei vel ideam corporis, nihil aliud intelligendum est quam nos habere ideam proprietatum vel attributorum Dei vel ideam proprietatum quibus corpora vel a Deo vel a seinvicem distinguuntur. Unde est quod de ideis substantiarum a proprietatibus abstractarum nullibi disputemus, conclusiones nullas ab eisdem deducamus.

Ideas habemus attributorum ejus sed quid sit rei alicujus substantia minime cognoscimus. Videmus tantum corporum figuras et colores, audimus tantum sonos, tangimus tantum superficies externas, olfacimus odores solos, & gustamus sapores; Intimas substantias nullo sensu nulla actione reflexa cognoscimus, & multo minus ideam habemus substantiae Dei. Hunc cognoscimus solummodo per proprietates et attributa [ends]

Prop. 1. Perparvas corporum particulas vel contiguas vel ad parvas ab invicem distantias se mutuo attrahere. Exper. 1. Vitrorum parallelorum. 2. Inclinatorum. 3. fistularum. 4. spongiarum. 5. Olei malorum citriorum.

Prop. 2. vel Schol. Attractionem esse electrici generis.

Prop. 3. Attractionem particularum ad minimas distantias esse longe fortissimam (Per Exp. 5) & ad cohaesionem corporum sufficere.

Prop. 4. Attractionem sine frictione ad parvas tantum distantias extendi ad majores distantias particulas se invicem fugere. Per exper. 5. Exper. 6. De solutione metallorum.

Prop. 5. Spiritum electricum esse medium maxime subtilem & corpora solida facillime permeare. Exp. 7. Vitrum permeat.

Prop. 6. Spiritum electricum esse medium maxime actuosum et lucem emittere. Exper. 8.

Prop. 7. Spiritum electricum a luce agitari idque motu vibratorio, & in hoc motu calorem consistere. Exper. 9. Corporum in luce solis.

Hic omnia regit...dici solet [*as printed, nearly*]. Aeternus est & Infinitus, seu durat ab aeterno in aeternum & adest ab infinito in infinitum. Duratio ejus non est nunc stans sine duratione neque praesentia ejus est nusquam. Cum unaquaque spatii particula sit semper, & unumquodque temporis momentum ubique certe Deus non erit nunquam nusquam. Semper durat et ubique adest. Omnipraesens est non per virtutem solam, sed etiam per substantiam: Nam virtus sine substantia subsistere non potest. Quod fingitur sine substantia subsistere, jam fingitur esse substantia. In ipso continentur & moventur universa...more minime corporeo, more nobis prorsus incognito. Quomodo Deus sentit & intelligit omnia non magis intelligere possumus quam caecus

357

intelligere quid sit videre. Vivit Deus sine corde et sanguine praesens praesentia sine organis sensuum et sine cerebro sentit & intelligit. Et cum corpore omni & fixura corporea prorsus destituatur videri non potest, nec audiri, nec tangi, nec sub specie rei alicujus corporei coli debet. Caetera omnia figuras corporeas induunt.

Jam vero si Deus Systema Solis et Planetarum in ordinem pulcherrimum redegit, si motum Planetis tali directione ac tali velocitate dedit ut in orbibus concentricis circum solem eodem ordine in eodem plano ferantur; si motum Lunis quatuor Jovialibus Jovi prorsus concentricum eodem ordine in eodem plano & motus consimiles Lunis Saturninis Lunaeque Terrestri dedit; & machinam tantorum corporum ad tantas distantias tam accurate constituere sit artis summae et summae potentiae: Si praeterea Cometae in orbibus valde excentricis moventur ut per orbes Planetarum citissime transeant et transitu suo quam minime perturbent Planetarum motus, & nullae sint in motibus Planetarum irregularitates nisi quae ex Cometarum attractionibus oriri potuerint; Si Cometarum aphelia in omnes caelorum regiones disponuntur ut haec corpora ubi tardissime moventur se mutuo quam minime trahant & mutuos motus quam minime perturbent; si fixarum translatio inter se per parallaxim annuam nondum observetur, observari autem posset si modo ad minuta quatuor secunda ascenderet, & inde sequatur distantias fixarum proximarum a Sole superare distantiam Terrae a Sole in ratione plusquam 100000 ad 1; tantae autem sint distantiae fixarum et a Sole et a se invicem ne Systemata eorum in se mutuo cadant; certe causae finales in Philosophia naturali locum habent, & quem in finem conditus est hic mundus, quo consilio orbes coelestes situm tam elegantem inter se habeant, qua potentia corpora coelestia motus suos obtinuerunt & ad distantias tantas locata sunt, uti et quos in fines membra animalium formata sunt & quoque authore situm & structuram tam commodum tam elegantem nacta sunt inter se jam certe licebit inquirere.

Deus est nomen relativum et refertur ad servos ejus. Dicimus enim Deum meum, Deum nostrum, Deum vestrum, Deum servorum ejus id est Dominum meum supremum, Dominum nostrum supremum, Dominum vestrum supremum,

Dominum servorum. At non dicimus Ens perfectum meum, Ens perfectum nostrum, Ens perfectum vestrum, Ens perfectum servorum. Non dicimus Aeternus noster, aeternus vester, infinitus noster, infinitus vester. Qui Ens perfectum dari demonstraverit, & Dominum seu παντοκράτορα universorum dari nondum demonstraverit, Deum dari nondum demonstraverit. Ens aeternum, infinitum, sapientissimum, summe perfectum sine dominio non est Deus sed natura solum. Haec nonnullis aeterna, infinita, sapientissima, et potentissima est, & rerum omnium author necessario existens. Dei autem dominium seu Deitas non ex ideis abstractis sed ex phaenomenis et eorum causis finalibus optime demonstratur.

Prop. 8. Lucem incidendo in fundum oculi vibrationes excitare quae per solida nervi optici capillamenta in cerebrum propagatae visionem excitant.

Schol. Omnem sensationem omnemque motum animalem mediante spiritu electrico peragi.

Prop. 9. Vibrationes spiritus electrici ipsa luce celeriores esse.

Prop. 10. Lucem a spiritu electrico emitti refringi reflecti et inflecti.

Prop. 11. Corpora homogenea per attractionem electricam congregari heterogenea segregari.

Prop. 12. Nutritionem per attractionem electricam peragi.

TRANSLATION[1]

And if the fixed stars be the centres of similar systems, all these are under the same one dominion. This Being rules all things not as the soul of the world (for he has no body)...He is Eternal and infinite. He endures for ever and is everywhere present: for what is never and nowhere is nothing. Can God be nowhere when the moment of time is everywhere? Certainly. He is omnipresent not only virtually but substantially, for virtue cannot subsist without substance, and of that which is imagined to subsist without substance, the substance is already imagined. In him are all things contained and moved, yet God and matter do not interfere. God suffers nothing

---

[1] The translation begins with the third paragraph of the MS.

from the motions of bodies, and these suffer no resistance from the omnipresence of God. By the same necessity he is always and everywhere the same. He is wholly like to himself; he is all eye, ear, brain, hand; all power of feeling, understanding and acting, but in a manner not at all human, in an incorporeal manner, in a manner utterly unknown to us. He lives without heart and blood, a present presence, he feels without organs of sensation, understands without brain, acts without hands and cannot possibly be seen in bodily form, but God is utterly invisible. Indeed, if God did reduce to order the System of the Sun and Planets, final causes will have a place in natural philosophy, and it will be legitimate to inquire to what end the world was founded, to what ends the limbs of animals were formed, and by what wisdom they have so elegant an arrangement. Hitherto we have explained the phenomena of the heavens...provided that those Aphelia are stationary, and it impels bodies not towards the axes of vortices but towards the centre of the Sun and the Planets, and acts everywhere equally at equal distances from the centre. Indeed, I could not discover the cause of these properties from its phenomena. For I avoid hypotheses whether mechanical or of occult qualities. They are harmful and do not engender science. And it is enough that gravity does really exist and acts according to the laws which we have explained, and serves to account for all the motions of the celestial bodies and of our sea.

We do not know the substances of things. We have no idea of them. We gather only their properties from the phenomena, and from the properties [we infer] what the substances may be. That bodies do not penetrate each other we gather from the phenomena alone; that substances of different kinds do not penetrate each other does not at all appear from the phenomena. And we ought not rashly to assert that which cannot be inferred from the phenomena.

We know the properties of things from phenomena, and from the properties we infer that the things themselves exist and we call them substances: but we do not have any more idea of substances than a blind man has of colours. From the phenomena alone we gather that bodies do not penetrate each other; that substances of different kinds do not penetrate each other is not even evident from the phenomena. All allow that a Supreme God exists necessarily and by the same necessity he exists always and everywhere. Whence also he is wholly like to himself; he is all eye, all ear, all brain, all arm, all force.

From phenomena we know the properties of things, and from the properties we infer that the things themselves exist and we call them substances: but we do not have any idea of substances. We see but the shapes and colours of bodies, we hear but sounds, we touch but external surfaces, we smell odours and taste flavours; but we know the substances or essences themselves by no sense, by no reflex action, and therefore we have no more idea of them than a blind man has of colours. And when it is said that we have an idea of God or an idea of body, nothing other is to be understood than that we have an idea of the properties or attributes of God, or an idea of the properties by which bodies are distinguished from God or from each other. Whence it is that we nowhere argue about the ideas of substances apart from properties, and deduce no conclusions from the same.

We have ideas of its attributes but we do not know at all what the substance of anything may be. We see but the shapes and colours of bodies, we hear but sounds, we touch but external surfaces, we smell only odours and taste flavours; by no sense, by no act of reflection do we know the innermost substances, and much less have we any idea of the substance of God. Him we know only by his properties and attributes.

Proposition 1. That very small particles of bodies, whether contiguous or at very small distances, attract one another. Experiment 1. Of parallel pieces of glass. 2. Of inclined [pieces of glass]. 3. Of tubes. 4. Of sponges. 5. Of oil of oranges.

Proposition 2. Or Scholium. That attraction is of the electric kind.

Proposition 3. That attraction of particles at very small distances is exceedingly strong (by Experiment 5) and suffices for the cohesion of bodies.

Proposition 4. That attraction without friction extends only to small distances and at greater distances particles repel one another. By Experiment 5. Experiment 6. On the solution of metals.

Proposition 5. That the electric spirit is a most subtle medium and very easily permeates solid bodies. Experiment 7. It permeates glass.

Proposition 6. That the electric spirit is a medium most active and emits light. Experiment 8.

Proposition 7. That the electric spirit is set in motion by light and this is a vibratory motion, and of this motion heat consists. Experiment 9. Of bodies in the light of the Sun.

This Being rules all things...He is eternal and infinite, or endures from eternity to eternity and is present from infinity to infinity. His duration is not now standing without duration,[1] nor is his presence nowhere. Since every particle of space is always, and every moment of time is everywhere, certainly God will not be *never* and *nowhere*. He endures always and is present everywhere. He is omnipresent not in virtue only, but also in substance, for virtue cannot subsist without substance. What is feigned to exist without substance, is already feigned to be substance. In him the universe is contained and moved...in a manner not at all corporeal, in a manner quite unknown to us. We can no more comprehend how God understands and feels everything than a blind man can comprehend what seeing is. God lives without heart and blood, a present presence he feels without organs of sensation and comprehends without a brain. And since he is utterly destitute of all body, and of bodily trappings, he cannot be seen nor heard nor touched nor should he be revered under any bodily semblance whatever. All other things assume corporeal forms.

If indeed God reduced the System of the Sun and Planets to a most beautiful order; if he gave motion to the Planets in such direction and of such velocity that they are borne in concentric orbits about the Sun, in the same order and in the same plane; if he gave motion to the four moons of Jupiter which is absolutely concentric with the Planet in the same order and in the same plane, and similar motion to the moons of Saturn and of the Earth—and to set up so accurately a machine of so many bodies at such great distances requires consummate art and consummate power; if moreover Comets move in exceedingly eccentric orbits so that they pass very quickly

---

[1] Does Newton mean that God's duration is never suspended?

through the orbs of the Planets and hardly disturb the motions of the Planets in their passage; and if there are no irregularities in the motions of the Planets other than those that may arise from the attractions of Comets; if the aphelia of comets are so disposed in all parts of the heavens that where these bodies move most slowly they attract each other least, and disturb each others' motions least; if a relative displacement of the fixed stars due to annual parallax has not yet been observed (though it could be observed if it amounted to but four seconds of arc) whence it should follow that the distances of the fixed stars nearest to the Sun exceed the distance of the Earth from the Sun in the ratio of more than 100,000 to 1— so great indeed are the distances of the fixed stars, from the Sun and from each other, lest their systems should fall towards one another:[1] certainly final causes have a place in natural philosophy; and now it certainly will be legitimate to inquire to what end this world was founded, by what wisdom the celestial orbs have such an elegant arrangement, from what power the celestial bodies obtain their motions, and are placed at such great distances, how and to what ends the limbs of animals are formed and from what author they received so commodious and elegant an arrangement and structure.

God is a relative name and refers to his servants. For we say my God, our God, your God, God of his servants, that is, my supreme Lord, our supreme Lord, your supreme Lord, Lord of servants. But we do not say my Perfect Being, our Perfect Being, your Perfect Being, the Perfect Being of servants. We do not say our Eternal, your Eternal, our Infinite, your Infinite. He who shall demonstrate that there is a Perfect Being, and does not at the same time demonstrate that he is Lord of the Universe or Pantokrator, will not yet have demonstrated that God exists. A Being eternal, infinite, all-wise and most perfect without dominion is not God but only Nature. This is in a manner eternal, infinite, all-wise, and most powerful, and the necessarily existing author of all things; yet the dominion or Deity of God is best demonstrated not from abstract ideas but from phenomena, by their final causes.

---

[1] In the last clause, here literally translated, Newton seems to have become lost in his construction, but the meaning is clear enough.

Prop. 8. That light falling on the bottom of the eye excites vibrations which, propagated to the brain through the solid fibres of the optic nerves excite vision.

Scholium. That all sensation and all animal motion are accomplished by means of the electric spirit.

Prop. 9. That the vibrations of the electric spirit are swifter than light itself.

Prop. 10. That light is emitted, refracted, reflected and inflected by the electric spirit.

Prop. 11. That homogeneous bodies are brought together by electric attraction, and heterogeneous ones disassociated.

Prop. 12. Nutrition is accomplished by electric attraction.

PART V
EDUCATION

INTRODUCTION

For almost the whole of his active scientific life, from 1669 to 1696, Newton was engaged in regular teaching. The number of lectures he gave in each academic year was small indeed— probably no more than eight in most years—but the effort in research that they represented was large. Everything that Newton published in mathematics and science was rehearsed in his lecture courses, the manuscript copies of which survive in the Cambridge University Library. Besides his lectures, Newton gave private instruction in his rooms at Trinity, as he was bound to do under the statutes governing his professorship. He was a Fellow of a College, and must have had considerable experience of the manner in which College teaching was conducted, and in which appointments to Fellowships and Lectureships were made.

It may be less obvious, for the fact seems to run counter to the harsher traits in Newton's character, that he also took considerable pains to expound science plainly and simply when a specific request was made to him. His correspondence with such near-equals as Locke and Bentley, in which he sought to facilitate their understanding of his celestial system, is well known; so is the patient exposition contained in some of the letters Newton wrote concerning his optical discoveries. With other more humble correspondents who on occasion wrote to him of their difficulties in understanding his own or others' writings Newton was not less patient. He was far from unwilling to take trouble to make himself better understood. And he wrote approvingly of attempts to form clubs for scientific study in provincial towns; indeed the flourishing Spalding Society in Lincolnshire, founded in 1712, regarded him as their direct patron.[1]

Newton's letters on the education of the 'mathematical boys' at Christ's Hospital have already been published, and these alone give a fair notion of the seriousness with which he

---

[1] Cf. C. R. Weld, *A History of the Royal Society* (London, 1848), I, 422–3. Even in 1684–5, while deeply engaged in writing the *Principia*, Newton expressed interest in helping to promote in Cambridge the formation of a 'Philosophical Meeting...so far as I can do it without engaging the loss of my own time in those things' (Weld, I, 305).

would examine questions of educational policy.[1] The first document printed below shows that he equally contemplated a reform of University teaching. The others show Newton in an unfamiliar guise, as a frankly popular expositor. The purpose of the elementary presentations is unknown; but from their form they can hardly be responses to the questions of some correspondent. Perhaps they are merely sketches—like many others—of works which Newton for a moment thought of writing, but never continued beyond the first few pages. One is reminded of his remark in the Preface to the *Principia* that he had originally 'composed the third Book in a popular method, that it might be read by many'. The treatise *De Systemate Mundi* (if this is the draft to which he refers)[2] is hardly the lightest of reading, but the sketches printed here do indicate that Newton was capable of a really popular approach to science.

[1] Edleston, 279–99; cf. L. T. More, *Isaac Newton*, 400–4.
[2] Cf. above, p. 233.

1

OF EDUCATING YOUTH IN THE
UNIVERSITIES

MS. Add. 4005, fols. 14–15

There is no direct indication of the date of composition of this piece, but it may be attributed to about 1690. Before the crisis of the Revolution of 1688 Newton seems to have taken little interest in University affairs; yet when the crisis came he is said to have been one of the leading opponents in the University of James II's Catholic policy. Later, he twice represented the University in Parliament, and was a candidate for the Provostship of King's College. It seems likely that his engagement in public affairs may have led him to develop a scheme for University reform; on the other hand, he may have set down his ideas for the benefit of another person—for example, Richard Bentley—after leaving Cambridge in 1696. In either case, the third paragraph confirms a late date, for its reference to 'gravity and its laws' makes it certain that the paper was written after 1687.

Newton's reforming urge was not restricted to his own College, for the new plan was intended to apply to all the Colleges in the University. At this time, of course, teaching was entirely in the hands of the Tutors and Lecturers appointed by the several Colleges: there were no University teaching offices apart from the small group of Professors, whose lectures were infrequent and probably beyond the capabilities of most undergraduates. (Certainly this was true of Newton himself!) Newton's proposed reforms touch upon the teaching and direction of studies of undergraduates, the selection of College Lecturers and Tutors, and the discipline of students. It is worth noting that they provide for a considerable scientific element in teaching, though probably not one exceeding the bounds of what was already offered in the better Colleges.

Newton's abhorrence of religious tests and oaths is typical. In his own case, a dispensation was procured from the Crown to enable him to hold his fellowship without taking Orders, so long as he remained Lucasian professor. One is also reminded of Boyle's similar scruples about oath-taking, which furnished his overt reason for declining the Presidency of the Royal Society.

Undergraduates to be instructed by a Tutor an Humanity Lecturer a Greek Lecturer a Philosophy Lecturer & a Mathematick

Lecturer. The Tutor to read Logicks, Ethicks, the Globes & principles of Geography & Chronology in order to understand History, unless the Lecturers have time for any of these things.

The Humanity & Greek Lecturers to set tasks in Latin & Greek Authors once a day to the first year [students] & once a week to the rest & to examine diligently & instruct briefly & to punish by exercises such faults as concern Lectures & to appoint the reading of ye best Historians.

The Philosophy Lecturer to read first of things introductory to natural philosophy, time, space, body, place, motion & its laws, force, mechanical powers, gravity & its laws, Hydrostaticks, Projectiles solid & fluid,[1] circular motions, & ye forces relating to them: & then to read natural Philosophy beginning wth ye general systeme of the world & thence proceeding to ye particular constitution of this earth & the things therein, Meteors, elements, minerals, vegetables, animals & ending wth Anatomy if he have skill therein. Also to examin in Logicks & Ethicks.

The Mathematick Lecturer to read first some easy & usefull practical things, then Euclid, Sphericks, the Projections of the Sphere, the construction of Mapps, Trigonometry, Astronomy, Opticks, Musick, Algebra, &c. Also to examin & (if ye Tutor be deficient) to instruct in the principles of Chronology & Geography.

Several sciences wch depend not on one another are all learnt in less time together then successively, the mind being diverted & recreated by the variety & put more upon the stretch. And therefore diverse of these Lectures may proceed together: suppose the Tutor's after morning chappel, the Greek or Philosophy Lecturers two hours after, & ye Humanity & Mathematick in the afternoon. The Tutor to accompany his pupills to the Philosophy & Mathematick Lectures & to examin them the next morning both in those Lectures & in his own & make them understand where they hesitate. These two Lecturers to read five days in the week & wth ye other two to examin the sixt. Each Lecturer to read the same day successively to two or three years under several Tutors. Their Lectures to begin with Michaelmas Term & continue till the Commencement, the Tutors to begin the Commencement

---

[1] I.e. jets of liquid, fountains, etc.

before. The Greek & Humanity Lecturers to set bigger tasks in ye vacations then in reading time proportionally to ye spare hours of ye Students.

A Monitor to note those who miss Lectures, & give their names to ye Humanity Lecturer who shall punish them not by pecuniary mulcts but by tasks of making verses, Themes, Epistles, or getting any thing wthout book. All pecuniary mulcts of Undergraduates to be abolished, & Exercises, Admonitions, Recantations, & expulsions (according to ye nature of the crime) to succeed in their room.

In the long Vacation between ye Commencement & Michaelmas, the Tutor shall take care that his pupills read over all the last years Lessons again by themselves, & at ye end of ye Vacation they shall be examined again & those who are at any time found not fit to go on, turned down to ye Lectures of the year below that they do not retard the Lecturer & be an ill example to others.

The Lecturers to be chosen every three years, & the elections after the first institution to be in this manner. All those who have at any time been Lecturers shall chuse four out of their number one for each office, & the Master & Seniors of the College shall chuse other four who have not yet executed ye office & those eight with the Master shall by ballotting chuse four out of their number. No regard to Seniority or any thing but merit. The Lecturers to chuse yearly a publick Tutor, & to reprehend or displace him if there be reason. This Tutor wthout a new election to take none but those admitted in his year, untill their course of Lectures be gone through. No private Tutor to take two yeares together. All Sizars, poor Scholars & scholars of ye house to be under publick Tutors, except Westminster Scholars in Trin. Coll. when ye Tutor is of another school.

For encouraging able & fit men to accept of ye Readers [Lecturers'] places, their Fellowships during their office shall be doubled by the addition of four other Fellowships kept vacant for that purpose, one for each, unless some other competent provision be made for any of them.[1] And because the

---

[1] Each Fellow was paid a 'dividend' consisting of so many 'shares'. The value of each share was determined by dividing the disposable income of the College by the total number of shares. Thus a lecturer holding a double Fellowship would receive a double dividend, or stipend.

Philosophy & Mathematick Lecturers' Office is laborious, for encouraging them to diligence, none shall be compelled to come to their Lectures, but all that will be auditors shall offer each of them a quarterly gratuity, suppose of 10s ye sizar, 12 or 15s ye Pensioner & 20 or 25s ye Fellowcommoner.[1] And to encourage Auditors those shall be preferred to Scholarships & Fellowships wch are best skilled in all sciences, caeteris paribus, & shall have seniority of those who come not to Lectures. This Institution to begin in ye greater Colleges & be carried on in ye rest as men qualified & revenues can be had. In smaller Colleges ye Mathematick Lecturer may be omitted, & only a power granted ye College of instituting one when they can. Also ye Greek Lecturers office may be supplied by the Humanity Lecturer where it shall be thought fit. A gratuity to be given by all the first year to ye Greek & Humanity Lecturers.

For securing the Tutor & making his office desirable by fit persons, every student at his admission to deposit caution money in the hands of the Bursar of the College, suppose 10 or 12 lbs ye Sizar, 16 or 20 lbs ye Pensioner & 30 or 40 lbs the Fellowcommoner. And in case any Pupill at the end of any Quarter be in his Tutor's debt & do not discharge it within six weeks after his receipt of ye Qter Bill the Bursar to discharge it & return back the residue upon demand, & the Tutor forthwith upon pain of forfeiting his office to send home the Pupill. Yet may the Pupill be received again wth a new supply of money. This institution to be universal. The Master & Seniors to regulate the expenses of all under Tuition by certain limits common to them all & the Senior Dean to read over & signe all their Qter Bills. Extravagant Pupills after one admonition to be sent away.

Fellow-Commoners to perform all exercises in their course & to be equally subject to their Tutors & Governours wth other Scholars & alike punishable by exercises & those who are resty or idle to be sent away least they spoile others by their example. They shall read Geography Chronology & Mathematicks ye first year. All students who will be admitted to Lectures in naturall Philosophy to learn first Geometry & Mechanicks. By mechanicks I mean here the demonstrative doctrine of

[1] A Fellow-commoner was (and is) a matriculated member of the College, not yet admitted to a degree, who lived with the Fellows though not a Fellow himself.

forces & motions including Hydrostaticks. For wthout a judgment in these things a man can have none in Philosophy.

Whenever the major part of all the Mechanick Lecturers in the University shall desire a Master to teach Fellow-commoners & others Arithmetick & Designing, the University shall allow him 10 lb yearly out of their common chest, & he shall observe the orders of the Mathematick Lecturers & be placed or displaced by the major part of them at pleasure.

All Graduates wthout exception found by the Proctors in Taverns or other drinking houses, unless wth Travellers at their Inns, shall at least have their names given in to ye Vice-Chancellor who shall summon them to answer it before the next Consistory. The Deans to visit the Chambers of all Undergraduates once at least every week upon pain of forfeiting 10s to the Lecturers for every omission.

Fasting nights have a shadow of religion wthout any substance. Tis only supping more plentifully out of the publick Hall. And this does great mischief by sending young students to find suppers abroad where they get into company & grow debauched. Whether would it not be better to licence Undergraduates to sup Together in such place or places as the Dean shall appoint, wth a Monitor to note ye names of the absentees?

All these Lectures to consist in extempory explications of books in such an easy short & clear manner as may be most profitable to the Auditors. And if any Lecturer or other person shall compose any Treatise wch shall be preferred & used by ye major part of the Mathematick or Philosophie Lecturers, the University shall give the Author either 20 lb or if those Lecturers request it, 30, 40, or 50 lb out of their common chest.

Commissioners to be appointed for some years to set on foot inspect & amend ye institution.

No oaths of office to be imposed on the Lecturers. I do not know a greater abuse of religion then that sort of oaths they being harder to be kept then ye Jewish Law, so that yearly absolutions have been instituted. The Papists who believed such absolutions, might be excused for instituting such oaths, but we have no such doctrine & yet continue their practises. Admonitions & pecuniary mulcts for neglect of duty are less cruel punishments then ye consequence of perjury, & may be as effectual.

2

COSMOGRAPHY

MS. Add. 4005, fols. 21–2

This is a brief, popular exposition of physical astronomy prepared for an unknown purpose and left unfinished. Newton did not even trouble to fill in the figures of the first chapter. Its chief interest lies in its exposition of the idea that the stars (including the Sun) are distributed in space in an approximately uniform manner, an idea later explored by Halley, and by Sir William Herschel.

The draft is holograph and is complete as it stands although it breaks off abruptly. The reference to the five satellites of Saturn proves that it was written after 1684, when the fourth and fifth were discovered by J. D. Cassini. It seems to be a shorter version of 'Phaenomena'.

Chap. 1

Of the Sun & Fixt Starrs

The Universe consists of three sorts of great bodies, Fixed Stars, Planets, & Comets, & all these have a gravitating power tending towards them by which their parts fall down to each of them after the same manner as stones & other parts of the Earth do here towards the earth & by means of this gravity it is that they are all sphericall. For ye ☉ & ☽ appear round to ye naked eye & all ye other Planets appear round through Telescopes & ye earth casts a round Shaddow upon the Moon in Eclipses & has been sailed about & by reason of its round figure the Pole star rises & falls as we sail north and south. And as in this earth so in all other great bodies the gravity of their parts makes them endeavour downwards or towards the center of those bodies untill they stand round about the bodies at an equal height or distance from the center, that is untill the bodies be round. The fixt Stars are very great round bodies shining strongly with their own heat & scattered at very great distances from one another throughout the whole heavens. Those wch are nearest to us appear biggest & those wch are further of appear less & less till they vanish out of sight & cannot be seen without a Telescope. For by a good Telescope are discovered very great numbers of stars wch appear not to ye naked eye & ye better ye Telescope is the greater number of starrs are discovered by it insomuch that their number seems to be without limit.

The stars rest in their several places throughout the whole heavens. For they keep always the same position amongst one another & are therefore called fixt. Yet they seem (together with ye whole heavens) to move round about us once in 24 hours, but this happens because the Earth turns round in that time the contrary way as shall be explained hereafter.

Our Sun is one of ye fixt Stars & every fixt star is a Sun in its proper region. For could we be removed as far from ye Sun as we are from ye fixt stars, the Sun by reason of its great distance would appear like one of ye fixt stars. And could we approach as neare to any of ye fixt Stars as we are to ye Sun, that Star by reason of it nearness would appear like our Sun.

The light of ye Sun is about $[9.10^8]$ times greater in our region then that of one of ye brightest of ye fixt stars called stars of ye first magnitude. And if we were twice as far from ye Sun his light would be four times if thrice as far it would be nine times less, if four or five or six or 7 times as far it would be 16, or 25, or 36 or 49 times less & if we were $[3.10^4]$ times as far his light would be $[9.10^8]$ times less & by consequence equal to that of a star of ye first magnitude. So then the nearest fixt stars are about $[3.10^4]$ times further from us then the Sun.[1] And so far as ye nearest fixt stars are from our Sun, so far we may account ye fixt stars distant from one another. Yet this is to be understood wth some liberty of recconning. For we are not to account all the fixt starrs exactly equal to one another, nor placed at distances exactly equal nor all regions of the heavens equally replenished with them.

For some parts of the heavens are more replenished wth fixt stars then as the constellation of Orion wth greater or nearer stars & the milky way wth smaller or remoter ones. For ye milky way being viewed through a good Telescope appears very full of very small fixt stars & is nothing else then ye confused light of these stars. And so ye fixt clouds & cloudy

[1] The figures in brackets are found in *Phaenomena*.

stars are nothing else then heaps of stars so small & close together that without a Telescope they are not seen appart, but appear blended together like a cloud.

Were all the fixt stars equal & placed at equal distances from one another, the number of the stars next about us would be 12 or 13, those next about them 50, those next about them 110, those next about them 200 [those next about them 300, those next about them 450] or thereabouts.[1] And tho their magnitudes & distances be not equal yet this affords ye true reason why the smallest starrs are the most numerous. For there are about 15 stars of ye first magnitude 50 or 60 of ye second, 200 of ye 3d, 300 or 400 of ye 4th.

The fixt stars are bodies subject to various changes. For there are seen frequently spots upon the Sun some of wch are darker some brighter then the rest of his body & these spots are generated & corrupted like scum upon a pot & seldome last above a month & while they last they move round with him once in about 26 days. And to the like mutations the fixt stars are subject: For some of them have grown brighter others darker, some have vanished others appeared anew & some have appeared & disappeared & appeared again by many vicissitudes.

Chap. 2

Of the Earth & Planets

The Earth & Planets are great round moving opake dark bodies illuminated by the Sun & shining only by his light reflected from them. They are all of them much less then the Sun & some of them much less then others & are called Planets because they move. For a Planet in Greek signifies a wandering body. They move in circles or circular Orbs about other bodies much bigger then themselves. The bigger of them called Saturn, Jupiter, Mars, The Earth, Venus & Mercury move about ye Sun & the less about ye bigger. For there are five small planets moving about Saturn four about Jupiter & one about ye Earth. This last we call ye Moon. It appears great because it is very neare to us. For whenever it passeth

---

[1] The number of points, separated at least a radius apart, that can be arranged on the surface of a sphere is 12. On the next sphere, of double radius, if the separation is the same the number of points is multiplied by 4, on the next largest again by 9, and so on.

over any other of ye Planets or Stars it covers & hides them from or sight by coming between them & us & therefor is nearer to us then any of the rest. The Planets about ye Sun are called Primary Planets & those about Saturn Jupiter & the Earth secondary ones. And as the secondary planet about the Earth is called our moon so the secondary Planets about Jupiter & Saturn are by Astronomers called the Moons of Jupiter & Saturn. All the Planets move about in their Orbs the same way & almost in the same Plane & the motions are slower wch are done in the bigger orbs.

3

PHAENOMENA

MS. Add. 4005, fols. 45–9

This is a holograph draft in English, written on the single sides of five sheets. The blanks left for the tabulated information in fol. 49 have never been filled in; the Table missing from Phaenomenon 11 is to be found in the *Principia*, p. 403; that for Phaenomenon 12 occurs first in the second edition of the *Principia*, p. 360; that for Phaenomenon 13 in the first edition, p. 404.

Phaenomena 7–10 are summarized in the *Principia*, Book III, Hypothesis VI. (This in turn became Phaenomenon III in the second and third editions.)

Phaenomenon 11 corresponds roughly with Hypothesis V, Phaenomenon 13 with Hypothesis VII; these became Phaenomena I and II in later editions of the *Principia*. Phaenomenon 12 has no counterpart in the first edition, but corresponds to Phaenomenon II in the second. Phaenomenon 14 is not enunciated as such in the *Principia*, but may be compared with Prop. XIII of Book III.

Since the figures are more complete in *Phaenomena* than in the corresponding passages of *Cosmography* it seems natural to suppose that the former piece was written later. Both drafts are obviously post 1685 (as the reference to Saturn's satellites proves). The use of the term *Phaenomena*, which in the second edition of the *Principia* replaced the word *Hypotheses* used in the first edition, suggests a date of composition for this draft considerably after 1687.

No occasion for the composition of such a document in English is known. Certainly Newton never drafted material for the *Principia* or its revisions in English. It may perhaps be compared with Section IV, No. 2, *On Motion in Ellipses*, addressed to Locke, also in English and also dealing with material covered in the *Principia*, though it does not resemble the original as closely as this does.

Phaenom. 1. The earth is a round body in the form of a globe. For men have often sailed round about it, & in sailing from north to south the stars about the north pole descend lower & lower & at length sink under the horizon & the stars about ye south pole ascend from under the horizon & rise higher & higher & the contrary happens in sailing from south to north. And by the Eclipses of the Moon the shadow of the earth is

378

observed to be round. For the Moon being eclipsed by passing into this shaddow, Astronomers by the round figure & bigness of this shadow know, how to predict the form & duration of the Eclipses of the Moon.

Phaenom. 2. The Sun is a round body in the form of a globe & turns round about an axis in the space of about 26 or 27 days. For there appear frequently upon his body dark spots wch sometimes last above a month & go round about him in the space of about 26 days, spending one half of the time on this side of his body & the other half beyond it. And the spots appeare broader & swifter when they are seen in the middle of the Sun's body then when they approach to the side of it, the edge of the spot being then turned towards the eye of the spectator.

Phaenom. 3. The Moon is a round body in the form of a globe & full of mountains, & turns round about an axis once in a month so as to keepe the same side towards the earth without much variation. For the Moon appears to us like a globe illuminated by the sun on that side wch is towards him & dark on the other side. When she is in the full, her illuminated side is towards us. When she appears new, her dark side is towards us & we see only a small edge of her illuminated side. When she is in the quarters one half of her illuminated side & one half of her dark side are towards us. And all her appearances may be represented by a globe placed at a little distance from us in various positions to the sun, so that we may see more or less of that side of the globe wch is illuminated by the Sun. The roundness of her body is manifested also by the shadows wch fall from her mountains. For through good Telescopes her body appears full of mountains with shadows falling from them opposite to ye Sun. And as the shadows of our mountains are longest at Sun-rise & grow shorter till noon & then lengthen again till Sun-set, so the shadows of the mountains in the moon are longest when the Sun begins first to illuminate ye mountains & grow shorter & shorter till it be full Moon & then grow longer & longer again till the Sun ceases to enlighten the mountains. This roundness is also manifested by the libration of her body. For when by librating or rolling to & fro she turns her body a little eastward, the spots on her eastern limb vanish out of sight & go behind her body & others come from behind her body at her western limb. And the contrary happens when she turns her body a little westward.

Phaenom. 4. The Planets Mercury Venus Mars & Jupiter are round bodies in the form of globes, & move about their several axes. For Mercury & Venus appeare like full moons when beyond the Sun, like new moons when on this side of him & like Quarter Moons when over against him. And Mars appears like a full Moon when either opposite to the Sun or almost beyond him, & gibbous when over against him. And Venus hath a lucid spot upon her body wch revolves in 23 hours. Mars hath spots upon his body wch revolve in 24 hours 40 minutes. And Jupiter hath spots upon his body wch revolve in 9 hours 56′. .

Corol. By these instances it may be concluded that the stars & Planets are round bodies in the form of globes & revolve about their proper axes.

Phaenom. 5. The Sun & fixt stars are lucid bodies shining with a very strong light: but the earth & Planets are dark bodies shining only by the reflected light of the Sun. When the Moon moves over a fixt star & eclipses it, the star vanishes at once in the twinkling of an eye as the light of a candle vanishes in a moment when the candle is covered wth an extinguisher. And in ye end of the Eclips the star reappeares wth its whole light in as short a time. The whole time in wch the star either vanishes or reappears is less than a quarter of a second of a minute of a degree. And in that time the Moon moves but an eighth part of a second minute of a degree. And therefore the apparent diameter of a fixt star is less than an eighth part of a second minute of a degree, whereas the apparent diameter of the Planet Saturn is about 29 second minutes. The apparent diameter of Saturn is therefore above 160 times greater then that of a fixt star. And therefore if Saturn shone with as strong a light in proportion to his bignes as the fixt stars do, he would appear above 160×160 times more lucid then ye biggest fixt stars do: whereas all his light (when he is destitute of his ring,) appears but equal to that of a star of ye first magnitude. The fixt stars therefore are exceedingly more lucid then the Planets & so are bodies of another kind & by the great strength of their light resemble the Sun. The whole light wch comes from a hemisphere of the Sun or falls upon a hemisphere of Saturn's orb is to so much of his light as falls upon the body of Saturn as the surface of the hemisphere of Saturn's Orb is to the surface of Saturn's apparent diameter, that is as the Radius of a circle to the versed sine of 10″, or as 10000000000 to 11-3/4 or about 900000000 to 1. And since in removing from a lucid body the light thereof decreases in a duplicate proportion of the distance, if we were 30000 thousand[1] times remoter from the Sun then we are at present, his light would appear equal to that of Saturn without his ring or to that of a star of the first magnitude. And at that distance his apparent diameter would be 30000 times less then it is at present that is, it would be about the 15th or 16th part of a second minute, or of about the same bigness wth the diameter of a fixed star of the first magnitude. For we have shewed that their diameters are less then the eighth part of a second minute. So then the fixt stars shine wth a light of their own wch is about as strong & vigorous in proportion to their apparent magnitudes as the light of the Sun is in proportion to his apparent magnitude.

Phaenom. 6. The Sun & fixt stars appear like great lights scattered through all the heavens at great distances from one another for illuminating all places. Let us imagin that great Lamps were scattered through all the heavens at the distance of a mile or two from one another & that a man was lifted up into the heavens to view them & placed neare one of them so that it might appeare large & fulgid like the Sun: the rest of them would appear round about him like stars & those next about him would appear bigger then the rest like stars of the first magnitude & would be about 12 or 14 in number. And the lamps next beyond those would appeare like stars of the second magnitude & be about 48 or 50 in number, & the lamps next beyond those would appeare still less like stars of the third magnitude & be about 115. But if the Lamps be distinguished into magnitudes according to the quantity of their light decreasing in a geometrick progression (wch is the case for the fixt stars) there may be 120 or 140 of the third

---

[1] The word 'thousand' results from a natural mistake in writing: Newton means $\sqrt{9.10^8} = 3.10^4$, not 3.10^7.

magnitude, 3 or 4 hundred of the fourth nine hundred or a thousand of the fift.

And so it is in the fixt stars. There are about 12 or 14 stars of the first magnitude, about 50 of the second magnitude about 160 of the third about 4 or 5 hundred of the fourth & so on. For the stars of the smaller magnitudes being distinguished by their light decreasing in a geometric proportion, there are recconed about three times as many stars of the third magnitude as of the second & about three times as many of the fourth magnitude as of the third, & so on. And if the heavens be viewed through Telescopes, great numbers of small stars are discovered wch by reason of their smalness or distance could not be seen by the naked eye, & the better are the Telescopes the greater is the number of small stars discovered by them so that the number of stars lying one beyond another seems to be endless.[1]

Corol. 2. The Sun therefore seems to be a body of the same kind with the fixed stars, but appears much bigger by reason of our nearness to him. And the fixed starrs seem to be Sunns in their several regions placed at great distances from one another for illuminating all places, & to be equal in magnitude to our Sun, more or less.

Corol. 3. The distances of the fixt stars from our Sun & from one another seem to be about 30000 times greater then that of the earth from the Sun.

Corol. 3. The fixt stars rest in their several places & the earth turns about its axis once in 24 hours. For we shewed above that the Sun Moon & Planets turn about their axes & nature is uniformable [*sic*; Newton originally wrote 'conformable to her self'] & it is not reasonable to beleive that the fixed stars being so very numerous & such great bodies & at so great distances from us & from one another should revolve

---

[1] On the back of folio 47, which ends after Corol. 3, is the following, which is obviously a recasting of this paragraph:

But if the stars bee distinguished into magnitudes according to the quantity of their light decreasing almost in a geometric progression (wch is the case of the fixt stars) the number of the lamps of every magnitude will be reciprocally proportional to the magnitude of the light. And so it is in the fixt stars. There are about 14 or 15 stars of the first magnitude, about 50 or 60 of the second, about 180 of the third, about four hundred of the fourth, about seven or 800 of the fift & so on. And if the heavens be viewed &c.

The Corollaries are obviously misnumbered; the second Corol. 3 occurs at the top of fol. 48, and there is no Corol. 1. All of fol. 47 is much crossed out and interlined.

about the earth & all of them conspire to do it in the same time without altering their situations to one another. Many are apt to fansy that if ye earth moved we should feel its motion, not considering that in a boate wch is carried down a river wth an eaven streame wthout reeling or jogging a man cannot feel the motion, & that in a systeme of bodies the motions of bodies amongst themselves are the same whether the systeme be at rest or in motion provided the motion of the systeme be uniform & eaven without acceleration retardation or jogging.

Phaenom. 7. The Planets Mercury & Venus revolve about the Sun according to the order of the twelve signes. For when They appeare full faced & round like the full Moon they must be beyond the Sun that they may be illuminated on that side wch is towards us. And when they appear horned like the new moon they must [be] on this side the Sun that their dark side may be towards us. And when they appear like the quarter Moon they must be over against the Sun, that one half of their illuminated side may be towards us. And when they are beyond the Sun they move according to the order of the sines & whey [*sic*] they are on this side of the Sun they move contrary to the order of the signes in respect of the earth, & by consequence according to the order of the signes in respect of the Sun.

Phaenom. 8. The Planet Mars revolves about the sun according to ye order of the signes & his distance from the sun is greater then that of the earth. For he is sometimes opposite to the Sun. And when he is seen neare the Sun he appears very small & full faced & therefore is beyond the Sun. And when he is seen in Quadrature wth the Sun he appears gibbous like the Moon about three days & an half before or after the full. And all his phases & apparent magnitudes Astronomers find to be such as argue that he moves about the Sun according to the order of the signes at a distance wch is to ye distance of the earth from the sun as about three to two.

Phaenom. 9. The Planet Jupiter revolves about the Sun according to the order of the signes & his distance from the Sun is about five times greater then that of the earth. For the times of the Eclipses of Jupiters Satellites & of their shadows falling upon him determin the angle conteined by two lines drawn from Jupiter the one to the Sun & the other to the earth, & by this angle the distance of Jupiter from ye sun is determined, & is found to be about five times greater then that of the earth from the Sun.

Phaenom. 10. The Planet Saturn revolves about the Sun according to the order of the signes & his distance from ye Sun is about nine or ten times greater then that of the earth from the sun. For this distance is deducible from the times of the apparent conjunctions of Saturns biggest Satellite wth Saturns body & with the anses[1] of his ring.

Phaenom. 11. The Planet Jupiter is accompanied with four small Planets or Moons moving about him according to the order of the signes: & the times of their revolutions are in proportion to one another reciprocally as the square roots of ye cubes of their distances from the center of Jupiter.[2] For the times of their revolutions & their distances from the center of Jupiter are as in the following Table:

The times of revolving

The distances by observation

The distances by the times

Phaenom. 12. The Planet Saturn is accompanied with four small Planets or Moons moving about him according to the order of the signes:[3] & the times of their revolutions are reciprocally proportional to the square roots of the cubes of their distances from the center of Saturn. For the times of their revolutions & their distances from the center of Saturn are as in the following Table:

Phaenom. 13. The times of the revolutions of the Planets Saturn Jupiter Mars Venus & Mercury about the Sun & either of the earth about the Sun or the Sun about the earth, are reciprocally as the square roots [of the cubes] of their mean distances from the center about wch they revolve.[4] For the distances of Saturn Jupiter & Mars from the center of the

---

[1] From the Latin for 'handle'.

[2] Newton has interlineated the word 'reciprocally' here and in Phaenom. 12, which is a mistake; the periodic times are directly as the 3/2 power of the distance.

[3] Cassini in 1684 discovered the 4th *and* 5th satellites of Saturn; Newton elsewhere refers to 5 satellites, but he was perhaps writing carelessly here.

[4] 'Reciprocally' is again a mistake and the words in square brackets need to be added.

Sun are determined by their retrograde motions & those of
Venus & Mercury by their greatest elongations from the Sun.
And their mean distances so found by Observation are as in
the following Table:
  According to Kepler
  According to Bulliadus
And their distances answering to the times of their periodical
revolutions are these

Phaenom. 14. The Planets Saturn Jupiter Mars Venus &
Mercury move in Ovals about the Sun placed in the inferior
node of the Oval, & every Planet with a right line drawn from
it to the center of the Sun describes equal areas in equal times.
Kepler by an elaborate discourse has proved this in the Planet
Mars & Astronomers find that it holds true in all the primary
Planets.

Phaenom. 15. The earth is about 50 times bigger then the
Moon & almost a million of times less then the sun & differs
not much in bigness from the Planets Mars Venus & Mercury.
This Astronomers know by comparing the Parallaxes of the
planets with their apparent diameters.

4

ASTRONOMIA

MS. Add. 4005, fol. 43

The most important feature of this short holograph paper on some technical aspects of astronomy, still treated in an elementary fashion, is the description of a new instrument introducing the principle of reflection to procure apparent coincidence of two objects at the eye. Newton was the inventor of this principle, later perfected in Hadley's octant. The instrument described here is a precursor of Newton's reflecting sextant, described in the *Philosophical Transactions*, no. 465, 1742.[1] As in all the later instruments of this type, two plane mirrors are used, one fixed and one rotating, in such a way that the images of two objects are reflected down a single path to the eye. The measure of the angular separation of the objects is the angle through which the rotating mirror has to be turned to bring their images to coincide, starting from the zero position where both mirrors reflect the same object to the eye. In this first instrument of Newton's the two light paths are not quite parallel, and from its construction the instrument would be awkward to calibrate. The later instrument was improved by attaching the telescope tube and the mirror F rigidly to the frame of the sectant; the mirror E was then rotated independently, the angle of rotation being measured by an index extending to the limb.

Cap. 1 Mensura temporis

Ad devitandas aequationes temporis & reductionem ejus ad dies solares, adhibendi sunt dies et anni siderei, numerando horas diei a termino aliquo constanti et immutabili, ut a meridiano qui per stellam polarem transit. Mensurandum est hoc tempus per duplex Horologium; Oscillatorium et Astronomicum. Prius quoties vel minimum erraverit corrigendum est per posterius. Posterius ex perspicillo constat angulum visorium permagnum admittente, circa axem axi mundi parallelum cui index affigitur movente & filum in umbilico seu foco vitrorum habente. Axis fiet axi mundi parallelus si specillum versus eandem stellam semper dirigi potest. Stella versabitur circa tropicum ♋ & circa meri-

---

[1] Reprinted in *Papers & Letters*, 236–8.

dianum observabitur et per horas 4 vel 5 ante et post, habita per tabulam correctione pro ratione refractionis. Vice fili melior erit regula tenuis recta quae stellam tegendo faciat ipso observationis momento evanescere. Cognita per hujusmodi instrumentum hora ad unum aliquod tempus, cognoscitur ad tempus omne adhibendo stellas quarum differenta ascentionis rectae cognoscitur. Melior erit sextans tripedalis aequatori parallela cum indice telescopico.

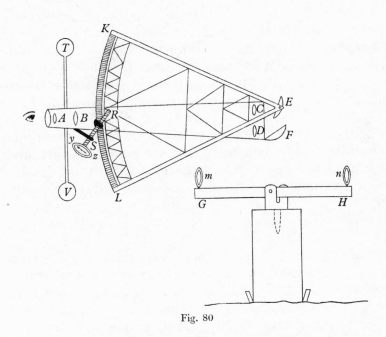

Fig. 80

Cap. 3. Instrumentum pro capiundis angulis et altitudinibus meridianis.

A vitrum oculare [Fig. 80]. *B* vitrum latum oculare ad ampliandum angulum visorium. *C, D* duo vitra objectiva aequalia et similia. *E, F* vitra plana specularia a tergo pice tecta ad tollendam reflexionem *F* fixum est ad tubum *ACD* & radios objecti reflectit per *D* ad *A*. *E* affigitur quadranti vel sextanti *KL*, & reflectit radios alterius objecti per *C* versus *A*. *EKL* quadrans vel sectans ex tenuibus & latis virgis ferreis conflata. *RS* cochlea cum indice quadrantem dirigens *TV* baculum ponderibus plumbeis *T, V* ad [term]inos onustum, quo ad dextram vel sinistram tracto, instrumentum stet in aequilibrio *GH* sustentaculum per cujus annulos *m*, *n* tubus transeat. Ubi objecta duo coeunt Divisiones vel dentes in limbo *KL* denotant partes majores & index *YZ* partes decimales et centesimales partium majorum, quibus anguli mensura ex tabula ad angulos constructa designantur.

Cap. 2. Linea meridiana. Et mensura temporis

De aedif[ic]io satis alto demittatur linea perpendicularis aequabilis cui pondus magnum aquae innatans appenditur. Ad distantiam stadij unius vel dimidij stadij boream versus mittatur servus cum candela quae per foranem [*sic*] tenue lucens appareat instar stellae primae vel secundae magnitudinis. Austrum versus recedat Astronomus donec videat stellam polarem juxta summitatem perpendiculi, et candelam simul jubeat transferri donec ea appareat itidem juxta latus perpendiculi. Dein motu lentissimo Astronomus moveat caput donec stella a perpendiculo occultatur et si candela eodem temporis momento occultatur ita ut simul videantur evanescere, habebitur Azimutha stellae vel digressio a meridiano.

Invenienda est maxima digressio stellae ab utraque parte poli et media distantia erit in meridiana. Illic collocetur foramen exiguum per quod applicatio stellarum ad perpendiculum & occultatio cernatur ad correctionem temporis ab Horologio oscillatorio indicati. Non est corrigendum horologium ipsum tempore observationum sed tempus tantum. Et hoc modo tempus ad 20''' haberi potest. Ubi haec correctio non in promptu est adhiberi potest Sextans praedicta quae tempus ad 1'' dabit.

Cap. 4. Altitudo meridiana

TRANSLATION

Chapter 1. The Measure of Time

To avoid the equations of time and its reduction to solar days, the sidereal days and years are to be used, enumerating the hours of the day from some constant and unchanging point such as the meridian which passes through the pole-star. This time is to be measured by a double clock: a pendulum clock and an astronomical one. The former, whenever it errs even by a very small amount, is to be corrected by the latter. The astronomical clock consists of a telescope allowing a very wide angle of vision, turning around an axis parallel to the axis of the Earth to which the index is fastened and having a thread at the centre or focus of the lenses. The axis is parallel to the axis of the Earth when the telescope can always be directed to the same star. The star will turn about the Tropic of Cancer and will be observed near the Meridian and for 4 or 5 hours before and after, if one has the correction for refraction from a table. Instead of a thread it will be preferable to use a thin straight rule which, covering the star, causes it to disappear at the very moment of its observation. If the hour is known at some certain time by such an instrument, it is known at all times by making use of stars whose different right ascensions are known. A three-foot sextant, parallel to the equator, and with a telescopic sight would be better.

Chapter 2. The Meridian Line.
And the Measure of Time

A uniform perpendicular line is let down from a sufficiently tall building, to which a heavy weight immersed in water is attached. An assistant is sent with a candle about 100 or 200 yards towards the north; the light of this candle is emitted through a little hole so that it resembles that of a star of the first or second magnitude. The Astronomer moves away towards the south until he can see the pole-star near the top of the perpendicular line, and he orders the candle to be moved until it also appears near the side of the perpendicular line. Then the Astronomer moves his head very slowly until the star is occulted by the perpendicular line, and if the candle is occulted at the same moment of time, so that they seem to vanish simultaneously, he will have the azimuth of the pole-star, or its digression from the Meridian. The greatest digression of the star on either side of the Pole is to be found, and the half of this distance will be the meridian. There a small hole is placed, through which the approach of stars to the perpendicular line and their occultations are observed, in order to correct the time indicated by the pendulum clock. The clock itself is not to be corrected by the time of the observations, but only the time [recorded]. And in this way it is possible to have the time to one-third of a second. When this method of correction is not feasible the aforesaid Sextant which gives the time to 1″ can be used.

Chapter 3. An Instrument for Taking Meridian Angles and Altitudes

A, the eye-lens [Fig. 80]. *B*, a broad eye-lens, to enlarge the angle of vision. *C*, *D*, two equal and similar objective lenses.

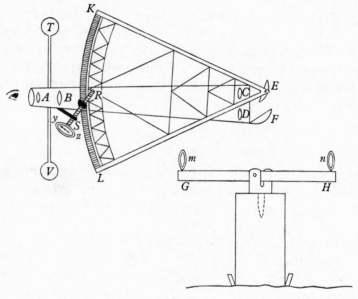

Fig. 80

E, *F*, plane looking-glasses covered with pitch on the back to increase reflection. *F* is fixed to the tube *ACD* and reflects the rays from the object through *D* to *A*. *E* is fastened to the quadrant or sextant *KL* and reflects rays from the other object through *C* towards *A*. *EKL*, a quadrant or sextant constructed of thin and wide strips of iron. *R*, *S*, a screw with an index directing the quadrant. *TV* a rod loaded with lead weights *T*, *V*, at the ends, by moving which to the right or left the instrument is kept in equilibrium. *GH*, a support, through the rings of which, *m* and *n*, the tube passes. When the two objects are brought to coincidence, the divisions or teeth of the limb *KL* denote the greater parts, and the index *YZ* the tenth and hundredth parts of these: by which the measure of the angle is determined from tables of angles.[1]

[1] Newton means that the instrument reads arbitrary units, tenths and hundredths; a conversion-table is used to give the angle in degrees and fractions thereof.

5

ASTRONOMY, GEOGRAPHY, NAVIGATION, ETC.

MS. Add. 4005, fols. 51–2

This document appears to be a series of rough notes of scientific observations that might be made by a person travelling abroad. Perhaps it may be compared with the lists of queries dispatched overseas in the early days of the Royal Society, and with Newton's own early letter to Francis Aston, 18 May 1669, O.S.

Astronomy

To be shewn the five satellites of Saturn & send over their apparent positions in some clear nights &c.

To get the time between the transit of Saturns & the Hygenian Satellit in its greatest elongation, observed in 3 or 4 several clear nights.

To get the Obs. of Comets

To send speedy notice of all new appearances of Comets.

Geography

To get observations of the Eclipses of the Moon & of 4 satellites in several places of the Mediterranean, the coasts of Afric Arabia foelix, the Persian Gulph, the East Indies the black sea, the coast of America &c, & to determine the times by the Altitudes of the stars taken with proper Instruments.

To observe wth magnetical needles the bendings turnings & direction of the Rivers of Euphrates, Tygris &c.

Navigation

To observe the longitudes & latitudes of places on all the sea coasts & the variation & dipping of the Needle & the currents of ye sea, & the winds & circumstances of ye weather in several regions.

To observe the tydes, their times their currents, their ascent & descent at all times of the year & in all positions of ye Moon.

Mechanicks

To get an account of all Mechanical engins celebrated abroad

Mineralogy

[blank]

PART VI
NOTES

INTRODUCTION

Newton's habit of taking extensive notes on his reading is well known: some hundreds of sheets of transcripts from alchemical authors were scattered over the world by the sale in 1936 of the portion of the Portsmouth Collection that was not deposited in the Cambridge University Library. No serious analysis of this strange habit has yet been made, so we include here Newton's annotations on Hooke's *Micrographia* (1665), both as a typical example of Newton's method of taking notes, and because these notes possess particular interest. It was in *Micrographia* that Hooke unfolded his theory of light, which was later to be a source of major controversy between himself and Newton. It is unfortunately impossible to date this manuscript (MS. Add. 3958, fols. 1–2), which was written on two folio sheets folded to make eight pages; the handwriting and the wildness of the spelling make it obvious that this is an early paper, but there is nothing to indicate precisely how soon after its publication Newton came to read Hooke's most important book.

Newton's predilection for taking notes was clearly little short of an obsession—like all obsessional habits it went beyond normal bounds—thereby causing endless trouble to later historians who have often been puzzled to understand its purpose and to differentiate between notes and original comments. The habit was one that Newton formed as a student. There are annotations in the earliest of Newton's notebooks,[1] while one which he used as an undergraduate, from about 1661 to 1665, contains at first mainly long extracts from the books he was currently reading—first Aristotle (in Greek and Latin), later moderns like Descartes and Boyle.[2] The habit continued until it was so engrained that Newton would copy out long passages from the later works of Boyle, for example, of which he possessed presentation copies from the author. Sometimes these passages relate to investigations that Newton had in hand at the time, but more often the chief reason for the note-taking seems to be no more than a general interest in the

[1] D. E. Smith, 'Two unpublished Documents of Sir Isaac Newton', in W. J. Greenstreet (ed.), *Isaac Newton, 1642–1727*, London, 1927, 16–34.
[2] A. R. Hall, 'Sir Isaac Newton's Note-Book, 1661–65', *Cambridge Historical Journal*, IX, 1948, 239–50.

subject. Apparently he could not read attentively without a pen in his hand, and he found the mechanical act of writing an aid to thought and memory rather than a burden. Yet it seems reasonable that one should not judge the depth of Newton's interest in any specific topic by the volume of notes he has left. For example, the mass of alchemical transcripts proves that Newton was widely read in alchemy, but gives no authority for the assumption that Newton accepted the pretensions of alchemists, or that his own interest in chemical science was modelled on that of the authors he transcribed.[1] On the other hand, curiously, we have discovered no surviving annotations from either Galileo or Kepler, whom Newton certainly read.

He seldom wrote his own comments in the form of critical notes; usually he seems to have been content merely to write down the facts or ideas that interested him more or less in the words of the book he was reading. It is for this reason that the notes from *Micrographia* are of such particular interest, for in these Newton did introduce his own comments. They are carefully marked off by square brackets; indeed, he was surprisingly careful to indicate minor variations of his own, especially at the beginning. Perhaps this may be partly due to the fact that here at least Newton took notes a chapter at a time, which enabled him to digest Hooke's work in his own mind, a process inevitably provocative of his comment and one which often sent him back to earlier sections of the *Micrographia*. Sometimes he did not write his notes till he had repeated Hooke's experiment—unless indeed Newton had independently performed that same experiment under the same stimulation as Hooke, that is, that of Descartes and Boyle.[2] Examples are his comment on Hooke's description of the colours of thin plates of 'Muscovy glass' (under p. 47), and on the formation of crystals in freezing experiments (under p. 88). A principal reason for Newton's care in indicating his variations was undoubtedly a genuine distrust of Hooke's

---

[1] Marie Boas and A. Rupert Hall, 'Newton's Chemical Experiments', *Archives Internationales d'Histoire des Sciences*, xi, 1958, 113–52.

[2] In this connexion it is unfortunate that one cannot be certain whether or not Newton had begun his own optical researches, or proceeded far in them, when he read *Micrographia*, which probably appeared at the booksellers' late in December 1664. One cannot assume that Newton saw the book as early as Pepys, who examined it on 2 January. Since it is very possible that Newton began his optical experiments before *Micrographia* was published, it is likely that Newton had ideas of his own before he read Hooke's.

theory of light. There is a note of acerbity in many of his comments foreign to his usual habit of note-taking. There is no doubt that although Newton regarded Hooke as a major scientist he found him intensely irritating; from these notes it would seem that this was so long before the two men met or had the slightest direct contact. For the notes, one can be sure, were put down long before 1672; Newton would have composed his reflections on *Micrographia* very differently, and with fuller reference to his own experiments, had he been reading the book with a view to a formal refutation of Hooke's hypothesis of colour. He clearly (as under p. 83) enjoyed catching Hooke in a contradiction, or (as under p. 60) an error; and though Newton was no partisan of the Cartesian theory of light, he was ready enough to defend Descartes against what he took to be unjustified criticism. The comments on Hooke's views on light and colour are brief, but dryly antagonistic (cf. under pp. 56–62, 230). His objection to Hooke's pulse theory is the same as his life-long objection to all wave-theories: that according to their principles, light ought, like sound, to bend around corners. Yet rejection of Hooke's hypotheses did not prevent Newton's carefully noting the observations on which Hooke had based them; in one place, however, Newton remarks that Hooke had 'framed his observation to his erroneous Hypothesis'.

Only on the subject of light is Newton severely critical of Hooke. Elsewhere, though he sometimes enters minor criticisms, he seems to regard Hooke's work as essentially sound. (Newton has left no trace of the typical *virtuoso* pursuit of microscopy.) It is interesting that he copied out Hooke's table of experimental results relating to 'Mr. Townley's Hypothesis' (under p. 226), though he could have obtained similar data on Boyle's Law from Boyle's own second edition of the *Physico-Mechanical Experiments* (1662). This may explain the otherwise odd fact that years later in the semi-popular *De Systemate Mundi* he referred to the relation between the pressure and volume of air as having been 'proved by the experiments of *Hooke* and others'.[1]

---

[1] Cajori, 609; cf. *De Aere et Aethere*, p. 216. Since this volume first went to press this manuscript has been transcribed and printed by Sir Geoffrey Keynes in his *Bibliography of Dr Robert Hooke* (Oxford, 1960).

Pag. 4. Glass most exactly pollished is rough full of broken & scratched parts wch reflect light variously and variously coloured.

13. A pin may bee easily unscrewed [or screwed faster] out of a plate whilst grated on by a file yn otherwise. Whence tis no wonder yt heate [burnes &] loosens bodys so much [perhaps ye magnet will invigorate iron best, wch is made to jarr by a file. Or yn best attract it &c]

13. That is Aire is nothing (probably) but a tincture, solution or saline substance dissolved out of ye Earth & water by ye Aether & agitated by it, whence ye explication of its rarefaction is easy for Cocheneel or Logwood will tinge 10000 times more water & salt will dilate as much in praecipitations. That Raine is Aire praecipitated.

15. The reason why some bodys doe easily mix together others not some are congregated other segregated by motion, is ye agreement or disagreement in their motions (caused by theire various bulkes, densitys or figures) to comply one wth another or beate one another of, like concords or discords in musick.

18. A drop of liquor in an equally ponderous medium is sphaericall, in a lighter or heavier medium is Ellipticall (as at C) betwixt two mediums its figure is various (as at ab)[1] [hence perhaps may be caused halos]

16. Hence (from pag. 15) may bee explained ye rising of water at ye sides in a cleane glasse vessell, in ye middest in a greasy vessell; Of Mercury at ye sides in a gold, silver tin copper or leaden vessell, at ye midst in a glasse or Iron vessell, because ye vessel hath more or lesse congruity wth ye Aire yn wth those liquors, soe ye rising of water in a glasse pipe

21. if not greasy, of Mercury in a silver pipe &c. of Liquors in a filtre, of Oyle in the weeke of a Lamp (though made of wire, threds of Asbestus, Strings of Glasse &c) Of water in bread or a spung & perhaps of sap in plants (at least out of the earth into their roots), & ye cause of ye more or lesse refraction of light passing through mediums of more

---

[1] We have omitted the sketches which Newton made from Hooke's figures, and to which he occasionally refers in lettered points.

or lesse congruous motion. & of gravity, ye Æther (its motion being incongruous to all other bodys) forcing bodys to retire from those places where it is in greatest plenty, towards ye Earth where it is in least plenty.

(23. Put as much Auripigmentum poudered as will lye on a halfe crowne to 20 pound of leade more or lesse untill ye lead (melted but not too hot & well stired) will drop into water without a tayle, ye lead powered into water through a copper trencher perforated wth many small holes in ye middest will bee good shot.)

22. Hence perhaps ye Sunn, starrs, plannets, fruit, Pebbles flints (wch seeme to have once beene fluid) drops of water Glasse leade &c are round, from ye heterogeneous ambient

25. body; Springs have their force from the heterogeneous included ambient, fountaines are caused by ye rising of water as in Syphons (Or rather by ye lightnesse of fresh water more yn of salt water in ye proportion of 45 to 46, (though it may bee to brine as 12 to 13) hence ye ebbing & flowing of a

27. Spring a Kilken in Flintshire on ye top of a mountaine). Dissolutions & ye mixing of severall bodys & partly praecipitations may bee caused hereby, & perhaps this may bee

28. a coefficient in causes of light heat & (consequently) Rarefaction condensations Hardnesse fluidity perspicuity opacity, Refractions colours & almost all ye Phaenomena of

29. nature. Of ye floting of bodys towards ye sides of a vessell or one towards another (ye aire pressing more upon ye midst of ye water yn on its sides & so causing bodys to recede to ye sides) if ye water rise at ye sides, otherwise

31 from it. ye viscousnes of syrups pitch Turpentine &c May arise from ye congruity of ye parts one among another, & also ye suspension of ye Mercury somtimes some feete above 29 inches is from ye congruity of it wth water & of water wth glasse. Electricall & magneticall attraction may perhaps bee hence explained.

9. 10. Silkes are watered by folding ym in ye midst yt one halfe of each wale (as neare as may bee) ly upon ye other, yt ye protruberant turnes of ye threds of one wale may make impressions on ye wale it is contiguous to or it selfe made sharper being pressed into ye other, yn sprinkle ye silke, & wind it up wth pastbord twixt each fold & lay it in a strong

presse till it bee stiffe & dry. The reason of these colours as also why whitned silver burnished or water ruffled looks black is ye same Viz: ye reflection of ye rays either much from or towards ye eye.

35 Bubbles in a glasse drop as also ye pupill of an eye appeares much bigger yn it is [may not ye pupill seeme to contract or dilate when it moves nearer to or farther from ye cornea

41. Alabaster heated swells much & appears like a fluid body untill a certaine watry stuff in wch it parts are brought to swim be evaporated, when they returne to pouder againe.

39. And soe glasses are rarefied & swelled by heate as appeare by ye manner of cracking it by suddeine heat or cold. so a cilinder of brasse or glasse &c will not goe into yt hole hot wch they will doe cold. And being in water ye outmost side grows hard & rigid whilst ye inmost having more leasure to coole is more soft & endeavours to shrinke more yn ye outside but is kept distended by it, (leaving onely some little bubbles in ye center) soe yt this drop being like an arched roof dissolvs all into dust when ye aequilibrium of pressure towards ye center &c is destroyed by braking ye least snip of its tayle or scratching it. But if ye drop bee cooled by degrees it hath not yt springy tension in its parts is more

44 soft & easily broken will not break into dust. Soe the springgy tension of ye parts in other bodys as in Iron or steel (wch is acquired by being quenched whilst they have vitrifyed parts interspersed) or brasse or silver (acquired by striking) is loosened by fire.

45 The sparks struck from Tindar are some of ym shivers of Iron only vitrifyed, others only melted into Globules, others onely heated red hot wch yet may suffice to fire tinder. Thus ye filings of iron dust are easily fired & burne like sparks struck from steele, whence Iron seems to have much Sulphur in it for ye air to prey upon when a little agitated. Filings

47 of lead or Tin strewed in a crucible wth fine & well dryed pouder of quick lime by a gentle fire may be melted into Globules & washed from ye lime.

48. Thin flake of Muscovy glasse, aire, water, metalline scumme doe exhibit divers colour according to their thiknesse, if ye midst be thinest there will bee a broad spot of one

colour & coloured rings about it (outward in this order. (white perhaps in the midst) blew, purple, scarlet, yellow, greene, blew, purple, scarlet, yellow, greene, blew, &c untill somtimes 8 or 9 such circuits. & ye outmost limb of ye flaws of Muscovy Glasse appeares white because of its thinness the other flaws not extending so far [The more oblique position of ye eye to ye glasse makes ye coloured circles dilate.]

60 [Though Descartes may bee mistaken so is Mr Hook in confuting his 10 Sec. 38 Cap. Meteorum he says well yt this Phaenominon of Muscovy Glasse cannot bee explained theirby, or that ye turbinated motion of ye Globuli signifys nothing unlesse they did not only endeavour but also move to ye eye. &c]

56. 62. Light is a vibration of ye Æther, wch pulse is made oblique by refraction & ye motion of ye precedent part of ye pulse being deaded by ye adjacent quiet medium makes blew, but those extreame rays make red whose followings parts of their pulse is weakest being deaded by ye medium adjacent on ye other hand. And these two strokes of the pulse are effected by reflection on either side of the Muscovy Glasse ye weaker stroke being effected by ye farther side of the glasse. [Why yn may not light deflect from streight lines as well as sounds &c? How doth ye formost weake pulse keepe pace wth ye following stronger & can it bee then sufficiently weaker.

52 Ye colours on steele heated, quenched & then again warmed are White Yellow Orange Minium scarlet purple blew watchet by reason of a vitrifyed scum melted & thrust out.

51 Those & onely those Mettalls & substances wch vitrify (that is wch by being coroded by a saline substance are converted into Scoria (yt is calcined) wch Scoria is evidently Glasse by its easy fusion & uniting wth Glasse its hardnesse & brittlenesse when cold &c) by fire (wch is a saline substance) (as all doe but gold & silver & those too being mixed wth saline substances as also most other vitrifications or calcinations are made by salts) are hardened by sudden cooling.

52. Iron is turned into steele by mixing wth certaine salts wth wch tis kept in the fire some time & Iron may bee

suddenly case hardened being daubed about wth a mixture of Urin soote sea Salt & Horse hoofes (wch are saline bodys) & inclosed wth clay & laid a good while in a strong fire, then soundly heated & suddenly quenched. Whence ye making & hardening of steele may bee from ye vitrification of some of its parts wch being rarefied by heate keepe of ye parts & open the pore of iron in wch distended posture ye Iron is kept by suddane cooling & so becoms hard & brittle (as was explained in ye Glasse dropp) but if ye steele bee kept in some convenient heate ye parts may thereby relax themselves complicate & joyne closer (protruding some of ye interspersed vitrum wch according to its thicknesse causes divers colours white, yellow, orange, minium, scarlet, purple, blew, watchet) & soe ye steele grows safter & tougher.

68. Noe colour can bee made wthout some refraction (& indeed ye least visible particles of all bodys if strictly examined by a microscope are transparent, & apt for refraction some perhaps being flawed by grinding appeare opace (& were it not for such flaws & pores to let in ye reflecting aire [or rather denser aether] noe body would bee opace for wee see there may bee made of any metall christalls or vitriolls, wch are transparent & variously coloured (as vitrioll of Gold is yellow of silver blew &c) wch variety of colours must proceede from ye metalline & not intermixed saline particles. The like effect follows from solutions of Metalls into Menstruums & their praecipitations & calcinations by fire wch is a kind of praecipitation by a salt proceeding from ye fire. Also calcined mettalls melted wth glasse tinge it wth transparent colours. & leafe gold is a transparent blew): Soe yt ye disseminated tinging parts of any liquor will transmit the rays but impede or promote their motion & so cause varyety of strong & weake pulses, besides other disorders.

73 Two concave glasse prismes (made by cementing glasse edges to woodden sides) filled with blew & red tinctures & variously applyed did exhibit all colours (though together they were almost wholy opace & yet singly very transparent) ye one wedge according to ye liquors thicknesse affording all sorts of reds & yellows ye other of blews.

Object: ye Thicknesse of some liquors varys not their colours only makes them more or lesse strong. nor are all redds diluted to yellows by more water nor can yellow layd upon yellow make a red. Resp: Tis to bee considered yt ye parts of those bodys are (perhaps) generally opacous by flakes & cracks, yt colours are onely whitened by setting ye tinging particles at a greater distance & not diluted unlesse those parts be divided into smaller parts. Thus (Blew) smalt & Bise & (yellow) Masticut are easily diluted by grinding by spreading thin (for every particle is as blew or yellow as ye whole masse untill it be ground smaller).

Now therefore most ordinary colours being generall Metal-line cannot bee diluted because their particles are already so small & can never scarce bee ground into so small parts as they are distinguished by ye flaw nor can they bee deepned (many particles being united into one) inlesse those flaws bee first shut up. [Yet in ye Experiment pag 73 all or none yellows may bee thikened into reds, though not all reds into yellows unlesse ye least particles of ye tincture bee already yellow.]

69. Purple is made of a deepe red & a deepe blew soe yt the blew being diluted ye red becomes praedominant & ye blew if ye red bee diluted. Boyle of Colours Exper 20. & it may bee produced by mixing vermilion & bise dry.

83. Whitnesse of bodys proceeds from their flaws & other pores filled wth aire [or some more reflecting Æther] & therefore is diminished by admitting water oyle turpentine & into its pores as in Paper linnen silke ye Occulus mundi stone, White marble &c. Hence may bee explained many Phaeno-mena of ye aire as about Mists clouds Meteors Halos, & of many other coloured bodys as liquors. [Is not $2 + : 1 :: d : e$ ye cause of quicsilvers opaknesse?

85. All the naturall regular figures of bodys may be imitated by composions of equilaterall triangles. & perhaps caused by globules wch doe naturally lye in that forme perhaps some may make a 4 square posture [or they may lye in a six square wch in effect varys not from a 3 square] & sollid are made by conveniently laying one such [?] surface on another.

88. The Ice on ye surface of frozen Urin is insipid, & branched from a center (a) wth six branches (of various lengths from 1/4 inch to 4 foot) & from those proceeded others in ye same angles of 60gr like hering bones or feathers

& from those others &c most prominent above ye Urin at ye center *a* & where the stemms are thickest ice, but ye watry interstitia when frozen swell as high. some of ye six great stems were knotty others conically eaven & so were ye secondary tertiany quartany quintany &c branches some eaven others knotty. The stems, & their primary branches crossed not, but their secondary & 3any &c did, not joyning but lying one over another in this order, viz: those always lying upmost wch were parallell to ye last biggest (Grandfather) stem from wch they sprung [yt is wch were nearest ye center being first frozen]. A most admirably curious figure (this figure is much imitated by Ferne or Brake of both wch perhaps there may bee the same cause) its six branches are not all exactly alike or of a length as they are in snowy figures. of wch in ye same snow ther is greate varyety.

81 Gravell in Urin are flat like fragments of muscovy glasse & perhaps alike divisible, they are dissolvable by oyle of Vitriol spirit of Urin & other saline menstrums, their diameter is about 1/128 of an inch.

94. Kettering stone (wch is harder yn free stone) consists of nothing but small globules (a little flatted into Ellipses by unequall pressure one upon another whilst they were soft liquor, & but just touching in small surfaces, yet very firmely united so as not to bee parted wthout breaking a peice one out of the other wch sometimes made a hole through ye outmost thin crusty shell & discovered two substance wthin like ye white & yelke of an Egge) whose intestitia appeare empty & therefore ye stone may be blowne through like a Cane. The least globule not exceeding ye greatest in diameter more yn 3 or 4 times & this curious stone appeares but rude to ye naked eye. Hence may some conjectures be raised of ye porousnesse of bodys for ye progresse of light.

97. White Marble wch may bee pollished as smooth as glasse will bee imbued quite through wth oyle of turpentine or pitch in wch tis boyled. Some liquors will corrode Glasse if kept long in it. The Occulus mundj stone put into water makes it in time bubble a little becomes transparent & heavyer weighing 6 3/256 graine wn wet & 5 202/256 gr wn dry. wch argues their porousnesse.

101. Cole charred (by covering ye wood in a crucible wth

sand to keepe of ye aire from setting it on flame whilst it is heated red hot 1/2 hower &c) is black because very porous like a honey come (ye pores in order radiating in rows streight from ye pith to ye barke & tending from one end of ye stick to ye other (for in a thin shiver of cole they may bee seene through wth ye microscope. a 1/18 inch line contains 150 of ym soe that in a stick of an inch diameter there are 5725350 such pores.) it shines because of its smoothnesse, & is brittle because ye liquor wch toughned it is expelled by the fire leaving ye sollid parts like soe much stone. Which liquor if collected in a Retort will bee found almost wholly incombustible yet makes uncharred wood flame more in burning ye liquor (wch rusheth out of each in ye forme of a vapor shattering & opening the wood & making each pore like soe

103. many Æolipiles to blow the fire. From ye manner of charing cole may be concluded yt the aire is the Menstruum or universall disolvent of all sulphurious bodys, it selfe being (as it were) a flegmatick body but abounding wth parts such as are fixed in salt peeter wch parts are true dissolving bodys & being but few in the aire a fresh supply must bee continually made by bellows &c (hence ye use of respiration) or otherwise made least ye flame goe out for ye aire like other menstruums is quickly glutted & will dissolve noe more wthout a fresh suply & then it consumes as quick as melted niter it selfe. Heate is caused by many dissolutions of other bodys by menstruums, as also many other menstruums like this of aire performe not their action of dissolving till ye body dissolvend bee sufficiently heated. Soe yt Flam is nothing but a mixture of aire & ye volatile or sulphurious parts of a body wch by their vehement acting one upon the other cause a vibration in the aether, rarify one another & soe ascend, some part of wch sulphur mixing & coagulating wth aire is againe praecipited becoming a certain salt in soot, the rest of ye soot being either fixed parts or such parts as had not time to dissolve but were carried up by ye streame of ye asscending flame. ye rest of ye body wch is not volatile (yt is those parts wch are so compact as yt by fire they cannot bee easily rarifyed into the forme of aire) remaine fixed in ashes.

107. Petrifyed wood seems to bee rotten wood whose

(Microscopicall & perhaps far smaller) pores (wch are somwhat bigger then in charred wood are crammed full of stony & earthy particles wch it separated from the water impregnated wth them either by straining & filtration or perhaps by praecipitation cohaesion or coagulation whereby it becoms 3 1/4 times heavyer yn water, as hard as flint endures ye fire better & will not cracking in it like flint, because ye stony parts are bound together wth the wood, onely becomes more darke by the woods being charred. Nor is it flexible because ye small parts of ye wood have noe places or pores to slide into. the like of petrifyed shells.

113. Corke (wch is ye excrescence of the barke of a tree like a 3d barke but all chapped) conteins in a cubick inch 1259712000 cells made by dividing a long pore which passeth through ye corks thicknesse into parts by many diaphragmas (as in ye figure the corke being cut cros to its thicknes) tis light because so porous as any hony come, springgy because of the includ aire (& perhaps ye matter of ye corke is springy too) but not perspirable becaus its pores reach not through it. Elder pith is such another body but yt its pores run parallell to the branch. pith of feathers is the same but yt its blebs or cells are wholly irregular.

118 ye leves of plant animalls are set in round sockets of ye stem (as som bones are joyned in animalls) full of juice if open, juceless if shut, more apt to move in warme then cold weather, ye lower leaves more spedily acting on the upmost then those on them &c.

121 On rose or brier leaves &c grow little codded plants like a black spot the length of ye cod being 1/500 or 1/1000 parte of an inch, ye stalke as long &c.

126. Blew mold & other mushroms (probably) have noe seede (for they may bee made artificially) & therefore their groth is mechanicall & ye simplest of vegetables, perhaps somwhat like petrifyed drops of water, wch are made by ye avolation of ye volatile water from ye stony particles whereby the drop is incrusted & still increases by ye descent of more water &c.

135. Mosse is a perfect plant. Sponge seemes to have beene a plant animall adhaering to rocks, ye stringy substance of it inveloped wth a naucious mucous juce or gelly seeming like

the animall vessels incompassed wth ye Parenchima. In each knot meete 3 fibers, had there met 4 I should have thought it was at first a froth whose blebs breaking left onely their stringgy juncture to harden. & yt ye larger furrows in it was made by the eruption of aire collected from many of the bubbles breaking as I have seene done in a froth on rosemary wch inveloped a little animall

142. On ye Fibres of Nettles *A*, grow severall baggs full of juice *B* wch by pression is trajected through ye conicall hollow pricks *C* wch grow upon them. For by ye Microscope ye juice may bee seene upon pressure to come into ye prick wch before was empty. (ye like of cowage horse haires, &c) wch corroding juice dispersed in ye body by ye moist vehicle it finds there creates a painfull hard swelling by ebullition, wch is ye usuall effect of two differing saline liquors mixed. Thus Toads froggs effs many Fishes &c are killed by throwing salt on their backs wch finding a moist vehicle there is dissolved & dispersed through ye whole body of ye Animall.

151. The spungy substance wthin the beard of a wild oate swelling wth moisture makes ye outward twisted barke wch will not swell to untist. Hence ye Hygroscope.

158. Some haire is sollid as mans, hoggs; others pithy as Horses, Cats others porous long wise others transversly as Indian Deares. Skin or leather consists of nothing but fibers like Sponge, all wch perhaps were once muscles (Or other vessells) soe that there is rome inough for the avolation of vapors. (wch avolation in plants perhaps may cause ye rising of new juice to suply ye reliquished pore.) The shells of all Fishes are curiously wrought. A bee can move ye spikes on her sting as also pluck in & out a little sword at ye end of it whereby (perhaps) she takes faster hold & pumps out her poysonous liquor.

166. ye downe on the spriggs of feathers is made prettily wth small sprigs having hookes on it to catch hold of ye opposite spriggs of ye neighbouring downe so composing a net of square holes two small for ye aire passe through.

168. The reason of coulors in feathers more then haire is onely their thinnesse &c for ye colors vanish by dipping the feathers in water. Thus flax may bee dressed so fine as to be as much coloured & glossy as silke wch colours disappeare in

the weaving because many laminae of flax then close into one wch cannot happen in ye cilindricall threds of silke.

The Feete of most flys taking hold on any body wth two claws draws forwards whereby wth a kind of patten indented wth many small bristles (like a Carde) it takes firmer hold &c Their wings are finely adorned wth haires & feathers whose motion create ye sound.

176. The eyes of a grey drone fly wch in two clusters almost cover the head are 14000. 6000 whereof looked upwards & backwards & were larger yn ye 8000 wch looked downwards. they conteine an humor & have a darke skin distant ye diameter of a buble of the cornea from ye outside of ye cornea & somthing more wch may bee ye retina of all the eys. The like of many other insects.

A snaile eats wth a round bone fastened in her upper jaw.

187. 189. Gnats were once watry insects & afterwards transformed. Most of those buttons wch are every where found on leaves roots water &c are ye Matrixes of Vermin. If any insect cast its seed on a leafe ye leafe often pits there & perhaps at last incloses ye seede. The vessels of Insects are as various exact & curious as those of larger animalls. they have a pulse & (some at least) breath.

199. The Eys of ye Sheepherd (long-legged) Spider are fixed on a hill or Pillar on ye Crowne of its head. 200. The hunting spider is very expert &c & disciplines her young.

202. Some threds of a Spiders webbs are so small as to bee coloured. Perhaps ye greate white clouds that appeare all summer time consist of such white thredds as fly up & downe after a Fog in the aire.

204. Insects made drunke wth well rectifyed spirits of vinegar &c will ly still & plump fit to bee looked on wth a Microscope.

206. Perhaps ye wandring mite is the parent of those in cheese, meale, seede, Corne, musty places &c The eggs of Insects receiving much augmentation & therefore (perhaps) alteration too from ye place where they are layd.

209. By working & tossing (as twere) a parcell of pure cleare christall glasse whilst kept red hot in the blowne flame of a lamp it becomes shattered into an infinite number of thin plates or string wth interposd aeriall plates or fibers whereby

it appeares like mother of Pearle or ye curious glittering white scaly back of the booke worme & retaines its colour & opacity though drawn into wyer. The like may bee don by some other glutinous bodys, or by thin glasse bubbles broken into an heape of shivers.

213. A louse hath veins & arterys &c The blood of insects is white

217. Vinegar stopped close from aire kills ye little Eeles in it.

218. The verge of ye sun & moone in ye Horizon appeare indented like saws espetially on ye right & left sides wch inequalitys continue not always ye same but $\begin{Bmatrix} \text{fluctuate} \\ \text{change} \end{Bmatrix}$ like ye waves of ye sea. & ye body appeas flatted ye lower parts being somwhat nerer to ye center yn ye upper. & ye flatter it is ye Redder it is though not always ye contrary. Stars ye nearer ye Horizon ye more red & dull they looke & the more they twinkle though not wth so quick consecutions as when in ye meridian. they twinkle wth severall colours somtimes blew, red, yellow; though a pretty way from ye Horizon. Sometimes for a while appearing more then usually bright & somtimes for a while disappearing. Now higher now lower, now throne on this side now on that.

219. The gravity of water to yt of spirit of wine is as 21 to 19 but for ye same refracted angle in both 30d, 00′. The angle of incidence in water was 41d 35′, & spirit of wine 42d, 45′. Water refracts more then ice.

222. If cd is a glasse buble emptied of aire ye light coming from a & b to e will bee refracted to f. soe yt a ray coming out of a vacuum into aire is a little refracted towards ye perpendicular as if it came out of aire into water.

And since ye aire from ye earth grows continually more thin, hence in ye atmosphere is an inflection of ye rays.

226. Mr Townlys Hypothesis is ye dimension (or expansion) of ye aire is reciprocall proportional to its spring (or force required to compresse it). By Mr Hookes experience

(*A Table follows, on page 412*)

The dimension of ye aire	The perpendicular height of a Mercuriall cilinder compressing it	The Mercuriall cilinder counterpoised by ye Atmosphaere	The whole incumbent weight pressing ye aire	By Mr. Townlys Hypothesis ye incumbent weight should bee
12	———29—	29 ———	58 ———	58
13	———$24\frac{11}{16}$	29 ———	$53\frac{11}{16}$ ———	$53\frac{7}{13}$
14	$20\frac{3}{16}$ ———	29 ———	$49\frac{3}{16}$ ———	$49\frac{5}{7}$
16	——— 14 ———	29 ———	43 ———	$43\frac{1}{2}$
18	——— $9\frac{1}{8}$ —	29 ———	$38\frac{1}{8}$ ———	$38\frac{2}{3}$
20	———$5\frac{3}{16}$	29 ———	$34\frac{3}{16}$ ———	$34\frac{4}{5}$
24	0———	29 ———	29 ———	29
48	$-14\frac{5}{8}$ ———	29 ———	$14\frac{3}{8}$ ———	$14\frac{1}{2}$
96	$-22\frac{1}{8}$ ———	29 ———	$6\frac{7}{8}$ ———	$7\frac{1}{4}$
192	$-25\frac{5}{8}$ —	29 ———	$3\frac{3}{8}$ ———	$3\frac{5}{8}$
384	$-27\frac{1}{8}$ —	29 ———	$1\frac{3}{4}$ ———	$1\frac{7}{16}$
576	$--27\frac{7}{8}$ ———	29 ———	$1\frac{1}{8}$ ———	$1\frac{5}{24}$
768	$--28\frac{1}{8}$ ———	29 ———	$0\frac{7}{8}$ ———	$0\frac{1/74}{8}$ *
960	$--28\frac{3}{8}$ ———	29 ———	$0\frac{5}{8}$ ———	$0\frac{4/55}{8}$ †
1152	$--28\frac{7}{16}$ —	29 ———	$0\frac{9}{16}$ ———	$0\frac{10}{16}$

* Newton has misread Hooke's $\dfrac{7\frac{1}{4}}{8}$.

† Hooke's $\dfrac{4/5}{8}$ was incorrect and Newton's misreading makes it worse.

228. If Venice paper bee held or cleamed[1] on yt side of a Glasse globe full of water there appeares a red ring like an Halo on ye paper whence Mr Hooke endeavours to explicate ye rednesse of Starrs in ye Horizon.

230. Mr Hooke sayth yt ye Horizontall Diameter of ye Sunn is made longer a well as its perpendicular diameter shorter, wch hee would explicate by ye Solar rays being refracted towards ye center of ye Atmosphere [but he framed his observation to his erroneous Hypothesis.

231. The aire hath parts (& veines) unequally dense by reason of heate & vapors unequally dispersed (wch is ye cause of a stars wavering) ye like wavering will appeare by the sides of a red hot glasse suspended in the aire wch has noe steames. The ascending vapors being made globular by the ambient aire, & their heate & rarity being gradually greater towards

[1] Cleamed means stuck, made to adhere (O.E.D.).

the midst, they have ye effect of a concave lens of Glasse every way. Those wch descend are more cold & dense towards ye mids & have ye effect of a convex lens.

234 Quaere 1. May there not bee made a globe gradually denser towards ye center so as to refract all ye rays to one point wch come // out from another. this would perfect dioptricks 2. May not ye new appearance of Starrs bee deduced hence 3 & ye height of ye Atmosphaere. 4 May not ye disparity of ye upper & under aire bee so greate as somtimes to make a reflecting surface (as I have seene salt & fresh water in most places blended, yet here & there had a seperating surface) if so this may bee ye cause of ye equall limits of clouds & of Parelij either by reflection or refraction. 5. Whither clouds may not bee suspended twixt ye heavier & lighter aire, as 6 I have suspended a glasse buble in ye midst of water whose top has beene rarefied by heate more yn ye bottome. & I could make ye bubble to rise or fall at pleasure by heating ye water more or lesse. 7 The reason of remote appearance of Mountaines. 8. The uncertainty of ye Horizontall parallax of ye Planets, but it will bee better to delineate their motions wth good Telescopes in Iconismes of small starrs from an hower before to an hower after noone in two distant places under or neare ye same Meridian. as at London & St Hellens on ye coast of Africa. Probably the Planets are nearer us then Astronomers yet esteeme & ye Moone especially, ye Earth shaddow being shortened by inflection; nor will Keplers Hypothesis of ye Penumbra doe.

INDEX

The page-numbers of passages in Latin are given in italic type. Numerals in brackets indicate dates. Book-titles are in italics.

INDEX